Cavitation and Bubble Dynamics

Images in the Front Cover

Background: Representation of a bubble collapse near a wall, with the characteristic asymmetric collapse and jet formation, which pierces the bubble forming toroidal structures.

Top row, image at left: Representation of instantaneous turbulent structures in a cavitating channel.

Top row, image at centre: Simulation of bubble cluster collapsing near a wall. Note the shielding effect induced by the external bubbles, which collapse first with the characteristic jet formation.

Top row, image at right: Shock waves emitted during a bubble collapse, represented with numerical Schlieren. During bubble collapse, local hotspots may be created reaching luminescence conditions.

Cavitation and Bubble Dynamics

Fundamentals and Applications

Edited by

Phoevos Koukouvinis

Department of Mechanical Engineering and Aeronautics, School of Mathematics, Computer Science and Engineering, CITY, University of London, London, United Kingdom

Manolis Gavaises

Department of Mechanical Engineering and Aeronautics, School of Mathematics, Computer Science and Engineering, CITY, University of London, London, United Kingdom

ELSEVIER

ACADEMIC PRESS

An imprint of Elsevier

Academic Press is an imprint of Elsevier
125 London Wall, London EC2Y 5AS, United Kingdom
525 B Street, Suite 1650, San Diego, CA 92101, United States
50 Hampshire Street, 5th Floor, Cambridge, MA 02139, United States
The Boulevard, Langford Lane, Kidlington, Oxford OX5 1GB, United Kingdom

British Library Cataloguing-in-Publication Data
A catalogue record for this book is available from the British Library

Library of Congress Cataloging-in-Publication Data
A catalog record for this book is available from the Library of Congress

ISBN: 978-0-12-823388-7

For Information on all Academic Press publications
visit our website at https://www.elsevier.com/books-and-journals

Publisher: Matthew Deans
Acquisitions Editor: Brian Guerin
Editorial Project Manager: Rachel Pomery
Production Project Manager: Kamesh Ramajogi
Cover Designer: Greg Harris

Typeset by MPS Limited, Chennai, India

Contents

List of contributors

Ioannis Anagnostopoulos Laboratory of Hydraulic Turbomachines, School of Mechanical Engineering, National Technical University of Athens, Athens, Greece

Rickard E. Bensow Mechanics and Maritime Sciences, Chalmers University of Technology, Gothenburg, Sweden

Johan Bosschers MARIN/TU Delft, Faculty 3ME, Delft, The Netherlands

Chris Conrad Institute of Engineering Thermodynamics, Friedrich-Alexander-University Erlangen-Nuremberg, Erlangen, Germany

Manolis Gavaises Department of Mechanical Engineering and Aeronautics, School of Mathematics, Computer Science and Engineering, CITY, University of London, London, United Kingdom

Milad Heidari-Koochi School of Mathematics, Computer Science and Engineering, CITY, University of London, London, United Kingdom

Joonsik Hwang Department of Mechanical Engineering, Mississippi State University, Starkville, MS, United States; Center for Advanced Vehicular Systems (CAVS), Mississippi State University, Starkville, MS, United States

Hesamaldin Jadidbonab Department of Mechanical Engineering and Aeronautics, School of Mathematics, Computer Science and Engineering, CITY, University of London, London, United Kingdom

Ioannis K. Karathanassis Department of Mechanical Engineering and Aeronautics, School of Mathematics, Computer Science and Engineering, CITY, University of London, London, United Kingdom

Ioannis Kassanos Laboratory of Hydraulic Turbomachines, School of Mechanical Engineering, National Technical University of Athens, Athens, Greece

Phoevos Koukouvinis Department of Mechanical Engineering and Aeronautics, School of Mathematics, Computer Science and Engineering, CITY, University of London, London, United Kingdom

Nikolaos Kyriazis Department of Mechanical Engineering and Aeronautics, School of Mathematics, Computer Science and Engineering, CITY, University of London, London, United Kingdom

Dmitrii Mamaikin Institute of Engineering Thermodynamics, Friedrich-Alexander-University Erlangen-Nuremberg, Erlangen, Germany

Themis Melissaris MARIN/TU Delft, Faculty 3ME, Delft, The Netherlands

Georgios Mousmoulis Laboratory of Hydraulic Turbomachines, School of Mechanical Engineering, National Technical University of Athens, Athens, Greece

Mithun Murali-Girija Department of Mechanical Engineering and Aeronautics, School of Mathematics, Computer Science and Engineering, CITY, University of London, London, United Kingdom

Alexander Neubauer Institute of Engineering Thermodynamics, Friedrich-Alexander-University Erlangen-Nuremberg, Erlangen, Germany

Claus-Dieter Ohl Department for Soft Matter, Institute for Physics, Otto-von-Guericke University, Magdeburg, Germany

Fabian Reuter Department for Soft Matter, Institute for Physics, Otto-von-Guericke University, Magdeburg, Germany

Sebastian Riess Institute of Engineering Thermodynamics, Friedrich-Alexander-University Erlangen-Nuremberg, Erlangen, Germany

Sören Schenke MARIN/TU Delft, Faculty 3ME, Delft, The Netherlands

Evangelos Stavropoulos-Vasilakis Department of Mechanical Engineering and Aeronautics, School of Mathematics, Computer Science and Engineering, CITY, University of London, London, United Kingdom

Martijn van Rijsbergen MARIN/TU Delft, Faculty 3ME, Delft, The Netherlands

Tom J.C. van Terwisga MARIN/TU Delft, Faculty 3ME, Delft, The Netherlands

Erik van Wijngaarden MARIN/TU Delft, Faculty 3ME, Delft, The Netherlands

Michael Wensing Institute of Engineering Thermodynamics, Friedrich-Alexander-University Erlangen-Nuremberg, Erlangen, Germany

Preface

The aim of the book is to summarize the existing literature in the field of hydrodynamic cavitation investigations in the industrial field and to expand the current understanding by presenting new investigations covering both fundamentals and applications in a wide spectrum of industrial fields. Hydrodynamic cavitation occurs in flowing liquids when the local pressure drops below the saturation pressure threshold, forming vaporous cavities. Generally speaking, there are deviations from this description, as in exotic conditions (ultrapure environments or very fast dynamics), liquid tension may develop without the formation of cavities; however, in most cases, the fluid relaxes, forming a bubbly mixture well before reaching the tension regime.

The intriguing character of these gaseous bubbles stems from the fact that in practical flows, once pressure recovers, the bubbles collapse violently, focusing energy at very localized hot spots. Collapses can occur in a variety of ways, either symmetrically or asymmetrically with a characteristic jet formation, depending on the local pressure gradient. The energy concentration is so immense that during collapse, solid surfaces may be eroded, and luminescence may even be produced from the bubble contents as a result of compression heating.

Topics discussed in the book involve practical applications in which cavitation may be encountered. Such applications include turbomachinery (pumps and hydraulic turbines), marine applications and ship propulsion, fuel systems and positive displacement pumps, and microfluidics. The book also includes a literature review of modern methods of cavitation models and advanced experimental diagnostics for the quantification of density distribution in cavitating flows.

This book is intended for both new and experienced researchers who want to get up to speed with new developments in the field of cavitation in different disciplines and applied fields (marine and ship engineering, hydraulic machines, and fuel systems) and fundamental investigations of bubble dynamics (shock wave emission, jetting phenomena), with both experimental and numerical techniques. Further, it may appeal to engineers who are dealing with maritime applications, hydropower, and hydraulic/fuel systems and are interested in the advances in these fields.

The book contents summarize the research conducted in several institutions across the world with experience in multiphase flows and the respective application. Among these are City, University of London; Otto-von-Guericke University of Magdeburg; Technical University of Delft; Chalmers University; National Technical University of Athens; University of Mississippi; and Friedrich-Alexander-University. Part of the research was largely supported by the European Union, either in form of Innovative Training Network or Global Fellowship projects, which are gratefully acknowledged.

Chapter 1, Review of Numerical Methodologies for Modeling Cavitation, presents a short overview of cavitation phenomena and physical mechanisms on cavitation generation. It also discusses different methodologies that can be applied to capture cavitation effects for either industrial and fundamental applications and provides a literature review on the recent developments in the field of cavitation simulation.

Chapter 2, Nonspherical Collapse of Single Bubbles Near Boundaries and in Confined Spaces, discusses single-bubble formation and collapse in microfluidics. Modes of bubble collapse are extensively discussed in terms of spherical and asymmetric collapse with jetting, reviewing classical work, and bringing it into context with recent scientific findings that may lead to applications in biomedical and engineering in the very near future.

On more practical applications, Chapter 3, Cavitation Nuisance in Ship Propulsion: A Review of Developments, gives a description of the cavitation problem in marine applications and reviews the understanding of physical mechanisms underlying the hindrance. It presents the state of the art in experimental research and reviews the latest numerical models for the assessment of cavitation hindrance.

Similarly, Chapter 4, Numerical Prediction of Cavitation and Related Nuisance in Marine Propulsion Systems, aims to give an overview of methodologies for numerical predictions of cavitation in marine engineering, relevant to both research and industry, with a discussion of the limitations and possibilities of different methods. Topics covered include the determination of cavitation inception points, hull pressure pulse predictions, and radiated noise levels. Finally, the chapter examines proposed methods for assessing the risk of cavitation erosion.

Chapter 5, Study and Detection of Cavitation in Centrifugal Pumps, discusses the interaction between cavitation and centrifugal pump operation. Topics that are covered include different types of large-scale vapor formation that appear in centrifugal pumps, the effect of the establishment design on the development of cavitation, and a summary of the state-of-the-art experimental and numerical works on the topic.

Focusing on turbomachinery, Chapter 6, Cavitation in Hydraulic Turbines, covers the effects of cavitation on the operation of hydraulic turbines. Experimental measurement techniques that can be incorporated into the study of the various cavitation phenomena observed in hydraulic turbines are discussed.

Also, the available numerical tools that have been developed and applied toward an in-depth understanding of the physical processes involved in cavitation in hydraulic turbines are reviewed.

Chapter 7, X-ray Measurement Techniques Used for Wall-Bounded Cavitating Flows, focuses on state-of-the-art experimental methods that can be applied in practical industrial cavitating flows. X-ray diagnostics are extensively analyzed and classified according to their potential to provide quantitative information about the volumetric distribution of cavitation in the form of fluid density, from densitometry to X-ray microcomputed tomography.

Chapters 8 and 9 discuss cavitation phenomena in fuel systems and, in particular, the effect cavitation has on fuel injectors. Chapter 8, Interaction of Cavitation with Sprays in High-Pressure Diesel Injection Systems, discusses mainly diesel injection systems and Chapter 9, Experimental Diagnostics and Investigations on Cavitation in Engine Injector Nozzles, discusses gasoline systems. Both chapters provide an overview of recent studies, including modern optical diagnostics for understanding the interaction between cavitation and spray formation.

With regard to fuel systems, Chapter 10, Cavitation in Positive Displacement Pumps, discusses cavitation in high-pressure positive displacement pumps. Theoretical details of these pumps are provided along with a discussion of the thermodynamic phenomena that take place, such as compression heating. Special focus is placed on gear pumps and piston pumps, and indicative examples are discussed.

Closing, we would like to thank all of the colleagues who devoted their time and were involved in the preparation of this book. Finally, our publisher, Elsevier, deserves special acknowledgment. We are particularly grateful to Brian Guerin for taking the initiative in this venture and to Rachel Pomery for her continuous motivation.

Phoevos Koukouvinis and Manolis Gavaises

Chapter 1

Review of Numerical Methodologies for Modeling Cavitation

Evangelos Stavropoulos-Vasilakis, Nikolaos Kyriazis, Hesamaldin Jadidbonab, Phoevos Koukouvinis and Manolis Gavaises

Department of Mechanical Engineering and Aeronautics, School of Mathematics, Computer Science and Engineering, CITY, University of London, London, United Kingdom

1.1 Introduction

Cavitation is the phenomenon of vapor formation in liquids due to pressure drop [1]. It is relevant to many engineering applications, as it may occur during the operation of hydraulic machines, ship propellers, injection systems of internal combustion engines, and advanced rocket propulsion systems. The flow in hydraulic turbines [2,3] or diesel injector nozzles [4–7] is severely and rapidly accelerated; hence static pressure may drop locally below the saturation value, and vapor cavities may be initiated, grow to a critical size, and subsequently, when pressure recovers, collapse violently. In these cases, cavitation is a side effect and may be the source of erosion damage, noise, or low efficiency of the mechanical devices. On the other hand, cavitation may be harnessed for useful purposes, as when it is employed in noninvasive medical procedures for therapeutic purposes, such as high-intensity focused ultrasound [8–11], which is used for lithotripsy.

Particular complexities in understanding cavitation arise from its initiation and its extent in vastly different temporal and spatial scales. It is intriguing that liquids actually exhibit attractive intermolecular forces apart from repulsive,

Cavitation and Bubble Dynamics. DOI: https://doi.org/10.1016/B978-0-12-823388-7.00004-7

hence they exhibit an elastic behavior, more similar to solids rather than gases [12]. To illustrate this effect in a clearer manner, it is useful to understand the thermodynamic diagram of a fluid undergoing expansion at constant temperature (see Fig. 1.1). Assuming a cubic equation of state (EoS), such as van der Waals (other cubics, e.g., Peng-Robinson, behave in a similar way), the saturation dome can be identified on the basis of the Maxwell criterion, that is, pressure and Gibbs free energy equilibria. Under a thermodynamic equilibrium assumption, below the saturation dome, only a mixture of the liquid and vapor phases can exist with averaged properties of the two phases. However, the thermodynamic nonequilibrium model indicates that a fluid can exist as a pure phase under the saturation dome in a metastable state; "metastable" implies that it is short-lived and will decay progressively to a stable state. It is notable that under such conditions, liquid can actually exist in a tension, that is, a negative pressure regime.

Theoretical estimations indicate that the liquid tensile stress should be of the order of thousands of atmospheres, yet practical applications and careful experiments [13–15] indicate that voids in liquids are formed under less extreme conditions. It is widely accepted that this effect is caused by microscopic impurities, which can be submicrometric particles, bubbles, or even wall crevices, which act as nucleation points, hence lowering the tensile stress of liquids, thus seeding the

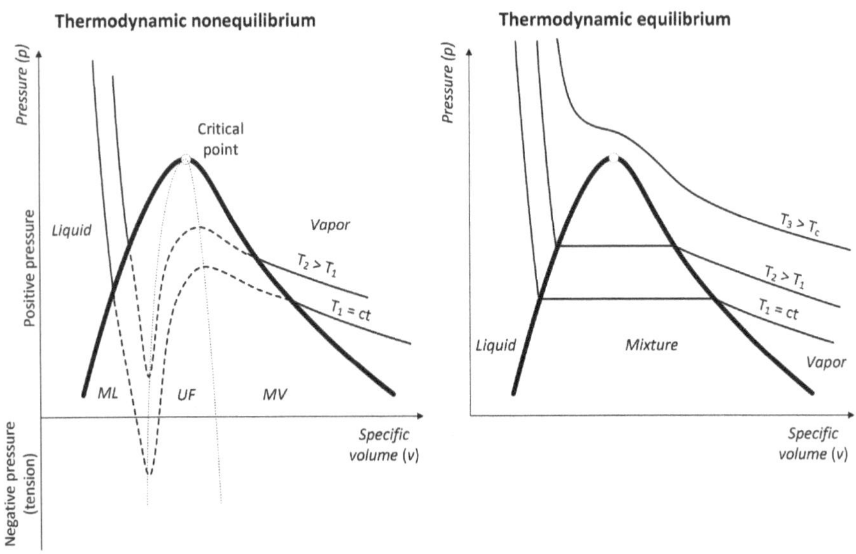

Figure 1.1 Illustration of thermodynamic diagrams of a real fluid under thermodynamic nonequilibrium (*left*) and thermodynamic equilibrium (*right*). Thin lines indicate different isotherms. Thick lines indicate the saturation dome, as identified by the Maxwell criterion. Thin dashed lines indicate the spinodal curves, denoting the border of the metastable liquid (ML) and metastable vapor (MV) regions with the unstable fluid (UF) regime.

formation of large cavities and a macroscopic behavior more like thermodynamic equilibrium. It becomes apparent that whereas nucleation may occur on the submicrometric scale, cavitation may extend to the scale of millimeters or centimeters, hence growing over five orders of magnitude. Moreover, cavitation time scales may range from microseconds, considering the Rayleigh collapse of a micrometric bubble, to even hours over which cavitation erosion develops. Since capturing all these scales in the same simulation framework is unfeasible from computational resources standpoint, different submodels and/or assumptions are employed in different methodologies, as will be further analyzed.

1.2 Classification of Multiphase Methods

A basic classification of the multiphase flow-modeling strategies can be made by the assumptions about each computational cell. Methods that assume that each cell is occupied by a single phase fall into the category of interface-tracking methods. Such methods track the interface by a set of markers or elements and dynamically split cells or deform the computational mesh in such a way that the part of the mesh is always aligned with the interface (Lagrangian perspective). However, the interface-tracking approach poses computational constraints regarding topological changes on the interface [16–18], and thus it complicates the numerical modeling of cavitation, which usually involves the violent collapse of vapor cavities and the fragmentation of the vapor-liquid interface.

On the other hand, methods that assume mixing of the two phases within the computational element can be further classified as nonhomogeneous or homogeneous in terms of velocity, temperature, and pressure fields. In the first case the different fluids are described by separate set of equations, which model the mass, momentum, and energy conservation and account for phase-to-phase interactions by appropriate source terms. In the latter case the different fluids are assumed to be in mechanical and thermal equilibrium and thus can be described by a single set of conservation equations of a pseudofluid of average properties. The nonhomogeneous mixture is a more general formulation of the physics of two-phase flow, as it stems from the large differences in state and transport properties between the gas and the liquid and considers the finite rate of interfacial transfer processes. The precise description of the interface is required to correctly estimate the interfacial interactions. This is achieved either by correlating empirically the interface with local flow conditions or by explicitly advecting or tracking the interface. On the other hand, the homogeneous mixture approach relies on averaged flow properties and no longer needs the interface coupling. The mechanical and thermal equilibrium means that the two fluids share the same velocity, pressure, and temperature field. However, the transfer processes between the phases can be conducted with either an infinite or a finite rate. In-depth analysis of modeling strategies and numerical methods for multiphase flows can be found in the books of Städtke [19] and Prosperetti [16]. For simulations of cavitating flows, in which liquid and gas phases coexist, further classifications can be made on the basis of interface capturing (sharp interface or diffuse interface), the description of the gaseous phase (continuous or dispersed, in Eulerian or Lagrangian frameworks), and the thermodynamic assumptions (thermodynamic equilibrium or non-equilibrium); see also Fig. 1.2.

Nevertheless, the choice of modeling strategy for the cavitating flows is highly application dependent. A crucial parameter for the choice of methods is the relationship between the physical length scales and the numerical modeling length scales [20]. The latter refers to the computational method and the available computational resources that dictate the spatial resolution. If the physical scales are adequately resolved by the available numerical resolution, sharp interface techniques are applicable, while if they are marginally or completely unresolved, diffuse interface methods are preferred.

In addition, the flow features of interest are of primary concern in making a choice of the methodology. With regards to hydrodynamic cavitation over hydrofoils, mass-transfer models might be preferred [21,22] over barotropic models, as the influence of baroclinic torque on vapor dynamics may be important. Barotropic models are commonly combined with explicit density-based solvers, which have the capability to capture strong dynamics of bubble collapse but impose tighter time-step restrictions [22], rendering the computational cost of such a lengthy simulation significant. On the other hand, mass-transfer models are susceptible to closure issues; indeed, a comparison with extensive available experimental data shows that the numerical results may depend greatly on the mass-transfer model used or the empirical tunable model constants [21,22]. Barotropic models are free of such tuning parameters and hence are preferred in some applications [23].

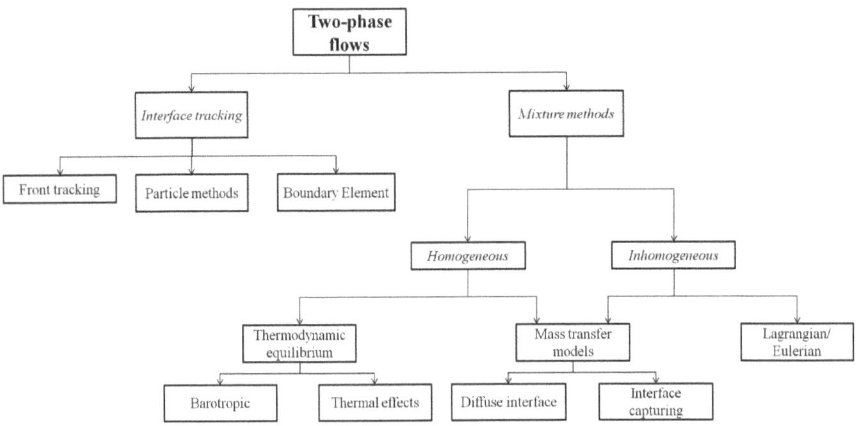

Figure 1.2 Classification of multiphase methods for cavitation.

In this work, the different methodologies employed for cavitation modeling are reviewed. The applicability of the methods to different flow regimes and physical scales is assessed, along with their relevance to the physical processes under examination. Along these lines, it is believed that the best way to present the various models is through the characteristic length scales of the problems they are applied to. This will shed light on the features, advantages, and the disadvantages of the different methodologies and on their range of applicability. Therefore the methods are categorized into three classes: the methods applied in simulations of individual bubbles, the methods applied in simulations of larger clusters of bubbles, and the methods applied in simulations at engineering scales.

1.3 Governing Equations

Let us consider the fluid domain Ω enclosing the flow field of n different media; the interface Γ between two of these media divides the domain into two subdomains Ω_i and Ω_j, occupied by phases or fluids i and j, respectively, as Fig. 1.3 illustrates. If the mixture is composed of only two media, then $\Omega_j = \overline{\Omega}_i$ and $\Gamma = \partial\Omega_i$ if i is the phase of primarily interest.

Without loss of generality, each fluid is governed by the Navier-Stokes (NS) equations, which express momentum conservation; the continuity equation, which expresses mass conservation; and the energy conservation equation. In differential form, for each phase k the system of equations is written as follows:

$$\begin{cases} \dfrac{\partial \rho_k}{\partial t} + \nabla \rho_k \boldsymbol{u}_k = 0 \\ \dfrac{\partial \rho \boldsymbol{u}_k}{\partial t} + \nabla \cdot \left(\rho_k \boldsymbol{u}_k \boldsymbol{u}_k\right) = -\nabla p_k - \nabla \cdot \boldsymbol{\tau} + \boldsymbol{S} \\ \dfrac{\partial \rho e_k}{\partial t} + \nabla \cdot \left(\rho_k e_k \boldsymbol{u}_k\right) = -\nabla p_k \boldsymbol{u}_k - \nabla \cdot (\boldsymbol{\tau} \boldsymbol{u}_k) + \nabla q + \boldsymbol{S} \cdot \boldsymbol{u}_k \end{cases} \tag{1.1}$$

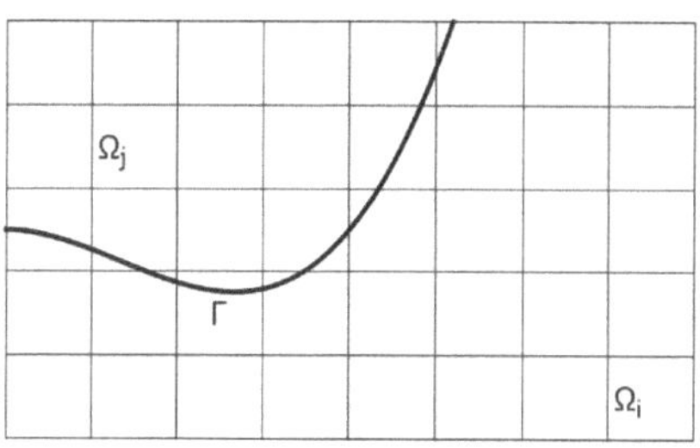

Figure 1.3 The interface between two phases or fluids for a generalized multiphase flow case.

The vectors of velocity and momentum sources are denoted with bold. In the case of incompressible isothermal flow, the energy conservation equation can be decoupled and omitted. The equations can be further simplified if the fluids are considered inviscid or other conditions exist. The system of differential equations is closed by an appropriate EoS linking the thermodynamic variables.

As has already been mentioned, the basic classification of the multiphase flow models is in nonhomogeneous and homogeneous models. In the first category the motion of each phase k is described by a separate set of NS equations, with the interface Γ representing a boundary of jump conditions in the flow variables. In the second category the flow of the mixture can be described by as single set of equations for a pseudofluid with variable density and properties according to the mixture's composition. In this case the phase volume fractions are defined as $\alpha_k = v_k/dv$ for each elementary volume (cell) of the discretized computational domain Ω. Depending on the approach that is followed, additional equations and relations may be used. In the following sections the various methodologies will be presented.

1.4 Single-Bubble Dynamics

Numerous studies deal with the simulation of bubble dynamics, not only because bubble collapse is a benchmark case for validating the developed numerical schemes, but also owing to their implications in many technological applications. The collapse of a single cavitation bubble represents a showcase of the various fundamental physical processes and complex phenomena taking place during phase change between the liquid and vapor phases of a fluid. Therefore, by studying them, the basis of modeling cavitating flows is solidified. The different approaches can be divided into two-fluid and single-fluid models. The latter can be further categorized into interface-tracking and interface-capturing algorithms as well as homogeneous equilibrium mixture (HEM) models.

1.4.1 Two-Fluid Models

The most accurate way to model multiphase flows is to consider the mixture of different fluids as nonhomogeneous and treat each fluid or phase separately with its own set of conservation equations for mass, momentum, and energy. In this way the interfaces are explicitly described and remain sharp, and they function as boundaries between media that have not only different properties but also different dynamics, as each phase or fluid has its own velocity, temperature, and pressure fields. At the interface, jump conditions are applied between the various flow fields that could also affect the evolution of the interface and may account for the slip velocity between the media. The interaction of the different phases on or through the interface is modeled by appropriate source terms.

The model of Saurel and Abgrall [24] for compressible multiphase flows solves seven equations for a two-phase problem, while considering different velocity and pressure fields for each, and introduces pressure and velocity relaxation across the interface to couple the dynamics of the different mediums. It is flexible with regard to its use of thermodynamic EoS for the closure for each phase and can be used for cavitating flows with dynamic creation of interfaces. The method is further generalized for an arbitrary number of fluids and for multidimensional cases, incorporating the concept of infinite rate of relaxation and thus of instantaneous pressure and velocity equilibrium on the interface [25], making it suitable for a wider range of applications.

Based on the extension of AUSM+ (advection upwind splitting method) scheme to compressible two-fluid models by Paillere and Garcia Cascales [26], Chang et al. [27] developed a compressible two-fluid model for nonhomogeneous and nonequilibrium mixtures with different velocity and temperature fields for each phase but a common pressure field. Their model is derived from a stratified flow concept, which provides a straightforward way to link the physics of the problem with their mathematical representation, conserves the hyperbolic nature of the equations, and employs an efficient scheme to overcome numerical oscillations. The gas phase is governed by the ideal gas and the liquid by the stiffened gas EoS. They unveil rich physics from the interaction of shockwaves with material interfaces in fast-slow (liquid-gas) configurations for shocks traveling over gas bubbles in liquid or underwater explosions. Lauer et al. [28] utilized a two-fluid method that solves a separate set of equations for each phase, tracking the interface and accounting for interface-exchange processes along with nonequilibrium thermodynamic effects, where the finite mass transfer was based on the Hertz-Knudsen relation, to investigate single-bubble dynamics next to a rigid wall. Their method has similarities with cut-cell immersed boundary methods [29] in the way in which the fluid domain is partitioned into subdomains enclosing the different phases. They explored the effect of the wall distance on the bubble shape during collapse.

1.4.2 Interface-Tracking Methods

Cavitation bubble collapse and its interaction with a solid boundary have been studied since 1971 by Plesset and Chapman [30], who followed an interface-tracking approach. They employed a potential flow solver for the liquid phase and a marker-and-cell (MAC) technique for tracking the bubble interface. In the

MAC method, which was introduced by Harlow [31,32], marker particles were used to identify the different fluid regions on a stationary grid. In the same context, boundary element method (BEM) employed in the study of bubble deformation and collapse near a solid wall [33,34] or for simulating a single cavitation bubble and two cavitation bubbles below a free surface [35]. However, the flow is investigated only until the moment of the transformation of the bubble to a torus. Such methods are still being used for gas bubble simulations near wall [36] and interactions with deformable bodies [37,38]. In an extension of the BEM method, Wang [39] proposed a weakly compressible model for the first two cycles of the bubble oscillation in the vicinity of a wall and performed axisymmetric simulations. The suggested model considers the energy loss due to pressure waves created during the bubble collapse and were found to be in good agreement with previous experimental works [40]. BEMs significantly reduce the computational cost by transforming the three-dimensional (3D) problem of the bubble collapse to a two-dimensional (2D) one, and thus only the bubble interface is discretized. However, this approach is subject to instabilities and cannot handle topological changes of the bubble interface [41], such as the jet formation during bubble collapse. Moreover, the potential solver is prone to singularities, owing to small-scale dissipative mechanisms [42]. In principle, BEM is an incompressible model, which may behave well in the initial stages of the collapse but cannot advance beyond the point of the jet impact. For that reason, Chahine [43] extended the incompressible BEM potential flow solver with a multiphase compressible flow solver for simulating the growth and collapse of a bubble near deformable walls. Depending on the stage of the bubble collapse, different methodologies were applied: the BEM in the initial stages, where flow velocities are small, and the compressible approach for the violent growth and collapse of the bubble, where compressibility effects are dominant.

On the other hand, the front-tracking methods were originally developed by Glimm et al. [44] and in a follow-up study by Unverdi and Tryggvason [45]. The moving boundary (interface) is explicitly described by a lower-dimension Lagrangian grid, accompanied by a 3D Eulerian grid for the hyperbolic solver. The latter grid is used for solving the bulk flow, away from the discontinuity. Because they offer high accuracy in resolving the bubble shape, front-tracking methods have been widely utilized for problems of bubble dynamics. For instance, Yu et al. [46] simulated the collapse of an initially spherical cavitation bubble in shear flow near a solid boundary, and more recently, Peng et al. [47] evaluated the motion of the bubble interfaces, in a collapsing cluster of seven bubbles by performing 2D direct numerical simulations (DNS). In another DNS study, Lu et al. [48] studied the shock wave propagation in bubbly fluids. The liquid pressure calculated by the DNS was used as an input in the Keller equation in order to find the collapse pressure of the bubbles. Furthermore, Hawker and

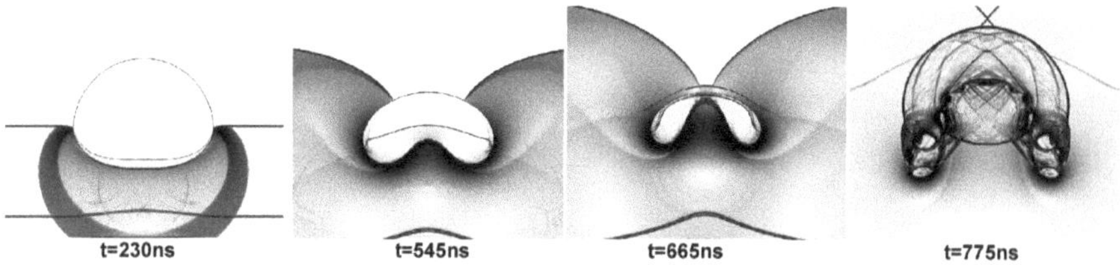

Figure 1.4 Indicative results of the interaction of a shock wave with a bubble handled with interface tracking. Shading indicates the density gradient. *Reproduced with permission from [52].*

Ventikos [49,50] simulated gas bubble collapse due its interaction with a shock wave by using the FrontTier code for tracking the liquid-gas interface; the computational grid was split into two regions, with different EoS applied for each phase (ideal gas EoS for the air and stiffened polytropic EoS for the water). They also expanded their initial configuration to multiple gas-filled bubbles [51,52], such as two identical bubbles placed one behind the other or a triangular array of three bubbles in which the central bubble was significantly smaller than the surrounding bubbles (see Fig. 1.4). In a similar study, Popinet and Zaleski [53,54] employed a finite volume formulation with a fixed grid and a front-tracking approach; free surface boundary conditions were imposed for simulating bubble flows near solid boundaries, and cubic splines were used for the surface representation. The effect of viscosity on jet formation during bubble collapse was studied and compared against experimental data. More recent works include the combination of the front-tracking approach with the immersed boundary method for simulating bubble dynamics in a microchannel [55] and the investigation of a toroidal bubble collapse near a solid wall [56]. The main attribute of the front-tracking methods is that they offer smear-free interfaces and are applicable to problems involving complex geometries. However, the location of the inner nodes is not described; therefore in cases of extreme deformation, degenerate cells are formed that need smoothing [45]. In addition, although they allow for large deformations of the interface surface, they cannot capture large topological changes, such as the transition from a spherical bubble to a torus.

1.4.3 Interface-Capturing Methods

Concerning interface-capturing methodologies, bubble collapse simulations have been performed utilizing either the level-set method (LSM) [57,58] or the volume of fluid (VOF) method [59,60]. Compared to the interface-tracking approach, in which the interface is explicitly tracked by Lagrangian marker points, in the interface-capturing schemes the interface is implicitly reconstructed by a continuous or discontinuous field variable.

1.4.3.1 Level-Set Methods

Osher and Sethian [57] proposed a continuous interface-capturing approach in which a level-set function ϕ is used for tracking the interface Γ, which is moving under velocity $\boldsymbol{u}$ and is bounding an open region Ω. At $t = 0$ the zero level-set of ϕ defines the material interface. The level-set function has the following properties:

$$\begin{cases} \phi(\boldsymbol{x}, t) = 0, & \text{for} x \in \vartheta\Omega_k = \Gamma(t) \\ \phi(\boldsymbol{x}, t) > 0, & \text{for} x \in \Omega_k \\ \phi(\boldsymbol{x}, t) < 0, & \text{for} x \in \overline{\Omega_k} \end{cases} \tag{1.2}$$

The value of ϕ is convected with the velocity field $\boldsymbol{u}$:

$$\frac{\partial \phi}{\partial t} + \boldsymbol{u} \cdot \nabla \phi = 0 \tag{1.3}$$

In general, the LSM is characterized by the advection step of the transport equation for identifying the phases, the reinitialization step to remove interfacial diffusion, and the coupling between the level set function φ and the conservative properties.

Several researchers have adopted the level-set approach for modeling bubble collapse problems, since it offers highly accurate calculation of the interface curvature and the surface tension. Furthermore, Hu et al. [61,62] developed a conservative interface method based on the level-set technique for solving compressible multiphase flows, maintaining a sharp liquid-gas interface. The methodology was validated against fundamental shock tube cases, bubble-shock wave interactions, underwater explosions, and bubble collapse simulations, among others. In addition, finite element method (FEM) and LSM were combined to perform DNS studies of spherical bubble collapse by Nagrath [63], in which both the air and the water were treated as compressible fluids. In a more recent work by Lee et al. [64] the ultrasound-driven bubble motion was modeled by a LSM in conjunction with a semiimplicit pressure correction solver. Compressibility and phase-change effects were considered, and the motion of a spherical bubble in a periodic acoustic field was compared against analytical methods.

More recently, researchers followed a hybrid approach, combining the LSM with interface-tracking methods, aiming to outweigh their disadvantages. Tan et al. [65] overcame the deficiency of BEM by combining them with a LSM and demonstrated complex topological changes in bubble simulations, such as axisymmetric bubble merging, bubble splitting, and bubble jet impact. The LSM was employed for capturing the bubble surface motion, and it was combined with the BEM to compute the surface velocities. In a similar work, Huang and Zhang [66] combined a MAC projection algorithm with a LSM of nonuniform grids for bubble collapse simulations near a rigid wall, whereas Sussman [67] coupled a LSM with a VOF for simulating vapor bubble oscillation and collapse.

1.4.3.2 Volume of Fluid

In the VOF method the treatment of the interface Γ is based on the transport equation for the volume fraction α, which is discontinuous across Γ:

$$\frac{\partial \alpha}{\partial t} + \boldsymbol{u} \cdot \nabla \alpha = 0 \tag{1.4}$$

Although VOF was originally introduced for incompressible flows, it was later expanded for compressible flow simulations as well [68], and therefore α is replaced by $\rho\alpha$ in the advection equation.

The VOF family of methods have two main steps: the reconstruction step and the advection step. In the reconstruction stage, an approximate interface is computed on the basis of the accuracy of the method, whereas in the advection stage, the boundary fluxes are approximated, correcting the interface. The main advantage of the VOF methods is that they ensure conservation. However, it is difficult to accurately reconstruct the curvature of the interface in complex geometries and unstructured grids, owing to the discontinuous nature of the volume fraction field. Moreover, to maintain interface sharpness, a convective Courant-Friedrichs-Lewy condition less than unity is required for the interface.

Considering works that adapted the VOF method for bubble collapse simulations, Li et al. [69] investigated the bubble collapse near a conical rigid boundary, while Johnsen [70] simulated nonspherical bubble collapse in water by deriving a high-order interface-capturing scheme [71] with WENO reconstruction and an HLLC solver, which was third-order accurate in time. Koukouvinis et al. [72,73] studied the effect of asymmetries on the collapse of the bubbles by employing the VOF method. They performed bubble collapse simulations subject to gravity [72] and close to a free surface [73] demonstrating jetting effects. In a more recent study by the same research group [74], the shear stress and the pressure pulse near the wall were quantified and parametric bubble collapse studies were performed, investigating the influence of the bubble size and the wall distance. The surface tension was included by the continuous surface force (CSF) model [75], and comparison with experiment was provided aiming to give an insight to the latter.

In connection with the previous work, Zeng et al. [76] employed a compressible VOF approach for axisymmetric simulations in OpenFOAM [77] for modeling the wall shear stress induced by cavitation bubbles. A polytropic EoS for the gas and the Tait equation for the liquid were used along with the CSF model, whereas heat transfer was neglected. Owing to experimental challenges, only a qualitative comparison between experiment and simulations was performed. Koch et al. [78] also followed a two-phase VOF approach in OpenFOAM to simulate laser generated cavitation bubbles. They compared their solution against the Gilmore model and experiments. The gas was modeled as noncondensable gas (no mass transfer) by the Noble-Abel EoS (covolume), and the Tait EoS was used for the liquid; viscosity and surface tension were also considered. The same methodology was employed for wedge simulations of collapsing bubbles [79]; the acoustic waves upon bubble growth, jet impact, collapse, rebound, and second collapse were demonstrated.

In conclusion, interface-capturing techniques can provide a decent alternative to interface tracking by reproducing a sharp interface with minimum smearing due to numerical diffusion. It is notable, though, that at specific conditions near the critical point of fluids, the a sharp interface assumption may not make sense anymore, as the interface becomes progressively more diffuse and even disappears (see also Fig. 1.5).

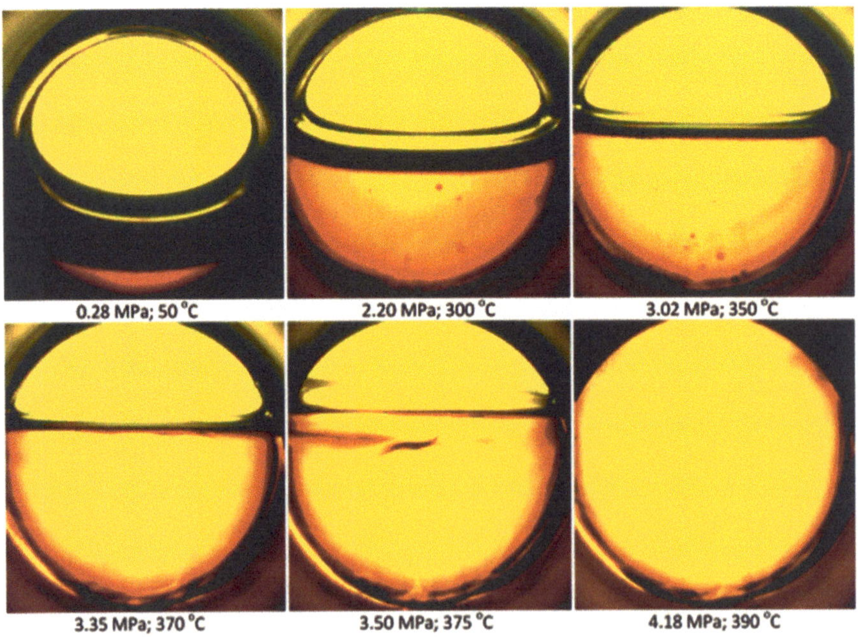

Figure 1.5 Changes of surface tension and interface appearance of the meniscus, shown for a dieseline mixture, as temperature and pressure progressively reach the critical point. *Reproduced with permission from [80].*

1.4.3.3 Homogeneous Equilibrium Mixture Models

In HEM models, all phases are in thermodynamic and mechanical equilibrium, as they share the same temperature and pressure (mixture). HEM is also known as a single-phase model, since only the compressible NS equations are solved without any additional transport equation. Therefore the phase change between the liquid and the vapor is predicted through the variation in the density from the corresponding EoS. The properties of the mixture are calculated as the weighted average of the pure liquid (l) and the liquid-vapor mixture (v) phases; for example, the density and internal energy of the mixture are calculated as follows:

$$\begin{aligned} \rho &= (1-\alpha_v)\rho_{\text{sat},l} + \alpha_v \rho_{\text{sat},v} \\ \rho e &= (1-\alpha_v) e_{\text{sat},l}\rho_{\text{sat},l} + \alpha_v e_{\text{sat},l}\rho_{\text{sat},v} \end{aligned} \tag{1.5}$$

In the relations, α_v and α_l denote the vapor and liquid volume fractions, respectively, and hence $\alpha_v + \alpha_l = 1$. The mixture speed of sound is determined by using the Wallis speed of sound formula [12]:

$$\frac{1}{\rho c^2} = \frac{\alpha_l}{\rho_{\text{sat},l} c^2_{\text{sat},l}} + \frac{\alpha_v}{\rho_{\text{sat},v} c^2_{\text{sat},v}} \tag{1.6}$$

The complexity of each EoS and the computational cost depend on the thermodynamic closure. The simplest approach is the barotropic EoS [81], in which temperature effects are not included. Note that temperature itself may change as the barotropic process can be assumed to follow an isentropic path. Aside from the barotropic model, more advanced models consider the energy equation [82] solved through iterative algorithms for finding the temperature. For instance, the Tait EoS or the stiffened gas equation for the liquid and the ideal gas EoS for the gas have been employed. Finally, high-order EoS, either cubic (Peng-Robinson [83]) or of an exponential form (Helmholtz EoS [84]), have also been utilized, increasing significantly the computational cost but offering a real fluid thermodynamic closure.

The main feature of HEM is that no tuning and empirical parameters are needed; hence they these methods have been widely used for modeling cavitation in engineering-scale problems (see Section 1.6). Previous work of the authors demonstrate that the HEM approach can also successfully capture the bubble collapse and the wave dynamics either by employing a barotropic EoS [85] or by employing real fluid thermodynamics [86], provided that the grid resolution is sufficient, and hence the diffuse interface is described within a reasonable amount of cells.

1.5 Bubble Clusters

Although benchmark cases for single-bubble dynamics that have already been presented are easily extended to arrays of a few distinct bubbles, investigation of the dynamics of larger clusters, or clouds, of randomly distributed bubbles poses significant numerical challenges, and according to the authors' best knowledge, few numerical works exist on that subject.

A first approach to modeling the dynamics of clusters of multiple bubbles would be the direct extension of the Rayleigh-Plesset equation to account for the interactions of neighboring bubbles. Modeling assumptions, such as considering the involved fluids inviscid or taking advantage of prominent symmetries, can simplify the governing equations and result in a one-dimensional formulation.

An example is the modeling strategy followed by Wang and Brennen [87], who investigated the collapse of a spherical cluster of uniformly distributed cavitation bubbles in which the governing equations were linearized and simplified into a single differential equation for the response of the bubbles' radius to a pressure perturbation, accounting for the bubble-bubble interactions. The flow was solved in a pure Lagrangian way for each individual bubble particle separately. However, such an approach has limited applicability to relatively simple or even idealized cases and is probably of mainly academic interest.

On the other hand, 3D modeling would consider the averaged conservation equations for the two-phase mixture in an Eulerian-Lagrangian manner [88–91]. The continuum phase would be described in a Eulerian frame of reference, while the dispersed bubbles would be tracked in a Lagrangian way. The dynamics of the individual bubble particles are governed by Rayleigh-Plesset equations, and the Eulerian-Lagrangian coupling is performed through the projected volume fraction of the disperse phase onto the Eulerian frame of reference. The void fraction can be considered simply in the definition of the averaged density [88–90] or can be accounted for in the governing equations in the form of source terms [91]. Interaction between the bubbles of the cluster can be considered when a local potential flow assumption is made, which extends in the vicinity of each bubble [91]. Although these models can treat more complex configurations than the one-dimensional models can, since the Rayleigh-Plesset equations assume inherently only a spherical bubble shape, only idealized applications can be considered.

Other approaches can account for asymmetries and nonspherical bubbles and capture the dynamics of more realistic configurations. Within the framework of front-tracking methods, Chahine and Duraiswami [17] employed a boundary integral method (BIM) to simulate the dynamics of a small cluster of 16 bubbles, capturing the inward propagation of collapse, the bubble deformations, the evolution of the nonspherical shape, and the reentrant jet. However, since the method cannot treat topological changes of the interface, the violent implosion was not modeled. In addition, a typical bubble cloud may comprise millions of discrete bubbles, which would be impossible to handle with such a method.

In one of the first steps toward large-scale bubble-detailed computations of bubble clusters, Tiwari et al. [18] simulated a cluster of 50 identical gas bubbles imploding next to a solid wall. They utilized a homogeneous mixture model in which the flow field was described by the averaged quantities, which obeyed the conservation equations, while a separate transport equation was solved for the advection of the secondary phase, corresponding to the gas bubbles, to capture the diffused interface and to account for the dynamics of the thin mixture region between the pure phases. Because the Weber number for the studied cases was estimated to be large enough that the inertia forces were dominant, viscosity and surface tension were neglected. In addition, thermal diffusion and mass diffusion were also not account for; therefore no phase change took place, while the liquid phase followed the stiffened gas EoS, and the gas inside the bubbles followed the ideal gas law. The simulations were performed over a grid of 20 million cells supported by adaptive mesh refinement (AMR) that ensured a resolution of 7.7 cells along the initial radius of the bubbles. The bubbles were initially randomly distributed in a hemispherical arrangement over the wall. The main focus of the study lay on the bubble-bubble interactions and the influence on the wave propagation; however, the simulations were able to go beyond the stage reached with BIM [17] and captured jetting and fragmentation of the bubbles.

Another modeling strategy is the use of the HEM approach, which was utilized by Adams and Schmidt [20] to simulate the collapse of a moderate-sized bubble cluster consisting of 125 bubbles on a 120 million cell grid. Although the HEM methodologies do not explicitly track or reconstruct the sharp interface of the bubbles, they implicitly capture it with a fine spatial resolution. According to their rationale, when physical and numerical characteristic length scales are equivalent, and therefore the flow structure could be partially or marginally resolved, the single-fluid formulations with additional local thermodynamic equilibrium assumption are more appropriate than mixture models or sharp interface methodologies. Thus the fully compressible inviscid conservation equations were considered for the mixture, and instead of the closure of the system by an EoS, the state variables were extracted by thermodynamic tables, considering the mass fractions' balance; surface tension was not accounted for. The simulation of a collapsing vapor bubble cluster near a rigid wall, comprising 125 bubbles randomly positioned with varying sizes and average radius of $\overline{R}_b = 0.95$mm, on a canonical grid with a resolution of at least 20 cells per initial radius, was carried out. Detailed information for the wave emission, propagation and focus, vapor volume. and pressure evolution as well as pressure loading on the wall was the focus of this study. More recently, the same research group used a similar HEM approach coupled with a barotropic model that neglects thermal effects to perform parametric studies of a similar cluster of 150 spherical vapor bubbles with regard to interaction parameter values and stand-off distance from the wall. The results were communicated in the publication of Ogloblina et al. [92].

A notable example of high-end simulations of bubble cloud collapse is found in the works of Rossinelli et al., Wermelinger et al., Rasthofer et al., and Rasthofer et al. [93–96], in which the authors performed simulations of resolved bubble clouds consisting of thousands of bubbles on incredibly large computational grids. A single-fluid formulation was followed, combined with the advection of a marker function, for tracking liquid and vapor and the thermodynamic characteristics of the mixture EoS. The simulations were carried out on large uniform Cartesian computational grids, comprising 1.2 billion [94,95], 232 billion [96], and even 13 trillion [93] cells, yielding a resolution of 10–15, 11–28, and 50 cells along the diameter of the smallest bubble. In this way the authors were able to investigate the symmetric collapse of spherical clusters of 2500 [94,95] and 12,500 [96] gas bubbles of average diameter $\overline{R}_b = 0.11$mm and $\overline{R}_b = 0.69$mm, respectively, and the asymmetric collapse of a 15,000-bubble cloud in the vicinity of a rigid wall [93]. Their most recent work [96] reported insightful findings on the rich physics of dynamics of the collapse and the subsequent mechanisms and processes, such as wave propagation and focusing, bubble oscillations, microjet formation, and pressure pulses. In addition, a brief comparison with a homogeneous mixture model was performed that demonstrated the promising applicability of these methods in such complex problems but also quantified their error with respect to the employed diffused-interface bubble-resolved computations. Nowadays, such simulations, even if demanding, are feasible with open source software such as OpenFOAM (see Fig. 1.6).

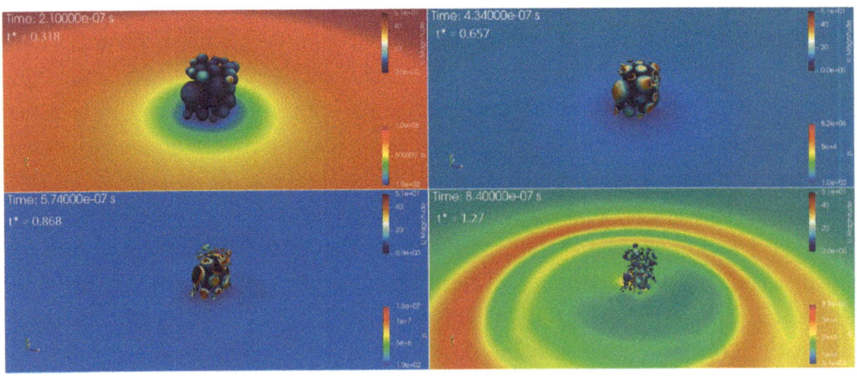

Figure 1.6 An example of a bubble cluster collapse near a wall for a population of 50 bubbles in water with a range of radii (1–15 μm), total equivalent radius 22.8 μm, at a pressure far field of 10 bar. The Rayleigh collapse time (used for the nondimensional time t^*) is 0.66 μs. Note the shielding effect induced by the outer layer of bubbles.

1.6 Engineering Scales

Apart from fundamental studies, cavitating flows are of practical and engineering interest, as they occur in a wide range of applications. Some of the most important are in the automotive industry (injection systems in IC engines), in naval engineering (propellers), in the energy industry (turbines and pumps), and in the biomedical sector (lithotripsy, drug delivery systems).

The cavitation models that have been applied to those length and time scales can be classified into the HEM approach, mass-transfer rate models, interface-capturing schemes, Eulerian-Lagrangian formulation, two-fluid models, and probability density functions (PDFs).

1.6.1 Homogeneous Equilibrium Mixture Models

The first approach invokes the thermodynamic equilibrium assumption, resulting in a single-fluid model for the mixture that predicts the vapor volume fraction directly from the cell-averaged fluid state [20]. Such models have been employed in an implicit large eddy simulation (LES) approach [23,97–99,98], since they compose a natural subgrid-scale model, by capturing individual bubbles for sufficient resolution. The equilibrium model has been proven to give correct predictions in terms of cavitation and wave dynamics for engineering applications [100–102], and in combination with its low computational cost, it has been widely used for simulating cases of industrial interest. Since the temperature variation of the liquid can be negligible in some cases, the energy equation has been omitted; therefore barotropic cavitation models have been utilized for predicting cavitation on macroscopic problems such as hydrofoils [103], Venturi nozzles [104], and high-pressure throttle flows [105]. Delannoy and Kuevy [81] proposed a barotropic law from which the phase change is predicted, and since then, different barotropic models have been developed and used [106,107]. Egerer et al. [23] demonstrated implicit LES of cavitating flow in a throttle, using the adaptive local deconvolution method (ALDM) for discretizing the convective terms in conjunction with a barotropic HEM for modeling the liquid and vapor phases. Their results were validated against experimental data for two different operating conditions. Örley et al. [98] further expanded this methodology to a two-phase model by adding a transport equation for the gas mass fraction to investigate the flow in a cavitating nozzle and to study the primary jet breakup. The LES results were compared against experiments, and it was determined that three mechanisms were responsible for the primary jet breakup: (1) turbulent fluctuations at the outlet region, (2) entrainment of gas in the nozzle, (3) bubble collapse close to the liquid-gas interface. Furthermore, Örley et al. [29,108] coupled the HEM approach with cut-cell immersed boundary method, with improved subcell resolution and small cut-cell treatment, to ease the simulation of cavitating flows through fuel injectors. In this way the fuel flow inside a nine-hole diesel injector with needle movement has been simulated [108], following a similar numerical approach as in [98]. The LES demonstrate that small-scale turbulence characterizes the opening and closing phases, whereas the main phase is dominated by large turbulent structures. Finally, Budich et al. [109] numerically studied cavity dynamics, periodic cavitation patterns, and instability mechanisms for a flow above a sharp convergent-divergent wedge and compared against experiments. A condensation shock was captured at the trailing part, which is in agreement with the experiment.

On the other hand, HEM with temperature effects has been employed by Saurel et al. [82] for studying cavitation in an underwater projectile. They modeled the liquid phase by the Tait EoS and the gas phase by the ideal gas relation. Later, Schnerr et al. [100] emphasized the detection of the shock formation and shock propagation in 3D cloud cavitation on a 3D twisted hydrofoil (see also Koop [110], Sezal [111]). Owing to the variation of the Mach number in the liquid and vapor regimes, a Mach consistent numerical flux was proposed and adopted for liquid flows with phase transition [101]. HEM models with temperature effects have also been used for diesel injector and nozzle simulations [101,102,112] and for estimating erosion [113] among other uses.

In the work of Dumbser et al. [84], high-order unstructured WENO schemes were implemented, and the HEM approach was expanded to real fluid EoS, based on IAPWS-IF97. Owing to the high computational cost of evaluating the thermodynamic properties compared to the closure models described in the previous two paragraphs, the EoS was projected onto the space of piecewise Lagrangian polynomials (in a FEM approximation). Since the properties drastically change across the saturation line, an AMR approach with Cartesian cut cells was employed. Apart from validation cases, 3D explosion and cavitating flow around a NACA0015 were simulated.

Implementing real fluid EoS in the discontinuous Galerkin framework can offer high-order spatial accuracy in comparison to traditional finite volume method (FVM) framework [114,115]. Under this prism a throttle and a jet flow were simulated at conditions that are realized in gas injectors for combustion engines. The pressure range in the throttle exceeds the applicability range of the ideal gas approximation; therefore a tabulated approach of the real fluid is required. A detector to identify regions where the Gibbs phenomenon prevails was employed [116], and then a finite volume subcell method was applied [117]. In a later work under the discontinuous Galerkin framework with a tabulated real fluid EoS, Hempert et al. [118] simulated subsonic and supersonic methane jets and resolved shocks and transient structures. They also assessed the real and ideal gas effects in the flow structure, density, pressure, and mass flow.

1.6.2 MASS-TRANSFER RATE MODELS

Cavitation has been also modeled by employing mass-transfer rate models in engineering-scale applications. In such models, a transport equation for the vapor volume fraction with source terms in order to model phase change is solved along the NS system of equations. Although the different phases may share the same velocity, pressure, and temperature as in HEM, the mass-transfer rate in this approach is finite and time dependent. Currently, diffuse interface models necessitate tuning for each case individually to predict realistic cavitation regimes.

A family of mass-transfer models is based on the kinetic theory of gases, such as the Hertz-Knudsen equation [119], in which only the evaporation term is present:

$$\begin{aligned} \frac{\partial \alpha \rho_v}{\partial t} + \nabla(\alpha \rho_v \boldsymbol{u}) &= \dot{R} \\ \dot{R} &= \frac{\lambda A_{\text{init}}(p_{\text{sat}} - p)}{\sqrt{2\pi R_g T_{\text{init}}}} \end{aligned} \tag{1.7}$$

In these relations, A_{init} is the overall bubble-cloud interphase surface area, R_g is the ideal gas constant, and T_{init} is the temperature at the bubble interphase. The coefficient λ denotes the deviation from the thermodynamic equilibrium conditions; $\lambda = 1$ for thermodynamic equilibrium, whereas if $\lambda = 0.1$ or lower, there is significant deviation from equilibrium. A similar formulation was proposed by Theofanous et al. [120]:

$$\dot{R} = \frac{\lambda A_{\text{init}}}{\sqrt{2\pi R_g}}\left(\frac{p_{\text{sat}}}{T_l} - \frac{p}{T_v}\right) \tag{1.8}$$

where T_l and T_v are the local liquid and vapor, respectively.

In another family of mass-transfer rate models, condensation and vaporization terms have been adopted, such as the models suggested by Kunz et al. [121], Merkle et al. [122], Schnerr-Sauer [123], and Zwart-Gerber-Belamri [124,125]. Therefore the transport equation for the vapor volume fraction takes the following form:

$$\frac{\partial \alpha \rho_v}{\partial t} + \nabla(\alpha \rho_v \boldsymbol{u}) = \dot{R}_e + \dot{R}_c \tag{1.9}$$

where $\dot{R}_e$ and $\dot{R}_c$ are the evaporation and condensation terms, respectively, which vary depending on the model.

Kunz et al. [121] developed an implicit methodology accounting for the liquid, vapor, and noncondensable gas phases. The phase-change between the liquid and the vapor was modeled by introducing the following evaporation and condensation terms:

$$\begin{aligned} \dot{R}_e &= \frac{C_{\text{dest}} \rho_v \alpha_l \cdot \min(0, p - p_{\text{sat}})}{\rho_l U_\infty^2 t_\infty / 2} \\ \dot{R}_c &= \frac{C_{\text{prod}} \rho_v \alpha_l^2 \cdot (1 - \alpha_l)}{t_\infty} \end{aligned} \tag{1.10}$$

where $C_{prod} = C_{dest} = 100$ are the production and destruction mass-transfer constants, respectively, and $t_\infty = L/U_\infty$ is the mean flow time scale for the characteristic length L. This model was incorporated in a dual-time, preconditioned algorithm with parallel capabilities. It was proposed for sheet cavitation, and supercavitation cases and 3D problems were simulated, such as cavitation on a one-caliber ogive. Senocak and Shyy [126] presented a pressure-based single-phase algorithm for turbulent cavitating flows. They modeled phase change by adapting the evaporation and condensation terms suggested by Kunz [121]. Merkle et al. [122] proposed an improved version of the Kunz model by incorporating a boundedness criterion for numerical stability:

$$\begin{aligned} \dot{R}_e &= k_v \frac{\rho_v \alpha_l}{t_\infty} \cdot \min\left(1, \max\left(\frac{p_{sat} - p}{k_p p_v}, 0\right)\right) \\ \dot{R}_c &= k_l \frac{\rho_v \alpha_v}{t_\infty} \cdot \min\left(1, \max\left(\frac{p_{sat} - p}{k_p p_v}, 0\right)\right) \end{aligned} \tag{1.11}$$

where k_v, k_l, and k_p are scaling constants.

Some other models utilize the asymptotic form of the Rayleigh-Plesset equation of bubble dynamics and are based on the assumption that there are plenty of nuclei in the mixture, sufficient for cavitation inception. Starting from nuclei, bubbles grow or collapse, depending on the flow conditions (pressure and temperature). Therefore information on the bubble number density and population that are present in the liquid prior to the onset of cavitation is required. Depending on their complexity and sophistication, they may include or ignore mass transfer between the liquid and the vapor phases and may consider or not consider the gas content in the liquid. Such models are currently available in commercial CFD models [127] and in open source CFD toolkits [128,129].

For instance, the Schnerr-Sauer model [123] is based on the Rayleigh-Plesset equation and the assumption that a large number of spherical bubbles are contained in the mixture. They utilized the model for simulating the flow in injection nozzles and for estimating the effect of injection pressure fluctuations on cavitation [130]. The evaporation and condensation terms become the following:

$$\begin{aligned} \dot{R}_e &= \frac{\rho_v \rho_l}{\rho} \alpha_v (1 - \alpha_v) \frac{3}{R_b} \sqrt{\frac{2}{3} \frac{(p_{sat} - p)}{\rho_l}} \\ \dot{R}_c &= \frac{\rho_v \rho_l}{\rho} \alpha_v (1 - \alpha_v) \frac{3}{R_b} \sqrt{\frac{2}{3} \frac{(p - p_{sat})}{\rho_l}} \end{aligned} \tag{1.12}$$

where R_b is the bubble radius, usually $R_b = 10^{-6}$ m. A similar model was developed by Zwart-Gerber-Belamri [124,125] under the assumption that all bubbles are of the same size, while the modified version considers the interaction of the cavitation bubbles as well:

$$\begin{aligned} \dot{R}_e &= F_{vap} \frac{3 r_{nuc} (1 - \alpha_v)}{R_b} \sqrt{\frac{2}{3} \frac{p_{sat} - p}{\rho_l}} \\ \dot{R}_c &= F_{cond} \frac{3 \alpha_v}{R_b} \rho_v \sqrt{\frac{2}{3} \frac{p - p_{sat}}{\rho_l}} \end{aligned} \tag{1.13}$$

where $R_b = 10^{-6}$ m is the bubble radius, $r_{\text{nuc}} = 5 \cdot 10^{-4}$ is the nucleation site volume fraction, $F_{\text{vap}} = 50$ is the evaporation coefficient, and $F_{\text{cond}} = 0.01$ is the condensation coefficient.

1.6.3 Interface-Capturing Schemes

LSMs are well known for offering sharp interface representation, and apart from being applied to single-bubble problems, as already discussed, they have been utilized in large-scale simulations irrespectively of their high computational cost.

A level-set approach that is applicable to both compressible and incompressible flows with a reinitialization step suitable for subgrid-scale interfacial breakup was developed by Kinzel et al. [131,132]. This approach was applied to cases of high-speed marine vehicles in cavitating flows around twisted wings [131] and developed supercavitation over moving vehicles [132] among others. The LSM has been also employed in combination with a mass-transfer model based on the Rayleigh-Plesset equation by Michael et al. [133,134] to predict the cavitation regime in hydrofoil geometries. Since the incompressible NS equations were solved, the phase-change source term was semiimplicitly included in the Poisson equation, thus increasing the stability of the method. In another study in which the LSM was used for capturing the interface and for modeling the surface tension, Dabiri et al. [135] identified locations that were subject to cavitation by proposing a total stress criterion instead of the conventional pressure criterion. Simulations of the flow in an orifice demonstrate that viscous stress has a significant effect on cavitation, while the impact of other parameters, such as the Reynolds and Weber numbers, as well as the curvature of the inlet corner on cavitation formation were investigated. Finally, Yue et al. [136] combined the LSM with LES to study the free surface motion in an open channel by employing two different subgrid-scale models and evaluating their effect on the free surface as well.

1.6.4 Eulerian-Lagrangian Formulation

The Eulerian-Lagrangian formulation couples the liquid, which is analyzed in the Eulerian coordinate system, with the vapor bubbles, which are described in the Lagrangian reference frame [137,138]. This approach has been successfully utilized for shock wave–bubble interaction, shock wave propagation in a bubbly medium [139], cavitation behind a blunt body [140], and propeller simulations [141,137]. In the latter studies, a Reynolds-averaged Navier-Stokes (RANS) solver for obtaining the flow field was combined with a Lagrangian multibubble-tracking approach, based on a modified Rayleigh-Plesset equation that accounts for a nonuniform flow field. Starting from nuclei distributions (micro-scale), bubbles grow and merge to form a cavity, or they shrink and break up from a larger cavity (inter-scale). The large bubbles and the cavities were modeled by utilizing the level-set approach in the Eulerian reference frame (macro-scale). Parametric investigations on the initial distribution of the nuclei, the dissolved gas concentration, and the cavitation number were performed in propeller simulations [141]. In a similar study by the same research group, sheet and tip vortex cavitation on a rotating propeller were simulated [137] by creating a blade-to-blade passage computational domain.

In another Eulerian-Lagrangian approach, the flow in a hole-type diesel injector was predicted and compared against experimental data by Giannadakis et al. [6] and Strotos et al. [142] (see Fig. 1.7). The Lagrangian cavitation model was based on the Monte Carlo approximation, similar to droplet discrete models that have been used in sprays. The total bubble population was described by a distribution function $f(\boldsymbol{x}, \boldsymbol{u}, R, t)$, which determines the probable number of bubbles per unit volume at each time t, in the spatial range of $(\boldsymbol{x}, \boldsymbol{x}+d\boldsymbol{x})$, with velocity in the range of $(\boldsymbol{u}, \boldsymbol{u}+d\boldsymbol{u})$, and with radius in the range of $(R, R+dR)$. The following equation must be satisfied and all physical subprocesses have been considered in order to achieve conservation of the total number of bubbles in any volume:

$$\frac{\partial f}{\partial t}+\nabla_x(f\cdot\boldsymbol{u})+\nabla_u(f\cdot\alpha)+\frac{\partial}{\partial R}\left(f\cdot\dot{R}\right)=\dot{f}_{\text{nucleation}}+\dot{f}_{\text{breakup}}+\dot{f}_{\text{coalescence}} \tag{1.14}$$

One of the most important models in this category is the Lagrangian cavitation model of Kubota [143], also employed by Giannadakis [6], which uses the Rayleigh-Plesset equation of bubble dynamics for estimating the cavitation volume fraction. Some more recent advances have proposed models that account for

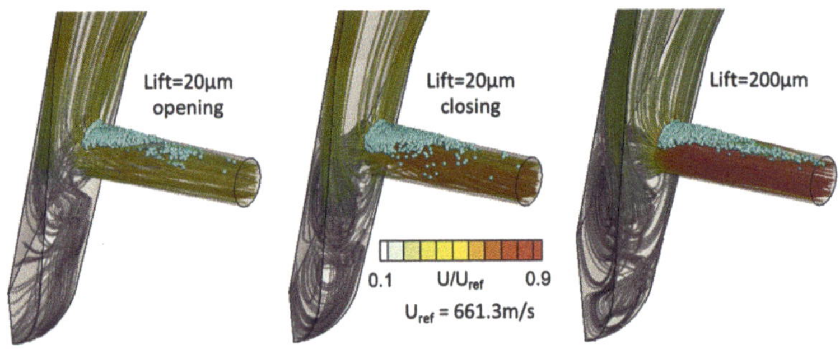

Figure 1.7 Indicative simulation using an Eulerian-Lagrangian assumption for cavitation. This simulation shows the flow inside the tip of a diesel injector. The streamlines are colored according to the velocity magnitude, whereas bubbles are resolved as parcels (*blue spheres*). *Reproduced with permission from [142].*

collective compressibility and shock wave interaction effects in polydispersed cavitating flows [91,144,145]. Finally, some models do exist for predicting the collapse process of individual vapor-air bubbles or bubble clouds within the bulk of the liquid or even near a wall surface [146–149].

1.6.5 Two-Fluid Models

Instead of treating cavitating flows as a homogeneous mixture governed by a single set of equations similar to an averaged single-fluid formulation, the two-fluid approach has been followed at engineering scales as well, despite the additional computational burden such an option may be thought to entail. In the model of Alajbegovic et al. [150] the two-fluid method was employed. Two sets of conservation equations are solved: one for the liquid phase and one for the vapor phase. In this approach. the two phases can have different velocities, and discontinuities on the flow fields are captured across the interface. Similarly, Cristofaro et al. [151] has resolved the cavitating flow, resembling a diesel injector orifice, within a square channel with a two-fluid approach, comparing the results with experimental results and obtaining details on erosion formation (see Fig. 1.8). An important observation of this investigation was that the slip velocity between the two phases was rather significant only for the shear layers, on the order of 75%.

In addition, based on their extension of the AUSM+ scheme to compressible two-fluid models [26], Paillere and Garcia Cascales [152] simulated the flow through a convergent-divergent nozzle, considering interfacial friction, heat, and mass transfer through the interface along with phase change with the aid of either

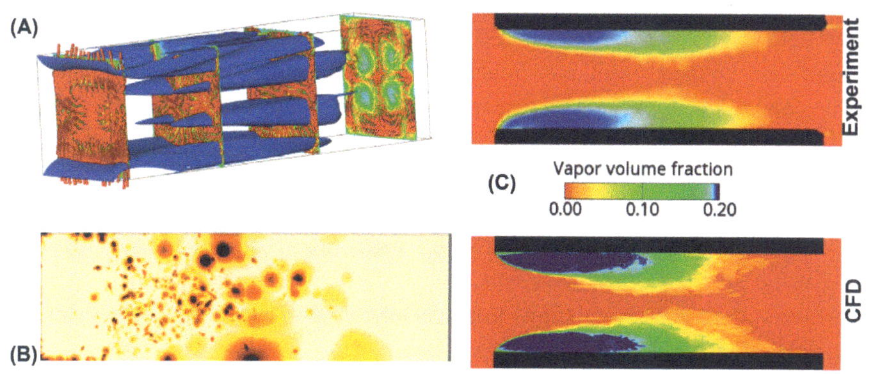

Figure 1.8 (A) Cavitating flow inside a square orifice (295 × 300 μm, length 993 μm). (B) Erosion pattern as expressed by the magnitude of pressure peaks in erosion collapse locations. (C) Comparison of vapor distribution between experiments and simulations. *Reproduced with permission from [151]. https://pubs.acs.org/doi/full/10.1021/acsomega.9b03623.*

analytical or tabulated equations of state. The increase in the computational load by using a separate set of equations for each distinct phase in the simulations of a multiphase flow at engineering scales could be significant. A rather extreme approach would be to apply a similar partitioning of the governing equations with regard to the different physical length scales of the flow instead of the species. An example could be the modeling of liquid jet atomization and spray formation, relevant to high-pressure fuel injection systems, through an Eulerian-Eulerian approach that considers a separate set of equations for each droplet size class [153] and accounts for all relevant physical processes of primary and secondary breakup, collision, and evaporation but disregards cavitation induction. On the other hand, the reduced five-equation two-fluid compressible model used by Saurel et al. [154] to simulate cavitating flows at engineering scales around a projectile or through a injector nozzle considered a single pressure field and a single velocity field, continuous along the interface, and two separate temperature and entropy fields for the two phases. This approach could be derived by the asymptotic limit of a nonequilibrium two-phase hyperbolic model, which accounts for discontinuities across the interface for all fields through stiff mechanical relaxation by employing infinite rate of relaxation and thus instantaneous equilibrium.

1.6.6 Probability Density Functions

Another approach for describing cavitation structures is by employing PDFs and related transport models. In the work of Dumond et al. [155] a PDF transport model has been used for the vapor fraction, based on the Boltzmann transport equation, to model the highly dynamic and stochastic interaction of the turbulent flow field with cavitation. An additional novelty of this work is the fact that solution of the PDF is done entirely in an Eulerian framework, avoiding the expensive computations and inaccuracies induced by coupling an Eulerian with a Lagrangian solver. It has also been shown that by coupling the PDF method with a compressible LES framework, good results may be obtained for a variety of Venturi-like tubes and shapes. The applicability of such models to engineering-scale problems has not yet been thoroughly assessed.

Finally, apart from the aforementioned models, which are based on the FVM framework, there have been efforts to model cavitation using alternative frameworks. Examples of such works include (1) simulation of cavities due to the entry of high speed objects, using meshless smoothed particle hydrodynamics (SPH) [156] and FEM [157]; (2) simulation of cavities at the wake of submerged bodies in liquid [158], using distributed particle methods and focusing on the SPH method in particular; and (3) simulations of forward step geometries, resembling the orifice of injectors, using lattice Boltzmann methodologies [159]. These examples are, of course, not exhaustive. There are many different approaches, most of them in an infancy stage, for attacking the phenomenon of cavitation, each having specific advantages and disadvantages on specific flow types. Nevertheless, the FVM framework is mature enough and offers better handling of the underlying flow phenomena with less uncertainty over the physics for general flow types.

1.6.7 Turbulence Modeling for Cavitating Flows

For a wide range of industrial applications and flows at engineering scales, the treatment of turbulence in the simulations is of high importance. The most widely used approach is to solve for the mean flow and model small-scale spatial or temporal turbulent features by conducting RANS computations. However, with the increase of computational means, advanced numerical computations become more affordable, and LES, which resolves the entire range of turbulent scales, becomes an attractive, realistic alternative. Therefore the influence of turbulence modeling on simulations of cavitating flows at these scales has been addressed by various researchers.

In their review of the state of the art on attached turbulent cavitating flows, Wang et al. [160], as well as Fuster [161] in their presentation of different numerical approaches on bubble clusters, apart from the physics and the alternative multiphase modeling methodologies, thoroughly discuss turbulence modeling. Within the framework of RANS the default eddy-viscosity models (EVMs), suitable for and widely used in engineering applications, seem to be lacking in the ability to capture the dynamics of turbulent cavitating flows. These models, compared to Reynolds stress models (RSMs), exhibit less sensitive connection between turbulent stresses and streamline curvature, body forces, and strong transport effects. Moreover, conflicts arise between turbulence modeling principles and some characteristics of the cavitating flows, such as the nonlinear behavior of the pressure-strain relation, the breakdown of turbulence scales in the liquid-vapor mixing region, and compressibility effects and reduced speed-of-sound values in the vapor regions. Because the EVMs, such as the popular two-equation $k-\omega$ and $k-\varepsilon$ models, rely on empirical constants, they need adjustment for nonequilibrium conditions found in such transient flows with swirl, adverse pressure gradients, and recirculation.

From numerical tests on sheet cavitation it was observed that standard EVMs underestimate the cavity length and have poor agreement with experiments regarding velocities and void ratio within the cavity [162]. The standard models tend to overpredict the turbulent eddy viscosity inside the vapor region. Especially for sheet cavitation the high viscosity values toward the rear end of the cavity yield higher stresses that restrict the reentrant jet [21,162–165], which is the principal mechanism of unsteady cavitation, initiating from the downstream end of the cavity, developing underneath it, and moving upstream to break the cavity cycle. The failure of the standard EVMs to capture the physics of the attached turbulent cavitating flows becomes more clear as the result of the numerical work of Reboud et al. [163], which demonstrated that inviscid simulations give results closer to experimental findings than viscous computations employing standard $k-\varepsilon$ or $k-\varepsilon$RNG models. To counteract this overprediction, Reboud et al. [163] and Coutier-Delgosha et al. [164] proposed the introduction of a scaling function that takes into account the compressibility of the mixture and reduces the turbulent viscosity in the mixture region. The Reboud correction has been widely adopted and has been applied to the well-known two-equation models $k-\varepsilon$, $k-\varepsilon$RNG, $k-\omega$, and $k-\omega$SST [21,103,165–167].

On the other hand, LES computations applied to cavitating flows do not encounter similar modeling issues because even the small-scale features are resolved. Koukouvinis et al. [168] performed a thorough assessment of turbulence modeling for the prediction of incipient and developed cavitation for flows inside diesel injector nozzles. They considered RANS simulations, using $k-\varepsilon$RNG, realizable$k-\varepsilon$, and $k-\omega$SST, linear pressure-strain RSM as well as LES computations with the WALE model. The Reboud correction was also taken into consideration. It was found that either EVMs or RSM may fail to capture cavitation inception, as low pressure existing in vortex cores was not resolved, while prediction of unsteady cavitation led to not physically stable cavities and proved highly dependent on the configuration. By contrast, the LES computations correctly predict the flow field, the vortical and vaporous structures, proper cavity shedding, and the associated instabilities. Following a different approach instead of the traditional explicit subgrid-scales model for LES, Egerer et al. [23] and Orley et al. [29,98,108] proposed the use of implicit LES, through ALDM, to simulate the turbulent cavitating flow in diesel injection systems, as has been already presented. Implicit LES introduces turbulence modeling directly into the numerical discretization of the conservation equations through ALDM, which is a nonlinear FVM capable of capturing pressure shocks and allowing pressure waves and turbulence to be propagated without numerical dissipation. Nevertheless, these findings cannot reduce the significance and relevance of RANS computations to industrial applications. Although the predictions may be found to be less accurate than those of LES, they offer reliable and efficient numerical tools for engineers.

1.7 Conclusions

Various techniques exist for modeling cavitating flows. The choice between the different methodologies and models is based mainly on the application under examination. Model choice is done according to the particular effect of interest, computational constraints, and available resources. When viewed through the prism of the length scales of the physical problem under examination, the attributes of the different approaches and their range of applicability are revealed. There is not a perfect method that is suitable for all the applications of cavitation.

In this chapter the available methodologies have been reviewed and a relative comparison has been presented with regard to a specific range of problems: single-bubble dynamics, bubble cluster dynamics, and cavitating flows at engineering scales. This assessment presents the state of the art, highlights the open questions, and pinpoints the outlook for current and feature research. While different numerical approaches have been utilized for modeling the collapse of single bubbles or bubble clusters, owing to either pressure difference or shock waves, there is a lack in the literature regarding bubble collapse under extreme conditions in which complex EoS or chemical reactions must be considered. Such cases are of scientific interest because of their implications in many applications of high industrial interest or of great potential owing to their relevance to uninvestigated physical phenomena.

For instance, cavitation-induced fusion can be a less costly alternative to nuclear fusion without producing radioactive waste, whereas pyrolysis in a diesel injector can reduce its efficiency. Regarding cavitation-induced fusion, light pulses have been noticed during bubble collapse, and extremely high temperatures have been measured [169]. Although the Moss EoS [170] describes such temperature increases and it has been combined with a front-tracking approach [171], it does not include dissociation and ionization of the air. It is necessary to derive a numerical methodology to model such phenomena precisely. Another field of study is numerical modeling of pyrolysis. A temperature increase in the surrounding liquid in a collapsing bubble is not negligible and can cause pyrolysis, which can contribute to deposit formation, reducing the performance of fuel injection systems, for instance. Modeling pyrolysis requires consideration of the chemical reactions that take place in conjunction with the multiphase flow solver.

Similar needs exist regarding cavitation modeling of engineering-scale flows, in which the effects of chemical composition have so far been ignored, yet they have been proven to happen, in cases of both sonoreactors [172] and hydrocarbon fuels [173]. In addition, it is worth mentioning that effects of dissolved gas, multicomponent fluids (such as fuels), and preexisting nucleation sites in the fluid have not been thoroughly investigated; hence these represent challenging fields for further research.

References

[1] J.-P. Franc, J.-M. Michel, Fundamentals of Cavitation, Kluwer Academic Publishers, 2005.

[2] P.A. Engel, Chapter 11—liquid erosion, in: P.A. Engel (Ed.), Impact Wear of Material, vol. 2, Elsevier, 1978, pp. 291–319.

[3] J.C. Marongiu, E. Parkinson, S. Lais, F. Leboeuf, J. Leduc, Application of SPH-ALE method to pelton hydraulic turbines, in: 5th Spheric Workshop, Manchester, 2010.

[4] M. Gavaises, C. Arcoumanis, F. Harminder, M. Badami, Cavitation in real-size multi-hole diesel injector nozzles, SAE International Journal of Engines 109 (2000). Available from: https://doi.org/10.4271/2000-01-1249.

[5] M. Gavaises, D. Papoulias, A. Andriotis, E. Giannadakis, Link between cavitation development and erosion damage in diesel injector nozzles, SAE Technical Paper 2007-01-0246, (2007). Available from: https://doi.org/10.4271/2007-01-0246.

[6] E. Giannadakis, M. Gavaises, C. Arcoumanis, Modelling of cavitation in diesel injector nozzles, Journal of Fluid Mechanics 616 (2008) 153–193.

[7] P. Koukouvinis, M. Gavaises, J. Li, L. Wang, Large Eddy simulation of diesel injector including cavitation effects and correlation to erosion damage, Fuel 175 (2016) 26–39.

[8] T. Kodama, K. Takayama, Dynamic behavior of bubbles during extracorporeal shock-wave lithotripsy, Ultrasound in Medicine and Biology 24 (1998) 723–738.

[9] A.J. Coleman, J.E. Saunders, L.A. Crum, M. Dyson, Acoustic cavitation generated by an extracorporeal shockwave lithotripter, Ultrasound in Medicine and Biology 13 (1987) 69–76.

[10] W. Mingxi, Y. Feng, G.T. Haar, Cavitation in Biomedicine: Principles and Techniques, Springer, 2015.

[11] S. Umemura, K. Kawabata, K. Fau Sasaki, K. Sasaki, In vivo acceleration of ultrasonic tissue heating by microbubble agent, IEEE Transactions on Ultrasonics, Ferroelectrics, and Frequency Control 52 (2005) 1690–1698.

[12] C.E. Brennen, Cavitation and Bubble Dynamics, Cambridge University Press, 1995. ISBN-139781107644762, ISBN-10: 1107644763.

[13] G.M. Lewis, The tensile strength of liquids in berthelot tubes, Proceedings of the Physical Society 78 (1961) 133–144.

[14] A.R. Imre, How to generate and measure negative pressure in liquids? in: Soft Matter under Exogenic Impacts, Dordrecht, 2007.

[15] V.P. Skripov, Metastable Liquids, Wiley, 1974.

[16] A. Prosperetti, G. Tryggvason, Computational Methods for Multiphase Flow, Cambridge University Press, Cambridge, 2007.

[17] G.L. Chahine, R. Duraiswami, Dynamical interactions in a multi-bubble cloud, Journal of Fluids Engineering 114 (1992) 680–686.

[18] A. Tiwari, C. Pantano, J.B. Freund, Growth-and-collapse dynamics of small bubble clusters near a wall, Journal of Fluid Mechanics 775 (2015) 1–23.

[19] H. Städtke, Gasdynamic Aspects of Two-Phase Flow: Hyperbolicity, Wave Propagation Phenomena and Related Numerical Methods, John Wiley & Sons, 2006.

[20] N.A. Adams, S.J. Schmidt, Shocks in cavitating flows, in: F.C. Delale (Ed.), Bubble Dynamics and Shock Waves, Springer Berlin Heidelberg, Berlin, Heidelberg, 2013, pp. 235–256.

[21] A. Ducoin, B. Huang, Y.L. Young, Numerical modeling of unsteady cavitating flows around a stationary hydrofoil, International Journal of Rotating Machinery (2012).

[22] T. Capurso, M. Lopez, M. Lorusso, M. Torresi, G. Pascazio, S.M. Camporeale, et al., Numerical investigation of cavitation on a NACA0015 hydrofoil by means of OpenFOAM, Energy Procedia 126 (2017) 794–801.

[23] C.P. Egerer, S. Hickel, S.J. Schmidt, N.A. Adams, Large-eddy simulation of turbulent cavitating flow in a micro channel, Physics of Fluids 26 (2014) 085102.

[24] R. Saurel, R. Abgrall, A multiphase Godunov method for compressible multifluid and multiphase flows, Journal of Computational Physics 150 (1999) 425–467.

[25] R. Saurel, O. Lemetayer, A multiphase model for compressible flows with interfaces, shocks, detonation waves and cavitation, Journal of Fluid Mechanics 431 (2001) 239–271.

[26] H. Paillère, C. Corre, J.R.G. Cascales, On the extension of the AUSM + scheme to compressible two-fluid models, Computers & Fluids 32 (2003) 891–916.

[27] C.-H. Chang, M.-S. Liou, A robust and accurate approach to computing compressible multiphase flow: stratified flow model and AUSM + -up scheme, Journal of Computational Physics 225 (2007) 840–873.

[28] E. Lauer, X.Y. Hu, S. Hickel, N.A. Adams, Numerical modelling and investigation of symmetric and asymmetric cavitation bubble dynamics, Computers & Fluids 69 (2012) 1–19.

[29] F. Örley, V. Pasquariello, S. Hickel, N.A. Adams, Cut-element based immersed boundary method for moving geometries in compressible liquid flows with cavitation, Journal of Computational Physics 283 (2015) 1–22.

[30] M.S. Plesset, R.B. Chapman, Collapse of an initially spherical vapour cavity in the neighbourhood of a solid boundary, Journal of Fluid Mechanics 47 (1971) 283–290.

[31] F.H. Harlow, The Particle-In-Cell Method for Numerical Solution of Problems in Fluid Dynamics, United States, 1962.

[32] F.H. Harlow, J.E. Welch, Numerical calculation of time-dependent viscous incompressible flow of fluid with free surface, The Physics of Fluids 8 (1965) 2182–2189.

[33] S. Zhang, J.H. Duncan, G.L. Chahine, The final stage of the collapse of a cavitation bubble near a rigid wall, Journal of Fluid Mechanics 257 (1993) 147–181.

[34] S. Zhang, J.H. Duncan, G.L. Chahine, The behavior of a cavitation bubble near a rigid wall, in: J.R. Blake, J.M. Boulton-Stone, N.H. Thomas (Eds.), Bubble Dynamics and Interface Phenomena: Proceedings of an IUTAM Symposium held in Birmingham, UK, September 6–9,1993, Springer Netherlands, Dordrecht, 1994, pp. 429–436.

[35] P.B. Robinson, J.R. Blake, T. Kodama, A. Shima, Y. Tomita, Interaction of cavitation bubbles with a free surface, Journal of Applied Physics 89 (2001) 8225–8237.

[36] A.M. Zhang, X.L. Yao, L.H. Feng, The dynamic behavior of a gas bubble near a wall, Ocean Engineering 36 (2009) 295–305.

[37] Z. Li, L. Sun, Z. Zong, Numerical analysis of gas bubbles in close proximity to a movable or deformable body, Archive of Applied Mechanics 83 (2013) 1715–1737.

[38] G. Curtiss, D. Leppinen, Q. Wang, J. Blake, Ultrasonic cavitation near a tissue layer, Journal of Fluid Mechanics 730 (2013) 245–272. Available from: https://doi.org/10.1017/jfm.2013.341.

[39] Q. Wang, Multi-oscillations of a bubble in a compressible liquid near a rigid boundary, Journal of Fluid Mechanics 745 (2014) 509–536.

[40] A. Philipp, W. Lauterborn, Cavitation erosion by single laser-produced bubbles, Journal of Fluid Mechanics 361 (1998) 75–116.

[41] Z.R. Li, L. Sun, Z. Zong, A boundary element method for the simulation of non-spherical bubbles and their interactions near a free surface, Acta Mechanica Sinica 28 (2012) 51–65. Available from: https://doi.org/10.1007/s10409-011-0539-4.

[42] G. Tryggvason, R. Scardovelli, S. Zaleski, Direct Numerical Simulations of Gas-Liquid Multiphase Flows, Cambridge University Press, 2011.

[43] G.L. Chahine, Modeling of cavitation dynamics and interaction with material, in: K. Kim, G. Chahine, J. Franc, A. Karimi (Eds.), Advanced Experimental and Numerical Techniques for Cavitation Erosion Prediction, Springer Netherlands, Dordrecht, 2014, pp. 123–161.

[44] J. Glimm, C. Klingenberg, O. McBryan, B. Plohr, D. Sharp, S. Yaniv, Front tracking and two-dimensional Riemann problems, Advances in Applied Mathematics 6 (1985) 259–290.

[45] S.O. Unverdi, G. Tryggvason, A front-tracking method for viscous, incompressible, multi-fluid flows, Journal of Computational Physics 100 (1992) 25–37.

[46] P. Yu, S.L. Ceccio, G. Tryggvason, The collapse of a cavitation bubble in shear flows—a numerical study, Physics of Fluids 7 (1995) 2608–2616.
[47] G. Peng, G. Tryggvason, S. Shimizu, Two-dimensional direct numerical simulation of bubble cloud cavitation by front-tracking method, in: IOP Conference Series: Materials Science and Engineering, vol. 72, 2015, p. 012001.
[48] T. Lu, R. Samulyak, J. Glimm, Direct numerical simulation of bubbly flows and application to cavitation mitigation, Journal of Fluids Engineering 129 (2006) 595–604.
[49] N.A. Hawker, Y. Ventikos, Shock/gas bubble interactions in infinite and finite volumes of liquid, in: 2nd Micro and Nano Flows Conference, 2009.
[50] N.A. Hawker, Y. Ventikos, Interaction of a strong shockwave with a gas bubble in a liquid medium: a numerical study, Journal of Fluid Mechanics 701 (2012) 59–97.
[51] M.R. Betney, B. Tully, N.A. Hawker, Y. Ventikos, Computational modelling of the interaction of shock waves with multiple gas-filled bubbles in a liquid, Physics of Fluids 27 (2015) 036101.
[52] N. Bempedelis, Y. Ventikos, A simplified approach for simulations of multidimensional compressible multicomponent flows: the grid-aligned ghost fluid method, Journal of Computational Physics 405 (2020) 109129.
[53] S. Popinet, S. Zaleski, A front-tracking algorithm for accurate representation of surface tension, International Journal for Numerical Methods in Fluids 30 (1999) 775–793.
[54] S.T. Popinet, S.T. Zaleski, Bubble collapse near a solid boundary: a numerical study of the influence of viscosity, Journal of Fluid Mechanics 464 (2002) 137–163.
[55] K.-L. Pan, Z.-J. Chen, Simulation of bubble dynamics in a microchannel using a front-tracking method, Computers & Mathematics With Applications 67 (2014) 290–306.
[56] L.T. Liu, X.L. Yao, A.M. Zhang, Y.Y. Chen, Numerical analysis of the jet stage of bubble near a solid wall using a front tracking method, Physics of Fluids 29 (2017) 012105.
[57] S. Osher, J.A. Sethian, Fronts propagating with curvature-dependent speed: algorithms based on Hamilton-Jacobi formulations, Journal of Computational Physics 79 (1988) 12–49.
[58] S. Osher, R. Fedkiw, Level Set Methods and Dynamic Implicit Surfaces, Springer, 2003.
[59] C.W. Hirt, B.D. Nichols, Volume of fluid (VOF) method for the dynamics of free boundaries, Journal of Computational Physics 39 (1981) 201–225.
[60] D.L. Youngs, An interface tracking method for a 3D Eulerian hydrodynamics code, Technical Report 44/92/ 35, AWRE, 1984.
[61] X.Y. Hu, B.C. Khoo, An interface interaction method for compressible multifluids, Journal of Computational Physics 198 (2004) 35–64.
[62] X.Y. Hu, B.C. Khoo, N.A. Adams, F.L. Huang, A conservative interface method for compressible flows, Journal of Computational Physics 219 (2006) 553–578.
[63] S. Nagrath, K. Jansen, R.T. Lahey, I. Akhatov, Hydrodynamic simulation of air bubble implosion using a level set approach, Journal of Computational Physics 215 (2006) 98–132.
[64] J. Lee, G. Son, A level-set method for ultrasound-driven bubble motion with a phase change, Numerical Heat Transfer, Part A: Applications 71 (2017) 928–943.
[65] K. Tan, B. Khoo, J. White, A level set–boundary element method for the simulation of underwater bubble dynamics, SIAM Journal on Scientific Computing 30 (2008) 549–571.
[66] J. Huang, H. Zhang, Level set method for numerical simulation of a cavitation bubble, its growth, collapse and rebound near a rigid wall, Acta Mechanica Sinica 23 (2007) 645–653.
[67] M. Sussman, A second order coupled level set and volume-of-fluid method for computing growth and collapse of vapor bubbles, Journal of Computational Physics 187 (2003) 110–136.
[68] M.L. Norman, K.-H.A. Winkler, 2-D Eulerian hydrodynamics with fluid interfaces, self-gravity and rotation, in: K.A. Winkler, M.L. Norman (Eds.), Astrophysical Radiation Hydrodynamics, Springer Netherlands, Dordrecht, 1986, pp. 187–221.

[69] B.B. Li, W. Jia, H.C. Zhang, J. Lu, Investigation on the collapse behavior of a cavitation bubble near a conical rigid boundary, Shock Waves 24 (2014) 317–324.

[70] E. Johnsen, T. Colonius, Implementation of WENO schemes in compressible multicomponent flow problems, Journal of Computational Physics 219 (2006) 715–732.

[71] E. Johnsen, T. Colonius, Numerical simulations of non-spherical bubble collapse, Journal of Fluid Mechanics 629 (2009) 231–262. Available from: https://doi.org/10.1017/S0022112009006351.

[72] P. Koukouvinis, M. Gavaises, O. Supponen, M. Farhat, Numerical simulation of a collapsing bubble subject to gravity, Physics of Fluids 28 (2016) 032110.

[73] P. Koukouvinis, M. Gavaises, O. Supponen, M. Farhat, Simulation of bubble expansion and collapse in the vicinity of a free surface, Physics of Fluids 28 (2016) 052103.

[74] P. Koukouvinis, G. Strotos, Q. Zeng, S.R. Gonzalez-Avila, A. Theodorakakos, M. Gavaises, et al., Parametric investigations of the induced shear stress by a laser-generated bubble, Langmuir: The ACS Journal of Surfaces and Colloids 34 (2018) 6428–6442.

[75] J.U. Brackbill, D.B. Kothe, C. Zemach, A continuum method for modeling surface tension, Journal of Computational Physics 100 (1992) 335–354.

[76] Q. Zeng, S.R. Gonzalez-Avila, R. Dijkink, P. Koukouvinis, M. Gavaises, C.-D. Ohl, Wall shear stress from jetting cavitation bubbles, Journal of Fluid Mechanics 846 (2018) 341–355.

[77] H.G. Weller, G. Tabor, H. Jasak, C. Fureby, A tensorial approach to computational continuum mechanics using object-oriented techniques, Computers in Physics 12 (1998) 620–631.

[78] M. Koch, C. Lechner, F. Reuter, K. Köhler, R. Mettin, W. Lauterborn, Numerical modeling of laser generated cavitation bubbles with the finite volume and volume of fluid method, using OpenFOAM, Computers & Fluids 126 (2016) 71–90.

[79] C. Lechner, M. Koch, W. Lauterborn, R. Mettin, Pressure and tension waves from bubble collapse near a solid boundary: a numerical approach, The Journal of the Acoustical Society of America 142 (2017) 3649–3659.

[80] G. Anitescu, T.J. Bruno, L.L. Tavlarides, Dieseline for supercritical injection and combustion in compression-ignition engines: volatility, phase transitions, spray/jet structure, and thermal stability, Energy & Fuels 26 (2012) 6247–6258.

[81] Y. Delannoy, J.L. Kueny, Two-phase flow approach in unsteady cavitation modeling, Cavitation and Multiphase Flow Forum, ASME-FED 98 (1990) 153–158.

[82] R. Saurel, J.P. Cocchi, P.B. Butler, Numerical study of cavitation in the wake of a hypervelocity underwater projectile, Journal of Propulsion and Power 15 (1999) 513–522.

[83] G. Lacaze, A. Misdariis, A. Ruiz, J.C. Oefelein, Analysis of high-pressure diesel fuel injection processes using LES with real-fluid thermodynamics and transport, Proceedings of the Combustion Institute 35 (2015) 1603–1611.

[84] M. Dumbser, U. Iben, C.-D. Munz, Efficient implementation of high order unstructured WENO schemes for cavitating flows, Computers & Fluids 86 (2013) 141–168.

[85] P. Koukouvinis, M. Gavaises, A. Georgoulas, M. Marengo, Compressible simulations of bubble dynamics with central-upwind schemes, International Journal of Computational Fluid Dynamics (2016) 1–12.

[86] N. Kyriazis, P. Koukouvinis, M. Gavaises, Numerical investigation of bubble dynamics using tabulated data, International Journal of Multiphase Flow 93 (2017) 158–177.

[87] Y.-C. Wang, C.E. Brennen, Numerical computation of shock waves in a spherical cloud of cavitation bubbles, Journal of Fluids Engineering 121 (1999) 872–880.

[88] J. Ma, C.-T. Hsiao, G.L. Chahine, Euler–lagrange simulations of bubble cloud dynamics near a wall, Journal of Fluids Engineering 137 (2015).

[89] J. Ma, C.-T. Hsiao, G.L. Chahine, Shared-memory parallelization for two-way coupled Euler–lagrange modeling of cavitating bubbly flows, Journal of Fluids Engineering 137 (2015).

[90] J. Ma, C.-T. Hsiao, G.L. Chahine, Numerical study of acoustically driven bubble cloud dynamics near a rigid wall, Ultrasonics Sonochemistry 40 (2018) 944–954.
[91] D. Fuster, T. Colonius, Modelling bubble clusters in compressible liquids, Journal of Fluid Mechanics 688 (2011) 352–389.
[92] D. Ogloblina, S.J. Schmidt, N.A. Adams, Numerical simulation of collapsing vapor bubble clusters close to a rigid wall, in: Proceedings of the 10th International Symposium on Cavitation (CAV2018), ASME Press, 2018.
[93] D. Rossinelli, B. Hejazialhosseini, P. Hadjidoukas, C. Bekas, A. Curioni, A. Bertsch, et al., 11 PFLOP/s Simulations of Cloud Cavitation Collapse, Denver, Colorado, 2013.
[94] F. Wermelinger, U. Rasthofer, P.E. Hadjidoukas, P. Koumoutsakos, Petascale simulations of compressible flows with interfaces, Journal of Computational Science 26 (2018) 217–225.
[95] U. Rasthofer, F. Wermelinger, P. Hadijdoukas, P. Koumoutsakos, Large scale simulation of cloud cavitation collapse, Procedia Computer Science 108 (2017) 1763–1772.
[96] U. Rasthofer, F. Wermelinger, P. Karnakov, J.Š. Šukys, P. Koumoutsakos, Computational study of the collapse of a cloud with 120.16em0ex500 gas bubbles in a liquid, Physical Review Fluids 4 (6) (2019) 063602.
[97] S. Hickel, M. Mihatsch, S.J. Schmidt, Implicit Large Eddy Simulation of Cavitation in Micro Channel Flows, University of Warwick, Coventry, 2011.
[98] F. Örley, T. Trummler, S. Hickel, M.S. Mihatsch, S.J. Schmidt, N.A. Adams, Large-eddy simulation of cavitating nozzle flow and primary jet break-up, Physics of Fluids 27 (2015) 086101.
[99] C.P. Egerer, S.J. Schmidt, S. Hickel, N.A. Adams, Efficient implicit LES method for the simulation of turbulent cavitating flows, Journal of Computational Physics 316 (2016) 453–469.
[100] G.H. Schnerr, I.H. Sezal, S.J. Schmidt, Numerical investigation of three-dimensional cloud cavitation with special emphasis on collapse induced shock dynamics, Physics of Fluids 20 (2008) 040703.
[101] S. Schmidt, I. Sezal, G. Schnerr, M. Talhamer, Riemann techniques for the simulation of compressible liquid flows with phase-transistion at all mach numbers - shock and wave dynamics in cavitating 3d micro and macro systems, in: 46th AIAA Aerospace Sciences Meeting and Exhibit, American Institute of Aeronautics and Astronautics, 2008.
[102] I.H. Sezal, S.J. Schmidt, G.H. Schnerr, M. Thalhamer, M. Förster, Shock and wave dynamics in cavitating compressible liquid flows in injection nozzles, Shock Waves 19 (2009) 49–58.
[103] M. Dular, O. Coutier-Delgosha, Numerical modelling of cavitation erosion, International Journal for Numerical Methods in Fluids 61 (2009) 1388–1410.
[104] J. Decaix, E. Goncalvès, Compressible effects modeling in turbulent cavitating flows, European Journal of Mechanics - B/Fluids 39 (2013) 11–31.
[105] P. Koukouvinis, M. Gavaises, Simulation of throttle flow with two phase and single phase homogenous equilibrium model, Journal of Physics: Conference Series 656 (2015) 012086.
[106] J.L. Reboud, Y. Delannoy, Two-phase flow modeling of unsteady cavitation, Tokyo, 1994.
[107] H.W.M. Hoeijmakers, M.E. Janssens, W. Kwan, Numerical simulation of sheet cavitation, in: 3rd International Symposium on Cavitation, CAV 1998, Grenoble, France, April 7–10 (1998) 257–262.
[108] F. Örley, S. Hickel, S.J. Schmidt, N.A. Adams, Large-eddy simulation of turbulent, cavitating fuel flow inside a 9-hole diesel injector including needle movement, International Journal of Engine Research 18 (2017) 195–211.
[109] B. Budich, S.J. Schmidt, N.A. Adams, Numerical simulation and analysis of condensation shocks in cavitating flow, Journal of Fluid Mechanics 838 (2018) 759–813.
[110] A.H. Koop, Numerical simulation of unsteady three-dimensional sheet cavitation, in: Proceedings of the 7th International Symposium on Cavitation CAV2009, Ann Arbor, Michigan, USA, August 17–22, 2009.

[111] I. Sezal, Compressible dynamics of cavitating 3D multi-phase flows, PhD Thesis, Technical University of Munich (TUM), 2009.

[112] S.J. Schmidt, I.H. Sezal, Schnerr, Compressible simulation of high-speed hydrodynamics with phase change, TU Delft, The Netherlands, 2006.

[113] M.S. Mihatsch, S.J. Schmidt, N.A. Adams, Cavitation erosion prediction based on analysis of flow dynamics and impact load spectra, Physics of Fluids 27 (2015) 103302.

[114] F. Hempert, S. Boblest, M. Hoffmann, P. Offenhäuser, F. Sadlo, C.W. Glass, et al., High-pressure real-gas jet and throttle flow as a simplified gas injector model using a discontinuous Galerkin method, in: High Performance Computing in Science and Engineering 16, Cham, 2016.

[115] M. Hoffmann, An explicit discontinuous galerkin method for parallel compressible two-phase flow simulations, PhD Thesis, Universität Stuttgart, 2017.

[116] P.-O. Persson, J. Peraire, Sub-cell shock capturing for discontinuous Galerkin methods, in: Aerospace Sciences Meetings, American Institute of Aeronautics and Astronautics, 2006.

[117] M. Sonntag, C.-D. Munz, Efficient parallelization of a shock capturing for discontinuous Galerkin methods using finite volume sub-cells, Journal of Scientific Computing 70 (2017) 1262–1289.

[118] F. Hempert, S. Boblest, T. Ertl, F. Sadlo, P. Offenhäuser, C.W. Glass, et al., Simulation of real gas effects in supersonic methane jets using a tabulated equation of state with a discontinuous Galerkin spectral element method, Computers & Fluids 145 (2017) 167–179.

[119] D. Fuster, G. Agbaglah, C. Josserand, S. Popinet, S. Zaleski, Numerical simulation of droplets, bubbles and waves: state of the art, Fluid Dynamics Research 41 (2009) 065001.

[120] T. Theofanous, L. Biasi, H.S. Isbin, H. Fauske, A theoretical study on bubble growth in constant and time-dependent pressure fields, Chemical Engineering Science 24 (1969) 885–897.

[121] R.F. Kunz, D.A. Boger, D.R. Stinebring, T.S. Chyczewski, J.W. Lindau, H.J. Gibeling, et al., A preconditioned Navier-Stokes method for two-phase flows with application to cavitation prediction, Computers & Fluids 29 (2000) 849–875.

[122] C.L. Merkle, J.Z. Feng, P.E.O. Buelow, Computational modeling of the dynamics of sheet cavitation, in: Proceedings of 3rd International Symposium on Cavitation, Grenoble, 1998.

[123] G.H. Schnerr, J. Sauer, Physical and numerical modeling of unsteady cavitation dynamics, in: ICMF-2001, 4th International Conference on Multiphase Flow, New Orleans, USA, 2001.

[124] P.J. Zwart, A.G. Gerber, T. Belamri, A two-phase flow model for predicting cavitation dynamics, in: Proceedings of the ICMF-2004, 5th International Conference on Multiphase Flow, ICMF, 2004.

[125] P.J. Zwart, Numerical Modelling of Free Surface Flows and Cavitating Flows, VKI Lucture Series, Ansys Canada Ltd, Canada, 2005, p. 25.

[126] W.S. Senocak, A pressure-based method for turbulent cavitating flow computations, Journal of Computational Physics 176 (2) (2002) 363–383.

[127] Y. Matsumoto, T. Kanbara, K. Sugiyama, Y. Tamura, Numerical study of cavitating flow structure on a hydrofoil, in: 4th KSME-JSME Fluids Eng Conf., Pusan, 1998.

[128] G. Cazzoli, S. Falfari, G.M. Bianchi, C. Forte, C. Catellani, Assessment of the cavitation models implemented in OpenFOAM® under DI-like conditions, Energy Procedia 101 (2016) 638–645.

[129] L. Hanimann, L. Mangani, E. Casartelli, M. Widmer, Cavitation modeling for steady-state CFD simulations, IOP Conference Series: Earth and Environmental Science 49 (2016) 092005.

[130] W. Yuan, G.H. Schnerr, Cavitation in injection nozzles - effect of injection pressure fluctuations, in: CAV2001, 4th Int. Symposium on Cavitation, Pasadena, 2001.

[131] M.P. Kinzel, J.W. Lindau, R.F. Kunz, A level-set approach for large scale cavitation, in: 2009 DoD High Performance Computing Modernization Program Users Group Conference, 2009.

[132] M.P. Kinzel, J.W. Lindau, R.F. Kunz, A multiphase level-set approach for all-Mach numbers, Computers & Fluids 167 (2018) 1–16.
[133] T. Michael, J. Yang, F. Stern, Sharp interface cavitation modeling using volume-of-fluid and level set methods, Proceedings of the ASME 2013 Fluids Engineering Division Summer Meeting, vol. 2, Incline Village, Nevada, USA, July 7–11, 2013, V002T06A008. ASME. Available from: https://doi.org/10.1115/FEDSM2013-16479.
[134] T. Michael, J. Yang, F. Stern, A sharp interface approach for cavitation modeling using volume-of-fluid and ghost-fluid methods, Journal of Hydrodynamics, Series B 29 (2017) 917–925.
[135] S. Dabiri, W.A. Sirignano, D.D. Joseph, Cavitation in an orifice flow, Physics of Fluids 19 (2007) 072112.
[136] W. Yue, C.-L. Lin, V.C. Patel, Large eddy simulation of turbulent open-channel flow with free surface simulated by level set method, Physics of Fluids 17 (2005) 025108.
[137] C.-T. Hsiao, J. Ma, G. Chahine, Simulation of sheet and tip vortex cavitation on a rotating propeller using a multiscale two-phase flow model, in: Fourth International Symposium on Marine Propulsors, Austin, Texas, June 2015.
[138] C.-T. Hsiao, J. Ma, G.L. Chahine, Multiscale tow-phase flow modeling of sheet and cloud cavitation, International Journal of Multiphase Flow 90 (2017) 102–117.
[139] A. Kapahi, C.-T. Hsiao, G.L. Chahine, A multi-material flow solver for high speed compressible flows, Computers & Fluids 115 (2015) 25–45.
[140] J. Ma, C.-T. Hsiao, G.L. Chahine, A physics based multiscale modeling of cavitating flows, Computers & Fluids 145 (2017) 68–84.
[141] C.-T. Hsiao, G. Chahine, Effect of a propeller and gas diffusion on bubble nuclei distribution in a liquid, Journal of Hydrodynamics, Series B 24 (2012) 809–822.
[142] G. Strotos, P. Koukouvinis, A. Theodorakakos, M. Gavaises, G. Bergeles, Transient heating effects in high pressure diesel injector nozzles, International Journal of Heat and Fluid Flow 51 (2015) 257–267.
[143] A. Kubota, H. Kato, H. Yamaguchi, A new modelling of cavitating flows: a numerical study of unsteady cavitation on a hydrofoil section, Journal of Fluid Mechanics 240 (1992) 59–96.
[144] A.R. Jamaluddin, G.J. Ball, C.K. Turangan, T.G. Leighton, The collapse of single bubbles and approximation of the far-field acoustic emissions for cavitation induced by shock wave lithotripsy, Journal of Fluid Mechanics 677 (2011) 305–341.
[145] K. Ando, T. Colonius, C.E. Brennen, Numerical simulation of shock propagation in a polydisperse bubbly liquid, International Journal of Multiphase Flow 37 (2011) 596–608.
[146] L. Yuan, Sonochemical effects on single-bubble sonoluminescence, Physical Review E 72 (2005) 1–14.
[147] W.C. Moss, J.L. Levantin, A.J. Szeri, A new damping mechanism in strongly collapsing bubbles, Proceedings of the Royal Society of London A: Mathematical, Physical and Engineering Sciences 456 (2000) 2983–2994.
[148] J. Buttenbender, P. Pelz, The influence of imposed strain rate and circulation on bubble and cloud dynamics, in: CAV2012, Singapore, 2012.
[149] Y. Lu, J. Katz, A. Prosperetti, Generation and transport of bubble clouds in high-intensity focused ultrasonic fields, in: CAV2012, Singapore, 2012.
[150] A. Alajbegovic, H.A. Grogger, H. Philipp, Calculation of transient cavitation in nozzle using the two-fluid model, in: 12th ILASS-America, United States, 1999.
[151] M. Cristofaro, W. Edelbauer, P. Koukouvinis, M. Gavaises, Influence of diesel fuel viscosity on cavitating throttle flow simulations under erosive operation conditions, ACS Omega 5 (2020) 7182–7192.

[152] J.R. García-Cascales, H. Paillère, Application of AUSM schemes to multi-dimensional compressible two-phase flow problems, Nuclear Engineering and Design 236 (2006) 1225–1239.
[153] M. Vujanović, Z. Petranović, W. Edelbauer, J. Baleta, N. Duić, Numerical modelling of diesel spray using the Eulerian multiphase approach, Energy Conversion and Management 104 (2015) 160–169.
[154] R. Saurel, F. Petitpas, R. Abgrall, Modelling phase transition in metastable liquids: application to cavitating and flashing flows, Journal of Fluid Mechanics 607 (2008) 313–350.
[155] J. Dumond, F. Magagnato, A. Class, Stochastic-field cavitation model, Physics of Fluids 25 (7) (2013).
[156] M.B. Liu, J.R. Shao, Z. Shang, SPH modeling of supercavity induced by underwater high speed objects, in: 8th International Symposium on Cavitation, Singapore, August 13–16, 2012.
[157] D. Varas, R. Zaera, J. López-Puente, Numerical modelling of the hydrodynamic ram phenomenon, International Journal of Impact Engineering 36 (3) (2009) 363–374.
[158] D.M. Bourg, Development of the distributed points method with application to cavitating flow, University of New Orleans Theses and Dissertations 904, 2008. Available from: https://scholarworks.uno.edu/td/9042008.
[159] G. Falcucci, S. Ubertini, G. Bella, S. Succi, Lattice Boltzmann simulation of cavitating flows, Communications in Computational Physics 13 (2013) 685–695.
[160] G. Wang, I. Senocak, W. Shyy, T. Ikohagi, S. Cao, Dynamics of attached turbulent cavitating flows, Progress in Aerospace Sciences 37 (2001) 551–581.
[161] D. Fuster, A review of models for bubble clusters in cavitating flows, Flow, Turbulence and Combustion 102 (2019) 497–536.
[162] O. Coutier-Delgosha, J.L. Reboud, Y. Delannoy, Numerical simulation of the unsteady behaviour of cavitating flows, International Journal for Numerical Methods in Fluids 42 (2003) 527–548.
[163] J.-L. Reboud, B. Stutz, O. Coutier, Two phase flow structure of cavitation: experiment and modeling of unsteady effects, in: 3rd International Symposium on Cavitation CAV1998, Grenoble, 1998.
[164] O. Coutier-Delgosha, R. Fortes-Patella, J.L. Reboud, Evaluation of the turbulence model influence on the numerical simulations of unsteady cavitation, Journal of Fluids Engineering 125 (2003) 38–45.
[165] J.-B. Leroux, O. Coutier-Delgosha, J.A. Astolfi, A joint experimental and numerical study of mechanisms associated to instability of partial cavitation on two-dimensional hydrofoil, Physics of Fluids 17 (2005) 052101.
[166] O. Coutier-Delgosha, R. Fortes-Patella, J.L. Reboud, Simulation of unsteady cavitation with a two-equation turbulence model including compressibility effects, Journal of Turbulence 3 (2002) N58.
[167] B. Huang, A. Ducoin, Y.L. Young, Physical and numerical investigation of cavitating flows around a pitching hydrofoil, Physics of Fluids 25 (2013) 102109.
[168] P. Koukouvinis, H. Naseri, M. Gavaises, Performance of turbulence and cavitation models in prediction of incipient and developed cavitation, International Journal of Engine Research 18 (2017) 333–350.
[169] D.F. Gaitan, L.A. Crum, C.C. Church, R.A. Roy, Sonoluminescence and bubble dynamics for a single, stable, cavitation bubble, The Journal of the Acoustical Society of America 91 (1992) 3166–3183.
[170] W.C. Moss, D.B. Clarke, J.W. White, D.A. Young, Hydrodynamic simulations of bubble collapse and picosecond sonoluminescence, Physics of Fluids 6 (1994).
[171] B. Tully, N. Hawker, M. Betney, Y. Ventikos, Simulation of warm dense matter in intense bubble collaps, in: Proceedings of Meetings on Acoustics, Canada, 2013.
[172] K.S. Suslick, Sonochemistry, Science (New York, N.Y.) 247 (1990) 1439–1445.
[173] R.D. Lockett, M. Jeshani, An experimental investigation into the effect of hydrodynamic cavitation on diesel, International Journal of Engine Research 14 (2013) 606–621.

Chapter 2

Nonspherical Collapse of Single Bubbles Near Boundaries and in Confined Spaces

Fabian Reuter and Claus-Dieter Ohl
Department for Soft Matter, Institute for Physics, Otto-von-Guericke University, Magdeburg, Germany

2.1 Introduction

In technical systems, cavitation most often appears in clouds of mutually interacting bubbles, resulting in rather complex phenomena, in particular adjacent to some boundaries, for example, from a container that confines the liquid carrying the cavitation. The significant effects of cavitation are basically the result of the formation, growth, and implosive collapse of the cavitation bubbles. Thus all action is produced at the active bubble sites. Therefore a systematic investigation of the fundamentals of cavitation requires the consideration of single collapsing bubbles, which is the topic of this chapter. In addition to their relevance as the smallest constituents of large-scale cavitation systems, single bubbles can be used for targeted manipulation in microfluidic environments. Because bubbles are ubiquitous and important to many applications, knowledge about them has grown over the years, and several reviews and books on the topic have been written, such as those of Cole [1], Plesset and Prosperetti [2], Blake and Gibson [3], Lauterborn and Kurz [4], Leigthon [5], and

Cavitation and Bubble Dynamics. DOI: https://doi.org/10.1016/B978-0-12-823388-7.00003-5

Yasui [6]. In this chapter we focus on the interaction of single bubbles with a single boundary and few boundaries to connect earlier work with recent findings.

2.2 Spherical Collapse

In general, a bubble can be considered to be a gas-filled void in a liquid, and at first it is helpful to examine the spherical case. In the simplest case of an empty, spherical bubble in static equilibrium in the absence of any external forces, the pressure in a bubble of radius R_{eq} reads according to the Young-Laplace equation:

$$p = 2\sigma \frac{1}{R_{eq}} + p_0 \tag{2.1}$$

where σ is the liquid's surface tension and p_0 is the ambient pressure. When the bubble is deflected from its equilibrium radius, it can start to oscillate. The corresponding equation of oscillation for an empty cavity in an incompressible liquid was first formulated by Rayleigh and can be written as follows [7]:

$$R\ddot{R} + \frac{3}{2}\dot{R}^2 = \frac{p(R) - p_0}{\rho} \tag{2.2}$$

where $R = R(t)$ and ρ are the radius and the density of the (incompressible) liquid, respectively. The evident assumption of isotropy implied by the sphericity of the problem allows us to describe the bubble oscillator entirely by its radius $R(t)$ as a function of time, reducing the problem to a one-dimensional one. Only recently, Kudryashov and Sinelshchikov [8] derived an analytical solution for Eq. (2.2), while Obreschkow et al. [9] gave an analytical, approximate solution of the time evolution of the normalized radius $r(t)$ as $r(t) = (1 - t^2)^{2/5}$. Here, the time t is normalized by $T_L/2$ [see Eq. (2.3)]. This simple empty-bubble model of Eq. (2.2), which neglects energy losses, gives many useful insights into bubble dynamics. For example, the bubble lifetime T_L, that is, the time including the expansion phase up to the maximum radius R_{max} and the contraction phase until to collapse, can be derived as follows:

$$T_L = 2 \cdot 0.915 R_{max} \sqrt{\rho / p_0}. \tag{2.3}$$

In Eq. (2.3) the expansion and contraction phases are assumed to have the same duration, resulting in the factor of 2.

A refined equation for the spherical bubble dynamics that proved useful in a wide range of applications and that today can be considered a standard model is the Rayleigh-Plesset equation (see Plesset [10]). It is derived from the Navier-Stokes equations for a free unbounded infinite-domain liquid (with the bubble center at rest) for a nonempty homogeneous bubble in an incompressible, viscous fluid, neglecting mass fluxes through the bubble wall. It can be written in the form of Eq. (2.4), illustrating individual pressure contributions from velocity, the external pressure, surface tension, viscosity, and gas pressure in the bubble (see Bogoyavlenskiy [11]):

$$\rho \ddot{R} R = (P_{\text{vel}} + P_{\text{surf}} + P_{\text{ext}}) - (P_{\text{vis}} + P_{\text{gas}}). \tag{2.4}$$

The terms are $P_{\text{vel}} = \frac{3}{2}\rho\dot{R}^2$, which results from the dynamic pressure plus the geometric effects from the spherical topology, and $P_{\text{surf}} = 2\sigma/R$ and $P_{\text{ext}} = p_0 + p_{\text{a}} + R\dot{p}_{\text{a}}/c$, where p_0 is the ambient (hydrostatic) pressure and p_{a} is a time-dependent (e.g., acoustic) driving pressure. The term contributed by viscosity is $P_{\text{vis}} = -4\eta\dot{R}/R$, where ρ is the mass density, η is the liquid's viscosity, and c the speed of sound in the liquid. The formulation of the total gas pressure $P_{\text{gas}} = P_{\text{g}} + R\dot{P}_{\text{g}}/c$ can be intricate and must be calculated via an appropriate equation of state for the gas, for which an adiabatic van der Waals model (at the bubble interface) is often used. In principle, the velocities and pressures anywhere in the liquid can be derived from the spherical bubble models.

A variety of bubble models have been developed to include further physical aspects. For example, for particularly violent collapses, that is, when strong shock waves are emitted during the collapse, which is the case for laser-induced or spark-induced bubbles, the Gilmore model [12] is appropriate, as it considers the enthalpy and compressibility of the surrounding liquid (via the Tait equation of state) and thus accounts for the emission of acoustic waves. In general, the energy of a spherical shock wave is given by Robert [1]

$$E_{\text{SW}} = \frac{4\pi r^2}{\rho_0 c_0} \int p_{\text{SW}}^2 \, dt. \tag{2.5}$$

Here, r^2 is the distance from the shock wave emission center (i.e., about the bubble center), and p_{SW}^2 is the shock wave pressure (as a function of time). Comparing E_{SW} to the total bubble energy E_{b} of Eq. (2.6) shows that during violent collapses, a bubble loses a large part of its energy by the emission of the shock wave. The bubble energy is determined mainly by the volume work done against the ambient pressure to expand the cavity, given by

$$E_{\text{b}} = \frac{4}{3}\pi R_{\max}^3 (p_0 - p_{\text{v}}) \tag{2.6}$$

where p_{v} denotes the vapor pressure. The contribution from surface tension is $4\pi\sigma R_{\max}^2$, which can be neglected most of the time.

A variety of models of bubble oscillation exist that include further phenomena, such as fluxes through the bubble interface (mass or heat), nonhomogeneous bubble interior, and chemical reaction. Overviews are given by Leighton [5], Brenner et al. [13], and Fuster [14].

2.3 Techniques for the Experimental Generation of Single Bubbles

In large-scale systems, cavitation is formed in low-pressure regions, that may stem from fluid flows or vibrations. It typically occurs in cavitation clouds, where the formation of the individual bubbles is hardly predictable. However, for the systematic investigation of cavitation it is necessary to produce single bubbles at controlled locations and adjustable times. The aim of this section is to elucidate the available techniques for single-bubble generation and discuss their differences and respective advantages from a practical point of view. The following considerations are made for water at atmospheric pressure unless otherwise noted, but they can be transferred to other configurations, as will be discussed later in the chapter.

Experimental techniques to produce single, violently collapsing cavitation bubbles of defined properties close to boundaries typically rely on either the local deposition of energy at the center of the nascent bubble or the production of a stable gaseous bubble that is subsequently collapsed by a sudden change of the ambient pressure. The energy deposition techniques utilize phase explosions induced by breakdown from alternating electromagnetic fields (optic cavitation) or induced by breakdown from static electric fields (spark-induced cavitation), alternatively by superheating the liquid through lasers, heaters, or wire contacts; or by the detonation of explosives (underwater explosion, or UNDEX). The latter techniques usually result in large and buoyant bubbles, and research is also motivated by military interests [1], but much smaller bubbles with sizes down to $R_{max} \approx 15$ cm [15] can also be obtained. Spark-induced cavitation was introduced by Naudé and Ellis [16] and used also in the pioneering work of Shutler and Mesler [17]. Thereafter, it became a standard method used by many groups. Between two electrodes submerged in liquid, a high voltage is applied such that by electrical breakdown a plasma spark is generated in the liquid, around which a cavitation bubble forms. It is a relatively easily accessible technique, as it requires few experimental resources; a capacitor bank can provide the current at high voltage, and wires can be employed as electrodes. In typical configurations their faces are aligned coaxially, and resistant materials such as tungsten are employed. Electrode diameters range between several hundreds of micrometers up to centimeters. Common disadvantages are a somehow limited repeatability of bubble size, as wear or even complete destruction of the electrode tips can occur, and spurious effects from electrolysis or Joule's heating cannot be ruled out. The sphericity of the bubbles is limited by the geometric interference of the electrodes with the growing bubble. Common bubble sizes obtained by this technique lie between $R_{max} = 10$ mm and $R_{max} = 40$ mm (at atmospheric pressure) and can be adjusted within some range by limiting the current, setting the discharge voltage or varying electrode and gap size.

In optic cavitation the energy is fed through the liquid with light. A pulsed laser is focused, often by a lens system or a parabolic mirror, to exceed the threshold for optical breakdown at the focal region. The dimensions of the focal region are typically in the micrometer range; the lower physical limit is set by the wavelength of light and the numerical aperture of the focusing optics. At the focal spot, as a result of the optical breakdown, a nonthermal plasma is generated. Owing to the short laser pulse duration of nanoseconds and below, enormous instantaneous powers of megawatts and above are acting on the liquid. As a consequence, thermal confinement is achieved, which leads to a phase explosion [18] because the rate of volumetric energy deposition provided by laser radiation is more rapid than the rate of energy consumed by vaporization and normal boiling, the details of which were carefully reviewed by Vogel and Venugopalan [19]. It should be emphasized that the involved processes are fundamentally different from producing a bubble with a continuous laser by superheating the liquid, sometimes called thermocavitation, a process that is described elsewhere [20]. Similarly to thermocavitation, superheating can be achieved with small heaters [21] and the very popular method of Joule's heating through thin electric wires submerged and in contact (see, e.g., Kannan et al. [22]). In contrast, for optical breakdown, there are fluence and irradiance thresholds, which range from 10^{13} to $10^{11}\,\mathrm{W/cm^2}$ and from 10^{-1} to $10^3\,\mathrm{J/cm^2}$, increasing with the duration of the laser pulse, that is, from femtosecond-long pulses to nanosecond-long pulses, as calculated by Vogel et al. [23].

The maximum radius of a laser-induced bubble depends on the focal spot size and the numerical aperture and can, for a given optics, be controlled mainly via the laser pulse duration and pulse energy. The maximum bubble radius roughly scales as $R_{\mathrm{max}} \sim (E_L - E_{\mathrm{th}})^{1/3}$, where E_L is the laser pulse energy and E_{th} is the threshold of optical breakdown below which no bubble is generated [24]. Typical sizes range from $R_{\mathrm{max}} = 400\ \mu m$ to $R_{\mathrm{max}} = 1200\ \mu m$ for nanosecond laser pulses, but even bubbles with maximum radii down to $R_{\mathrm{max}} \approx 200$ nm can be achieved with femtosecond laser pulses [25]. Since its introduction for the study of single bubbles by Lauterborn [26], the optic cavitation technique has become widely employed, since it is noninvasive and allows precise control of bubble seeding in space and time (micrometer and submicrosecond precision, respectively). However, to produce spherical bubbles, a spherical plasma has to be generated, which may be a nontrivial task. Yet, only in a spherical plasma can the nonsphericities induced, for example, by nearby boundaries be studied without interference from artefactual asphericities from imperfect bubble seeding. In our experience, a plasma of nonspherical shape can still result to a bubble that seems to be mostly spherical at maximum size, that is, at R_{max}; however, it seems to memorize the nonspherical seeding and thus collapses again aspherically. Therefore to reliably check for spherical bubble production, the sphericity of the plasma shape during bubble inception has to be observed. An example of a plasma with subsequent bubble formation is presented in Fig. 2.1. Here, the bubble dynamics can be nicely modeled with the Rayleigh-Plesset equation [Eq. (2.4)], but to also model its dynamics after the first collapse (not shown in the time series), the Gilmore model provides a more accurate description.

Sources [27] for aspherical plasma generation can be impurities in the liquid as well as optical aberrations that lead to plasma fragmentation. In addition, the plasma tends to grow within the light-focusing cone toward the focusing lens, resulting in a conical plasma shape. Its length depends critically on the focusing angle of the objective (i.e., the numerical aperture, or NA), and it is more elongated for low NAs. It further depends on the pulse energy: the higher the energy, the longer the plasma cone, and on the pulse duration it is more elongated for shorter pulses. These dependencies have been revealed in great detail by Vogel et al. [28]. Particularly conical plasmas were later studied by Tagakawa et al. [29] and Sinibaldi et al. [30].

A further technique for bubble generation consists of the initial placement of a sufficiently stable gaseous bubble that is subsequently collapsed by a sudden pressure change (e.g., from a shock wave) or an impulsive acceleration of the boundary to which the bubble adheres, as was done, for example, by Benjamin and Ellis [31]. The stable bubble can be generated at an electrode by electrolysis of water or with the help of a syringe to deposit some amount of gas. Alternatively, it can be grown by rectified diffusion in a sound field. However, these techniques do not allow for the same repeatability of bubble sizes and seeding loci.

In former times, it was debated whether the single bubbles generated with the aforementioned methods had the same behavior as cavitation bubbles in hydrodynamic or acoustic cavitation. It is now widely accepted that they do, as the bubble dynamics in all cases can be consistently described by the respective oscillation models, such as Rayleigh-Plesset or Gilmore. Still, there are some general differences regarding the collapse behavior for bubbles generated by the various methods. One important variation is the amount of (non)condensable gases in the

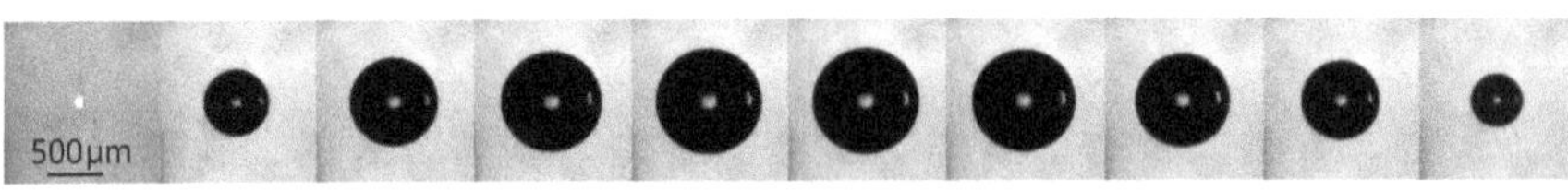

Figure 2.1 Time evolution of a laser-induced cavitation bubble in water (interframe time: 10 μs). In the first frame, a bright, spherical plasma is visible. It is produced by a nanosecond-long laser pulse focused by a custom-made focusing objective with a long working distance and numerical aperture >0.4. A spherical cavitation bubble develops around the plasma spot. *Image reproduced from F. Reuter. Collapsing Bubbles, Induced Flow Phenomena, and Their Application in Ultrasonic Cleaning (Ph.D. thesis), Georg-August-Universität Göttingen, Germany, 2017.*

bubble. This can be seen from the compression ratios of bubbles and subsequent shock wave peak pressures at generation and collapse or from comparison of the rebound radius with the maximum radius. A bubble with a large content of noncondensable gases will show a decreased shock wave peak pressure in comparison to the peak at seeding, because of damping effects but also sphericity may play a role. In addition, the rebound radius will be larger. In water, among the UNDEX, spark-induced, and laser-induced bubbles, UNDEX bubbles contain the most noncondensable gases, and laser-induced bubbles contain the least, while spark-induced bubbles lie in between (compare Refs. [15,32–36]). The noncondensable gases in the bubble may stem from electrolysis, vaporization of the electrodes, or explosive residues, respectively. Laser-induced bubbles in water are considered to contain the least amount of noncondensable gases or vapor as a result of thermal confinement during generation. Larger bubbles sizes can be achieved with the above-mentioned techniques *NULL* when generating the bubble at reduced ambient pressure, for example, in a vacuum chamber. Doing so will also increase the gas content in the bubble [37]. However, details of the mass transfer through the bubble interface during its dynamics, for example, by vaporization, condensation, and diffusion, present an intricate problem. More on that topic can be found, for example, in the work of Fujikawa [38].

2.4 Nonspherical Collapse

When a bubble oscillates and collapses in anisotropic surroundings, the spherical symmetry of the bubble is broken. Consequently, the bubble develops nonspherical flows, and a variety of effects and fluid dynamic phenomena emerge. The origins of anisotropies include pressure gradients, for example, due to hydrostatic pressure; shear flows or acoustic waves; and, very important, a confinement of the liquid domain by a nearby boundary. The number of possible boundaries is endless. A boundary can be a wall of any geometry and any kind, rigid or flexible, or the interface of a droplet or another bubble. The most extensively studied case resulting in nonspherical bubble dynamics at boundaries concerns the collapse near a rigid, infinitely extended plane wall. This case also has important model characteristics for cavitation erosion in technical applications, and we will discuss this configuration in the next section, together with some general aspects and phenomena that accompany the aspherical bubble collapse.

2.4.1 Infinitely Extended Plane Rigid Boundary

Our current understanding of cavitation and bubble dynamics is based on the early work of Plesset and Chapman [39], Lauterborn and Bolle [40], Blake et al. [41], and Tomita and Shima [42]. Since then, it has become known that bubble dynamics depend crucially on the maximum bubble radius and the standoff distance of the bubble from the boundary. It is common to describe the standoff in a dimensionless form as $\gamma = d/R_{\max}$ (see Fig. 2.2).

The presence of the boundary prolongs the Rayleigh lifetime [Eq. (2.3)]. Via perturbation theory, Rattray [43] (see also Plesset and Chapman [39]) derived a prolongation factor $k(\gamma) = 1 + \frac{0.205}{\gamma}$, so the collapse time at a rigid boundary is as follows:

$$T_L \approx \left(1 + \frac{0.205}{\gamma}\right) \cdot \left(2 \cdot 0.915 R_{\max} \sqrt{\rho/p_0}\right). \tag{2.7}$$

Experimental data from Vogel et al. [44] for $\gamma > 0.5$ reveal good agreement with Eq. (2.7).

In addition to the mere prolongation of the lifetime, the bubble dynamics reveals diverse phenomena and induces directed flows at and close to the boundary. An overview of the dynamics of the shape of a laser-induced bubble at a rigid (glass) boundary in water is presented in Fig. 2.3 for a standoff of $\gamma = 1.21$.

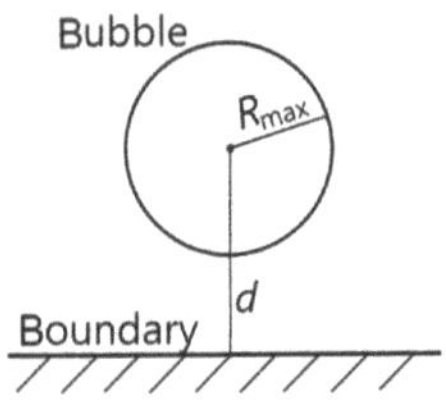

Figure 2.2 Definition of the nondimensionalized standoff distance $\gamma = d/R_{\max}$, where d is the distance from the bubble center to the boundary and $R_{\max}$ is the maximum bubble radius.

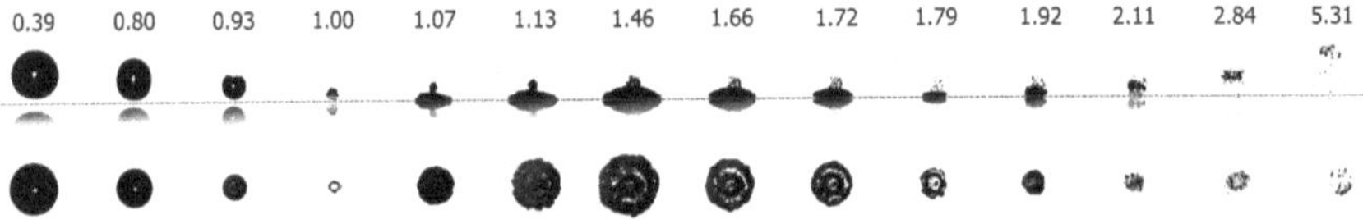

Figure 2.3 Synchronous recording of bubble shape dynamics in side view (top row) and bottom view (bottom row) through a glass substrate onto the laser-induced bubble ($R_{\max} = 414$ μm, $\gamma = 1.21$, image exposure time 360 ns, image frame dimensions about 1.7×1.5 mm^2). In the top view, the boundary runs horizontally at the bottom, indicated by the dashed line. Times above each pair of frames are given normalized on the bubble lifetime ($T_L = 85$ μs), where $t = 0$ corresponds to the instant of plasma generation. *Image adapted from F. Reuter, S.R. Gonzalez-Avila, R. Mettin, C.-D. Ohl, Flow fields and vortex dynamics of bubbles collapsing near a solid boundary, Physical Review Fluids 2 (6) (2017) 064202.*

A side view and a bottom view through the glass substrate onto the bubble were synchronously recorded. Despite the relative proximity to the boundary, the initial bubble oscillation is largely spherically, such that until the first collapse, the temporal evolution of the radius can be quite accurately described by the Rayleigh-Plesset or Gilmore equation. This allows estimation of the equilibrium radius of the bubble as well as determination of the bubble lifetime in case the bubble dynamics are temporarily undersampled, which may often be the case even with high-speed videocameras. Here, an equilibrium radius of $R_{eq} \approx 25\ \mu m$ is estimated with the help of the Gilmore model. At $t \simeq 0.50$ the bubble is at its maximum expansion of $R_{max} = 414\ \mu m$; thus it has significantly exceeded its equilibrium radius. At this standoff, only during the late collapse phase does the proximity of the boundary result in visual deformation of the bubble. In the side view, at $t = 0.80$, the bubble centroid is accelerating toward the boundary, and the bubble is a little flattened at its bottom. At $t = 0.93$ the bubble is indented on its top pole from a liquid jet toward the boundary that pierces the bubble axially and subsequently shapes it into a torus. It is produced from a directed liquid flow toward the bubble center that is caused by the boundary hindering the radial inflow. In this way the rigid boundary becomes effectively attractive for the shrinking bubble. Consequently, the bubble collapses as a torus onto the boundary. There, it rebounds with a flatter shape and shows a considerably rougher gas-liquid interface. From $t = 1.46$ onward, the second compression period begins, ending with another toroidal collapse (between $t = 1.79$ and $t = 1.92$) and a subsequent liftoff of the remnant bubbles from the boundary. Because of energy losses at the first collapse, mainly via shock wave emission, which is reported to amount to about 70%–84% of the bubble energy [42,44], the duration of the second oscillation period is reduced. Along with the bubble collapse, a vortex ring is generated. This long-lasting flow structure traps bubble fragments in its core and transports them away from the boundary into the bulk. For bubbles that are generated farther away than $\gamma = 1.33 \pm 0.03$, the ring vortex is generated with an opposite rotation direction, consequently spreading the bubble fragments over the solid [45].

In the imaging series shown in Fig. 2.3, collimated light was used to image the bubble outline. To better resolve the jet formation within the bubble and its piercing through the bubble, a more diffuse illumination is suitable (see Fig. 2.4). Recently, Koch et al. [47] studied the illumination of the jet inside the bubble through the highly curved bubble interface by a ray-tracing approach.

A particular violent but short-lived phenomenon escaped the imaging series above: the shock wave emission during bubble collapse. The shock wave can be made visible by using ultrashort exposure times (see Fig. 2.5) as pressure-induced changes in the refractive index from compression of the liquid allow for imaging of the shock front (shadowgraphy). Fig. 2.5 shows a laser-induced bubble similar to the one of Fig. 2.3 but at a slightly larger standoff distance and only around the very moment of collapse.

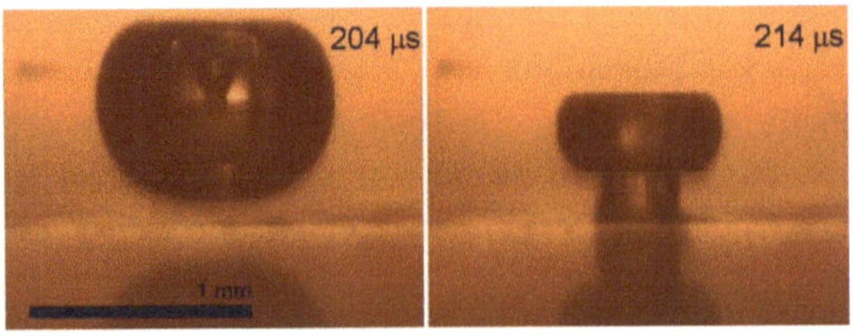

Figure 2.4 Picture of the axial liquid jet piercing the bubble during its collapse ($\gamma \approx 1.1$, exposure time 10 ns). *Image is reproduced from C.-D. Ohl, M. Arora, R. Dijkink, V. Janve, D. Lohse, Surface cleaning from laser-induced cavitation bubbles, Applied Physics Letters 89 (7) (2006) 074102.*

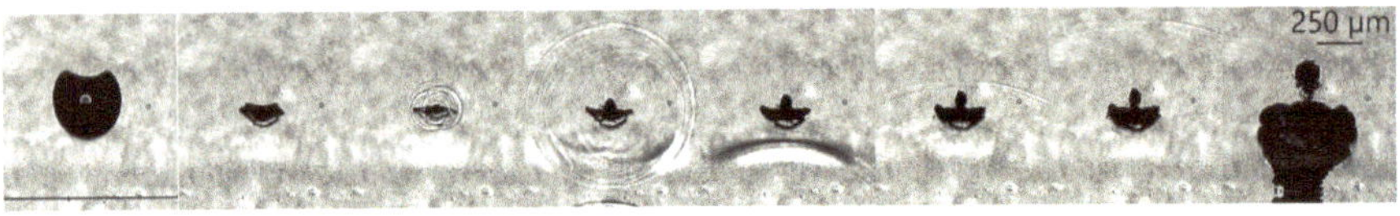

Figure 2.5 Shock wave emission from jet impact on the lower bubble wall and from the ring collapse in the vicinity of a rigid boundary (horizontally at the bottom), illuminated with a collimated femtosecond laser light (pulse duration: 220 fs). The first and the last frame are staggered in time while the center frames are separated at equidistant intervals of 200 ns, such that the frame times are −2.56 μs, −150 ns, 50 ns, 250 ns, 450 ns, 650 ns, 850 ns, and 9.25 μs with respect to the instant of collapse, $t = 0$.

In the first frame, the bubble is indented on the top pole by the axial jet. In the second frame, a shock wave from the jet impact onto the opposite bubble wall is visible. A complex emission of shock waves is produced by the torus collapse at $t = 50$ ns. The shock waves propagate in the successive frames, and after their reflection at the boundary they strike the bubble anew. With increasing propagation distance, the shock wave amplitudes decrease, thus the contrast of the shock front decreases. In the last frame, the jet is impacting onto the rigid boundary.

Thus as compared to the idealized case of a spherical bubble, here a shock wave is emitted from several loci. The shock wave impacts on other parts of the nonspherical bubble directly, again after reflection at the boundary. This can even lead to generation of secondary cavitation, in particular at the axis of symmetry, where shock waves from the toroidal collapse interfere. Recently, a numerical study of the microscopic details of shock wave emission was reported by Lechner et al. [48]. A systematic experimental parameter study of the peak pressures and energies of emitted shock waves as a function of ambient anisotropy was done by Supponen et al. [49]. Shock wave peak pressures are of the order of some hundreds of bars and are reported to scale linearly with $R_{\max}$ [4,44]. However, their exact value is hardly accessible from experiments because of the limited bandwidth of pressure sensors, confinement, and the number of loci of the shock wave emission while requiring a remote placement of the hydrophone to avoid geometric interference with the bubble dynamics [note the r^2-dependence in Eq. (2.5)]. The compression ratio of the gas phase that is achieved during collapse (as compared to the maximum bubble volume) should correlate with the peak pressures. Lindau and Lauterborn [50] found minimum compression ratios for $\gamma \rightarrow 0.6$, implying smaller shock energies. More detailed information on the shock wave from spherical and aspherical collapses can be found in Lauterborn and Kurz [4] and Lauterborn and Vogel [51] and references therein.

Since the axial jet was first hypothesized by Kornfeld and Suvorov [52] and later found by Naudé and Ellis [16] as well as by Shutler and Mesler [17], Benjamin and Ellis [31], and by Plesset and Chapman [39], a considerable amount of research studying the jet has been conducted (see, e.g., Lauterborn and Kurz [4], Lauterborn and Bolle [40], Pearson et al. [53]).

The radius of the axial jet can be assumed to scale linearly as $\frac{1}{10}R_{\max}$ [54] but can critically depend on γ. For a typical single bubble collapsing at ambient pressure, jet velocities are of the order of 80–175 ms^{-1}, also depending on γ.

From Plesset and Chapman's potential flow description, thus neglecting surface tension and viscosity, it follows that the Navier-Stokes equations for the bubble become self-similar, by scaling lengths with $R_{\max}$, pressures with Δp, and velocities with $\sqrt{\Delta p/\rho}$, where ρ_0 is the mass density of the (incompressible) liquid and Δp is the pressure difference that drives the bubble collapse. In particular, it follows that the jet velocity scales as $12\sqrt{\Delta p/\rho_0}$ [39] and thus is independent of $R_{\max}$. Consequently, faster jets can be produced by increasing Δp. Also, transient pressure pulses on a bubble, for example, from shock waves increase the jet velocity. Velocities of about 400 ms^{-1} were reported for bubbles being subjected to shock wave emission from neighboring bubbles [42,55]. For bubbles collapsed by sufficiently strong shock waves (e.g., from shock wave generators), even higher jet velocities of over 1000 ms^{-1} are reported, as, for example, in the detailed work by Lauer et al. [56]. Recently, Supponen et al. [57] studied the jet formation of aspherical bubbles subject to different sources of anisotropy, aiming for a dimensionless and universal description of the asymmetries. For a pressure field p they defined the anisotropy parameter $\vec{\zeta} = -\nabla p R_{\max} \Delta p^{-1}$ as the (normalized) liquid momentum gained around the bubble during its dynamics, which is synonymous with a dimensionless form of the Kelvin impulse [57]. For details on the Kelvin impulse in bubble dynamics that also describe the bubble's centroid motion, the reader is referred to [31,58,59]. To include into $\vec{\zeta}$ further anisotropies that are not only caused by pressure anisotropies, they derived an expression for the anisotropy caused by a nearby boundary, using their approximate solution [9] of the Rayleigh equation [Eq. (2.2)]. In this way, the rigid plane wall is realized as a special case of anisotropy, for which they obtain $\vec{\zeta} = \pm 0.195\gamma^{-2}\vec{n}$, where $\vec{n}$

is the normal vector of the surface. The sign is negative for a rigid boundary, indicating jetting toward the boundary surface, and positive for a free surface (jetting away from the surface, which will be discussed further later in the chapter). For large enough standoffs ($\gamma > \approx 1.4$), thus when the bubble is sufficiently spherical, with this model, jet velocity and centroid displacement in the first collapse can be predicted.

Lechner et al. [60] reported on another type of jet that is produced from an annular inflow toward the axis of symmetry for small γ-values very close to zero. According to their axisymmetric simulations, these jets are considerably thinner yet many more times faster (i.e., up to 1400 ms^{-1}) than the jets that were considered before. The geometry to induce these fast jets resembles the particularly fast jets from collapsing ellipsoidal bubbles that were reported by Voinov and Voinov [61] and Aganin et al. [62].

While only deformation and compression of the gas phase are visible in the above-presented high-speed imaging series of the shape dynamics, the energy of the system oscillates between the flow field (kinetic energy) and the gas phase. The flow fields around a collapsing bubble have been investigated experimentally with particle image velocimetry techniques (PIV) by Vogel et al. [44], Kröninger et al. [63], and Reuter et al. [45], and numerically, for example, by Brujan et al. [64]. The dynamics of the rebound and the subsequent toroidal bubble dynamics were studied by Best [65], Wang et al. [66], Wang [67], and Zhang et al. [68], among others.

There are many phenomena that accompany the nonspherical bubble collapse. Their respective strengths are known to depend crucially on the standoff γ. However, even though this scaling has often proved useful, one should not forget that it is strictly valid only for empty and spherical bubbles, neglecting viscosity and surface tension. For smaller γ-values, in particular $\gamma < 1$, viscosity can play a significant role, as part of the bubble dynamics is occurring close to the no-slip boundary and may be affected by the boundary layer. Also, an important consequence of the bubble growth at the boundary is that there resides a liquid film that is several micrometers thick between the bubble and the boundary. Thus the bubble does not directly collapse onto a "dry surface." Reuter and Kaiser [69] measured the temporal evolution of the layer thickness caused by bubbles with $R_{max} = 400$ μm near a glass substrate. The layer thickness can be expected to be rather independent on the particular wetting properties, as the rapid bubble dynamics process is inertia dominated; that is, the Weber number is large. The importance of viscosity is also indicated by the dependence of the layer thickness on numerical results of the toroidal dynamics (see Liu et al. [70]). Also, the high-speed imaging of the bubble shape in different viscous liquids done by Hupfeld et al. [71] points to the idea that the layer thickness is a function of the liquid viscosity.

Another implication is that with the separating liquid layer the bubble interface proximal to the boundary is flattened, that is, the bubble is not (hemi-)spherical, and R_{max} is not unambiguously defined anymore; obviously, this effect becomes more pronounced as γ approaches zero. In addition, the shock wave from bubble generation gets reflected at the boundary and subsequently acts on the growing bubble nucleus. Thus, owing to the viscous expansion dynamics and shock wave reflections, the idealized Rayleigh case of a spherical bubble, from which a part would be simply excised by the boundary for $\gamma < 1$, cannot be experimentally realized. Still, many numerical simulations initialize their flow field with a static hollow intersected by the boundary (for $\gamma < 1$). In recent numerical simulations, Lauterborn et al. [72] revealed the importance of including the expansion process for simulations and the consequences if that is omitted.

2.4.1.1 *Bubbles Collapse Near a Flat Boundary*

For good or ill, the rich bubble dynamics and resulting phenomena can affect a nearby boundary in various ways and at different stages of the bubble dynamics. It can result in the desired surface cleaning or unwanted material damage and erosion, or it can find application in material processing, for example, in cavitation peening. The mechanisms causing these effects are described next, and impact pressures and tangential shear stresses are discussed.

2.4.1.1.1 Impact Pressure

Earlier in the chapter, some light was shed on the complexity of shock wave emission. These, together with the water hammer pressures from the jet impact, are considered to be the main contributors to cavitation damage. When the shock waves are reflected at the solid boundary, their amplitude often can get substantially higher (ideally up to twofold), as solid materials that provide the rigid boundary usually have an acoustic impedance much larger than that of water. The compression of a boundary material subject to a collapsing bubble was imaged by Tomita and Shima [42]; in the same work they calculated the water hammer pressure of the jet according to $p_{WH} = \rho_l c v_j$, where ρ_l is the density of the (uncompressed) liquid, c is its speed of sound, and v_j is the jet velocity. Several experimental studies on the erosion from single bubbles on largely rigid boundaries have been conducted (Tomita and Shima [42], Lauterborn and Philippp [73], Isselin et al. [74], Dular et al. [75]). The breaking of ice during spark-induced single-bubble dynamics was studied by Cui et al. [76]. They find cracks on the ice plate from tension caused by the reflection of the shock waves. Detailed numerical studies on the peak pressures at the boundary have been conducted by Johnsen and Colonius [77], Chahine and Hsiao [78], and Chahine et al. [79], among others. The shock wave in the boundary material may contribute to material damage in certain cases [80]. More details on cavitation erosion and the role of jets and shock waves can be found in the book by Kim et al. [81].

2.4.1.1.2 Shear Stress

A collapsing bubble exerts also significant shear forces onto a boundary, that is, forces that are perpendicular to the flow direction and tangential to the boundary. Generally, the shear forces depend on the shear rate G, which is the gradient of the flow velocity perpendicular to the flow direction, defined as $G = \frac{du}{dz}$. In the simplest case, when the liquid is incompressible, isotropic, and Newtonian, which to a large extent is an adequate model for cavitation in water, the shear stress (which is the shear force per area) can be expressed as $\tau = \eta G = \eta \frac{du}{dz}$, where η is the (shear) viscosity. The wall shear stress on the boundary located at $z = 0$ is

$$\tau_{\mathrm{w}} = \eta \left.\frac{du}{dz}\right|_{z=0} . \tag{2.8}$$

An overview of models and further considerations concerning non-Newtonian liquids can be found in the work of Chhabra [82]. Among the many fluid dynamical effects brought about by bubble dynamics, several are associated with the generation of high shear rates. These include the axial jet that impacts on the boundary, the bubble wall motion directly at the boundary, which is followed by the respective fluid and the long-lived vortex that stretches over the boundary. Unlike many other flows, with a bubble sitting directly on the boundary, cavitation is able to produce its high-velocity flows directly in the near-wall liquid layer at the boundary, which leads to remarkably high wall shear rates taking a key role in several applications.

To measure the wall shear stresses from a single laser-induced bubble, Dijkink and Ohl [83] used a thin film resistor (dimensions: $750 \times 200\ \mu m^2$) embedded in a solid boundary as a constant-temperature anemometer. The resistor was maintained at an elevated constant temperature to infer the wall shear from the electrical current needed to maintain the temperature while advection is enhancing cooling of the sensor. The researchers conducted pointwise measurements and carried out synchronized high-speed imaging for a bubble with $R_{\max} = 750\ \mu m$ at a standoff of $\gamma = 1.07$. Some distances from the bubble center were probed, and the maximum peak of $\tau_{\mathrm{w,max}} = 3.5$ kPa was reached 17 μs after the collapse where the sensor was positioned about 200 μm off-axis. This shear could be attributed to the radially spreading jet after its impact on the boundary. It was also found that for a doubled wall standoff, $\tau_{\mathrm{w,max}}$ is about halved. However, the temporal bandwidth of the measurement is likely to have been limited by the internal frequency of the current control loop of 250 kHz of the anemometer.

In a similar work, but with the aim to measure the temporal evolution of the wall shear stresses from a laser-induced single bubble on a plane, Reuter and Mettin [84] built a wall shear scanning microscope synchronized with high-speed imaging of the bubble shape. It uses an electrochemical technique with a sensing microelectrode (radius: 12.5 μm) flush mounted into the boundary. The measurement principle is based on the advective perturbation of a chemical diffusion layer from a species that is oxidized at the electrode. Near-wall flows induce an additional oxidation current from which the wall shear rate can be deduced. In comparison to the earlier work [83], the sensor area was shrunk by a factor of 300, allowing for a more local measurement and, to some extent, faster rise times. The impact and spreading of the jet could be resolved temporally via the associated wall shear stresses. An excerpt of the wall shear planes is presented in Fig. 2.6. On the axis of symmetry a marked stagnation region was found, which

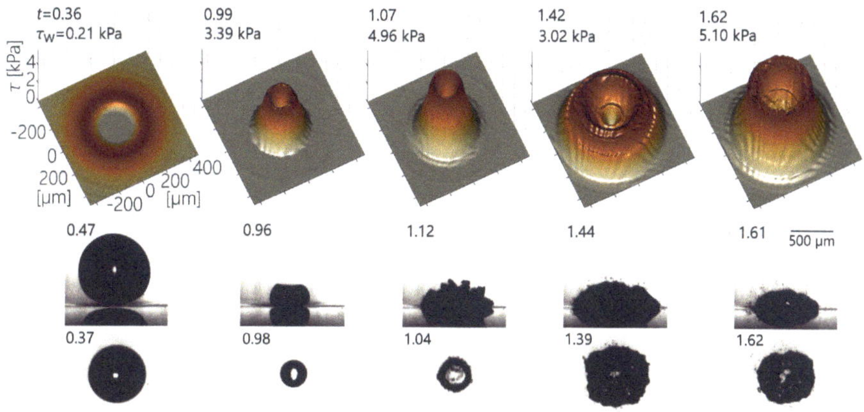

Figure 2.6 Time-resolved wall shear stress microscopy on a solid surface of 800×800 μm^2 of a single collapsing bubble ($\gamma \approx 1$, $R_{max} = 425$ μm). In the top row, the spatial distribution of the wall shear stress is height coded (same scale in all images) and color coded (scale adapted to maximum of each frame τ_w indicated). The normalized time ranges from $t = 0.36$ to $t = 1.62$, where $t = 0$ corresponds to the instance of plasma formation and $t = 1$ corresponds to the collapse. In the two bottom rows, the bubble shape dynamics are presented in side view (upper row, boundary extends horizontally at the bottom of the image) and bottom view through a transparent solid boundary. The staggered normalized times are indicated at the top left of each image. Some illustrative stages of the dynamics are chosen: the fairly spherical expansion phase ($t = 0.36$), the jet impact and spreading ($t = 0.99$, $t = 1.07$), and the collision of flows before the second collapse ($t = 1.42$, $t = 1.62$). In the first frame at expansion of the spherical bubble, a stagnation plane is formed below the bubble, and overall shear stresses are low. At times $t = 0.99$ and $t = 1.07$ the jet has already pierced the bubble. Its radial spreading over the substrate is imaged via the resulting shear stresses with a stagnation region at the axis of symmetry. The last two columns show the collapse phase of the toroidal bubble (instance of toroidal collapse at about $t = 1.75$). Maximum shear stresses are detected some time before the toroidal collapse on a larger area when flows following the inner torus wall and flows from the outer torus wall collide, which induces a liftoff of the collapsing gas phase. In the respective shape images, it can be seen how fast bubble dynamics take part directly at the boundary. *Image adapted from F. Reuter, R. Mettin, Electrochemical wall shear rate microscopy of collapsing bubbles, Physical Review Fluids 3 (6) (2018) 063601.*

explains the presence of the liquid layer that resides between bubble and boundary mentioned above [69]. Around the stagnation region, peak wall shear stresses are detected from the jet during its impact and early spreading at 6 μs after the collapse and about 12 μs after the first signs of jet impact. For the respective peak shear a value of $\tau_{w,max} \approx 5$ kPa is reported. These values were obtained for a somewhat smaller bubble ($R_{max} = 425$ μm) at a similar standoff distance of $\gamma = 1$. In contrast to [83], a further substantial contribution to the shear stress impulse with about the same $\tau_{w,max}$ as from jet impact was found just before the second collapse of the bubble. This is when the now toroidally shaped bubble is in the collapse phase and the flows collapsing the torus collide. The importance of this flow for surface cleaning was pointed out previously [85]. As the electrochemical reaction that is used takes place quasiinstantaneously at the electrode (large Damköhler number), a concentration of $c = 0$ at the boundary is maintained. Thus with a small sensor size, in principle this technique is well suited for transient flows. Still, the temporal bandwidth was limited by the sampling frequency of the potentiostat of 300 kHz.

Besides the two experimental works, there have been only a few numerical studies on the wall shear induced by the cavitation bubble collapse. Zeng et al. [86] made a careful convergence study of the mesh dimension in the direction perpendicular to the boundary, using a submerged jet as validation case, and eventually chose a cell length in that direction of 0.1 μm at the boundary, using the VoF framework OpenFOAM. They simulated the laser-induced bubble collapse and found a maximum shear stress at the jet rim a few microseconds after its impingement with shear stresses as high as $\tau_{w,max} \approx 100$ kPa (for a bubble of $R_{max} = 600$ μm). A detailed spatiotemporal map of the shear stresses is reproduced in Fig. 2.7. Denner et al. [87] and [88] obtained about the same maximum wall shear stress with their in-house code, applying a fully coupled pressure-based algorithm and finite-volume framework ($\tau_{w,max} \approx 105$ kPa for $\gamma = 0.88$ and $\tau_{w,max} \approx 67$ kPa for $\gamma = 0.4$). Koukouvinis et al. [89] simulated the shear stresses also using a volume of fluid method with a uniform grid sizing of 2.5 μm cell dimension. They probed the shear stresses on a large area to model the sensor size in the measurement procedure of [83]. A maximum wall shear stress of $\tau_{w,max} \approx 28$ kPa around the impact region of the jet was reported; however, stresses in the impact region itself ($< 2\,R_{jet}$) were not sampled. In accordance with the experimental results [84], in this work, wall shear stresses of magnitude similar to those from the jet impact are predicted for the toroidal bubble collapse and the collision of flows (for $R_{max} \approx 700$ μm). The authors also report that their data suggest that the maximum shear stress may be a logarithmic function of the bubble energy [i.e., of R^3_{max}; see Eq. (2.6)].

Currently, the peak shear stresses predicted from numerical simulations are about 5–10 times higher than the measured ones. It could be one reason that the

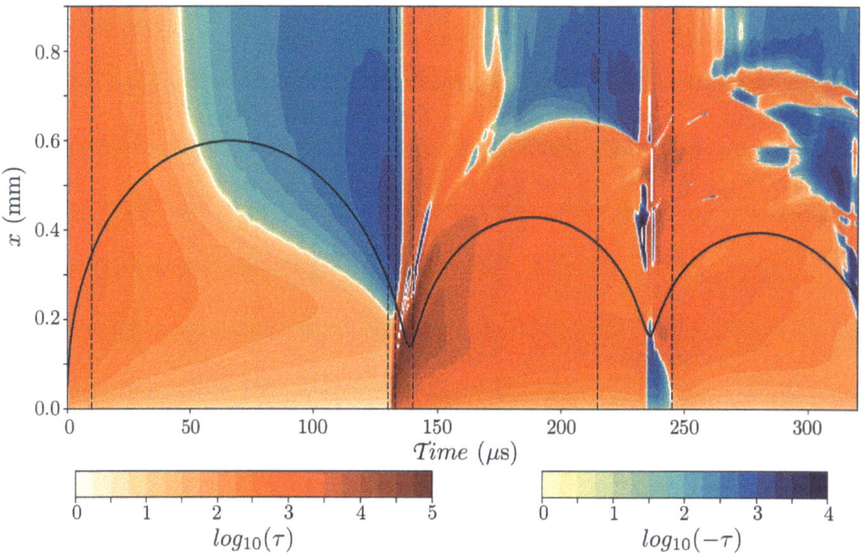

Figure 2.7 Mapping of the temporal (horizontal) and radial (vertical) dependent wall shear created by an expanding and collapsing bubble at $\gamma = 1.0$. The shear stress is plotted logarithmically with base 10, where red colors denote a shear stress in positive radial direction (away from the stagnation point $x = 0$) and blue colors denote a shear stress inwards. The bubble's volume dynamics is plotted with a black solid line; that is, the bubble reaches at $t \approx 70$ μs a maximum radius of $R_{max} \approx 600$ μm. Just before bubble collapse, the maximum shear stress of about $\tau \approx 100$ kPa is generated in a region close to the stagnation point that quickly moves in the positive x-direction. *Image is reproduced from Q. Zeng, S.R. Gonzalez-Avila, R. Dijkink, P. Koukouvinis, M. Gavaises, C.-D. Ohl, Wall shear stress from jetting cavitation bubbles, Journal of Fluid Mechanics 846 (2018) 341–355.*

predicted peaks occurring during jet impact are of transient nature with rise times below 1 μs, which would likely be smoothed by the currently available experimental techniques. The numerical works, on the other hand, predict shear stresses that are in accordance with wall stresses found in axisymmetric simulations of impinging submerged jets (assuming jet parameters estimated from collapsing bubbles). Nevertheless, the mentioned numerical works were undertaken assuming an axisymmetric geometry, such that instabilities, for example, of Rayleigh-Taylor type, splashing, and so, on are not truly modeled. The importance of these instabilities during the spreading of the jet have been demonstrated in Reuter and Kaiser [69] by imaging the rough and possibly partly fragmented bubble interface that faces and possibly partly touches the solid boundary during the spreading of the jet. Therefore the measurement of the magnitudes of the shear stresses are a topic of current research. This is deemed important because cavitation-induced shear has severe implications and important applications in technical flows.

In ultrasonic cleaning, for example, adherent contamination is removed by bubble activity at the boundary for which the wall shear stresses play a central role. Even though sometimes assisted by surfactants, ultrasonic cleaning is generally considered to be a physical cleaning method that invokes bubbles and avoids or reduces the usage of chemicals. While ultrasonic cleaning in the last 60 years has become a standard technique in a wide field of applications from laboratory use to industrial processes, the study of the actual cleaning mechanisms on a single-bubble level has gained attention only more recently, probably owing to the inherent challenges coming from the transient and microscopic nature. Also, the need to clean delicate microscopic (or nanoscopic) structures, for example, in wafer production, has stimulated research. Maisonhaute et al. [90], Kim et al. [91] and Okorn-Schmidt et al. [92] considered sonic cleaning of microparticulate or nanoparticulate contaminations on a wafer, and Chahine et al. [79] have simulated the collapse and interaction of a bubble near a mobile dirt particle with a focus on the forces exerted on the particle. In Okorn-Schmidt et al. [92] the necessary shear rates to set nanometer- to micrometer-sized particles adhered to a rigid boundary into sliding or rolling motion, that is, the necessary shear to clean, was calculated. Therefore smaller particles (radius ≈ 200 nm) in general require higher shear rates of the order of $2 \times 10^5 \mathrm{s}^{-1}$ for rolling and $2 \times 10^6 \mathrm{s}^{-1}$ for sliding, where the exact number depends on the material combination (here, an aqueous solution with a pH value of 7 and a silica/silica and silica/latex material combination for wall and particles was chosen). For both cases the threshold shear rates are exceeded by a cavitation bubble; therefore this kind of particle is expected to be cleaned by cavitation. In some scenarios a single bubble has been imaged with high-speed photography during its cleaning action on a surface: the cell detachment by a shock wave–induced bubble collapse [93], the cleaning of a grease layer [46] and of particulate matter [85] from a laser-induced bubble, and the cleaning of a cake layer from a membrane by an ultrasonically driven bubble [94]. In all cases the surface contamination was removed (cleaned) in the course of a bubble collapse in circular patterns with dimensions of the order of R_{max}. This is in accordance with the quantitative findings of high shear rates in the same area (as mentioned earlier in the chapter) that quickly decline beyond R_{max}. In the case of traveling cavitation bubbles, for example, in the acoustic fields of ultrasonic cavitation, cleaning is produced accordingly along lines of width $\approx R_{\mathrm{max}}$ from repeatedly collapsing bubbles [94]. The cleaned area in dependence of γ was studied experimentally in Reuter and Mettin [85].

Another subject area of cavitation-induced shear with a broad range of applications concerns shear flows interacting with biological boundaries, such as tissue and cell membranes. The cell permeability can be temporarily increased via cell poration, which can be exploited to selectively increase drug uptake into the cell [95]. Larger shear forces acting on a cell can rupture the membrane and consequently induce cell lysis [96]. These mechanisms are exploited in sonoporation and disinfection procedures of pathogen inactivation. Studies also allow conclusions to be drawn about the cavitation-induced shear. High-speed imaging studies on the interaction of a single laser-induced bubble with cells were done by Rau et al. [97] and Rau et al. [98], and of a shock wave–collapsed bubble by Ohl et al. [93]. They consistently identified fluid flow phenomena, in particular shear stresses, as a key factor for those processes rather than shock waves. After simplifying hydrodynamic considerations, Ohl et al. [93] modeled the shear stresses by a Glauert jet, while Rau et al. [98] regard the hydrodynamic boundary layer generated by the expanding bubble as the key factor. The later case was modeled as Stokes' first problem in an adapted and modified form. Indeed, in that work, the bubble was produced directly (i.e., at a distance of 10 μm) above the cell monolayer, such that it expanded with roughly a hemispherical shape. Helfield et al. [99] also studied the normal stresses from cavitation but did not find a correlation with cell poration. Instead, they found a shear stress threshold for cell poration, which confirms the above claims of the earlier works. Besides the absolute shear stress, the shear stress impulse (i.e. the shear stress integrated over time) is also reported to play a decisive role for cell lysis and poration (see Kodama et al. [100], Compton et al. [96]). Compton et al. coupled the Gilmore bubble model to a hydrodynamic model similar to the one used in Rau et al. [98] to estimate shear stresses and impulses on the cells. They found shear stress thresholds $\tau_{\mathrm{w,th}}$ for the permeabilization of their cells to be around 10 kPa and thresholds for shear stress impulses of $J = \int_0^{T_L} \tau_{\mathrm{W}}\, dt \approx 0.035$ Pa, where the wall shear stress is integrated over the bubble lifetime (one period). For cell necrosis they derived limit thresholds of $\tau_{\mathrm{w,th}} \approx 18 - 35$ kPa and concluded that for the (viscoelastic) cells, J may be a more meaningful criterion than $\tau_{\mathrm{w,th}}$. They carried out experiments with different bubble sizes between $R_{\mathrm{max}} = 44$ μm and 184 μm from varying laser pulse energies and durations and found poration in a region of radius $R_{\mathrm{perm}} \approx 1.02 R_{\mathrm{max}}$ and necrosis $R_{\mathrm{necr}} \approx 0.60 R_{\mathrm{max}}$ (averaged over all bubble sizes).

In general, shear flows can increase surface energies, for example, in emulsification, and accelerate diffusion limited processes. This is exploited for mixing purposes, enhanced cooling, or electroplating. Consequently, in mixing and emulsification procedures, cavitation is widely exploited, for example, in the cosmetic, pharmaceutical, and food industries [101], in which the applicability of physical means of emulsification by hydrodynamic or ultrasonic cavitation can help to reduce the usage of additional chemicals (stabilizers and emulgators) and in this way increase tolerability. The mass transfer enhancement from a single bubble to a rigid boundary was studied by Birkin et al. [102], who found a maximum mass transfer enhancement that is reported to likely stem from the bubble collapse. Reuter et al. [103] measured the mass transfer enhancement synchronized with high-speed imaging and identified the vortex that is generated at larger standoffs ($\gamma \geq 2.5$) and spreads over the boundary as a long-range and long-lasting mass transfer enhancement mechanism. Mass transfer toward the boundary and mixing of near-wall liquid layers have been visualized for a single bubble by Lagrangian ink maps (see Fig. 2.8). The ink maps are generated in silico from time integration of experimental flow field data obtained by PIV by Reuter et al. [45]. They allow tracking of the net fluid transport and mixing, intuitively accessible to the human eye imitating an experiment in which the bubble dynamics were carried out in a liquid of stratified ink layers. This visualization is achieved via the permanent assignment of a color to each fluid element that codes its initial proximity to the boundary.

The emulsification process on a single-bubble level for a water-oil boundary has been studied by Orthaber et al. [104]. Jetting from one phase into the other with subsequent shearing and Rayleigh-Taylor instabilities seems to be main factors.

In conclusion, the high shear rates (and small size of the boundary layer) produced by cavitation, their locality, and their transient behavior prohibit the application of standard measurement techniques. Also, the use of numerical techniques involves certain difficulties regarding cell refinement at the boundary, three-dimensionality, and flow instabilities as well as phase tracking of small gas entities that may get entrained by the jet. At the same time, cavitation-induced wall shear stresses and sometimes wall shear stress impulses play an important role in many applications that are based on cavitation. Because these applications deal with very different kinds of boundaries (rigid, viscoelastic, fluid-fluid) and different objects, we believe that results are somehow scattered in literature among different fields. However, consistent with the measured shear rates, in almost all

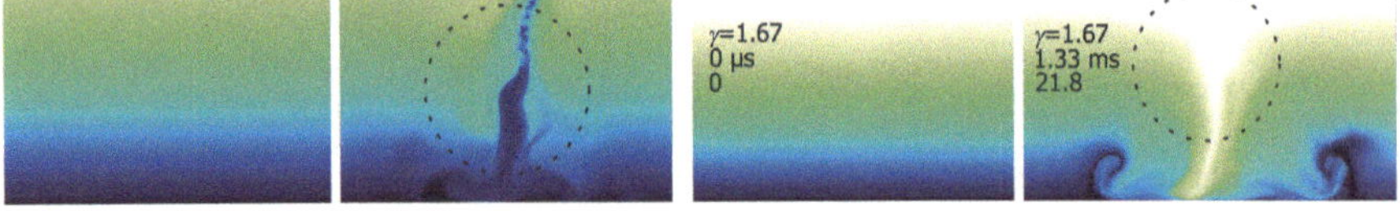

Figure 2.8 Lagrangian ink maps for a bubble generated at $\gamma = 1.21$ (left) and $\gamma = 1.67$ (right). From time-resolved flow field measurements, the net liquid displacement is calculated and plotted with color coding. The boundary is located horizontally at the bottom where the initial ink layering shows the dark blue colors. The bubble at maximum expansion is depicted in both cases by the dashed circle. Times are given in each frame in absolute form and normalized on T_L. In the left case, liquid from the near-wall layer is ejected into the bulk and sheared within the vortex. In the right case, bulk liquid is transported toward the boundary, where shear layers roll up. *For more details see F. Reuter, S.R. Gonzalez-Avila, R. Mettin, C.-D. Ohl, Flow fields and vortex dynamics of bubbles collapsing near a solid boundary, Physical Review Fluids, 2 (6) (2017) 064202.*

applications the range of bubble action along the boundary is about the maximum bubble radius R_{max} (when $\gamma <\approx 1.1$).

2.4.1.2 *Curved Rigid Boundaries*

The plane wall case discussed earlier in the chapter can be extended by allowing for some simple curvature of the boundary. Rigid boundaries of concave and convex geometry were studied by Best and Blake [105]. They were able to derive a generalized bubble lifetime [Eq. (2.7)] for these cases. It is modified by a prolongation factor: $(1 + 0.4065\mu)$, where $\mu = \frac{\xi^2}{\gamma(1+\xi)}$ is a function of the bubble-to-wall standoff γ and the curvature parameter as defined in Tomita et al. [106], where convex, flat, and concave geometries are indicated by $\xi < 1$, $\xi = 1$, and $\xi > 1$, respectively. Thus the bubble lifetime is increased for concave geometries and for smaller γ, that is, when the bubble is closer to the boundary. Tomita et al. [106] studied experimentally with laser-induced bubbles and numerically with the boundary integral method the bubble collapse near convexly and concavely curved, rigid boundaries. They consistently found that the velocity of the jet toward the boundary increases for convex boundary shapes and decreases for concave boundary shapes. For a highly convex boundary, that is, one that takes a needle-like shape, the formation of a second jet in the opposite direction can be observed. However, this phenomenon seems to rely on the rotational symmetry, which were assumed in the simulations by Tomita et al. [106]. While this jet was also observed in their experiments, it was also reported that this phenomenon is sensitive to small deviations from the rotational symmetry, for example, when the bubble is placed slightly off-axis to the needle tip–like boundary. Cui et al. [107] studied experimentally with spark bubbles the collapse at a convex, rigid, perforated surface and varied the bubble standoff and the perforation aperture radius. Saleki-Haselghoubi and Dadvand [108] simulated with the boundary element method a situation similar to that described in Cui et al. [107] (but naturally with the bubble being on axis). They varied the initial standoff distance and the aperture size of the perforation and identified four jetting regimes: no jetting, away from the boundary, toward the boundary, and both jets simultaneously. While the number of possible geometries is endless, some more and important geometries with concave and more confined geometries will be discussed in Section 2.4.3.

2.4.2 Elastic Boundaries and Flexible Membranes

Generally, a rigid boundary attracts a bubble during the collapse Section 2.4.1, while a free boundary repels the bubble (see Section 2.4.1). As a result the jet flow is directed either toward or away from the rigid boundary or the free boundary, respectively. For flexible boundaries, although they are blocking the flow, they may deflect the bubble from the boundary too. Gibson and Blake [109] tested these kinds of compliant boundaries that repel the bubble from the surface during collapse such that the damaging jet is directed away. Their composite surfaces were made of vulcanized rubber bonded to various backings ranging from air to natural foam. Duncan and coworkers [110,111] numerically investigated the bubble collapse at such compliant boundaries and found that the jet is directed away from the boundary when the displacement of the coating is about 15% of $R_{\max}$ and when the total energy transferred to the coating from the flow is about 10% of the total flow energy. The direction of jetting depends mainly on the inertia $m/(\rho R_{\max})$, where m is the mass area density of the compliant membrane boundary material and ρ is the density of the liquid. However, it does not depend on the relation $T_L L/T_m$, which is the ratio of the bubble lifetime in an infinite fluid to the spring-mass timescale of the compliant wall. Shima et al. [112] demonstrated that the kind of dynamics depends not only on the surface but also on the standoff distance and the maximum bubble size. Between the regimes of attractive and repulsive dynamics the bubbles may undergo a neutral collapse. That location is a nontrivial function of the composite layering and the mechanical properties [113], that is, the boundary inertia m^* and the boundary stiffness k^*. While the research goal initially was to find surfaces that are less prone to cavitation erosion by repelling the bubble, in later years the biomedical interest fueled the research on cavitation near elastic boundaries more and more.

Excellent work was done by Brujan et al. [114], who used instead of composite surfaces a thick slab made of transparent polyacrylamide (PAA) of 80% water content with an elastic modulus similar to that of tissue but having a considerably lower yield strength. For standoffs $\gamma \leq 0.76$ they found the formation of a mushroom-shaped bubble during collapse caused by the annular flow toward the bubble center. Later, this gave rise to two oppositely directed jets, which eventually split the bubble, see Fig. 2.9.

Interestingly, maximum velocities of $800\ \mathrm{ms}^{-1}$ and above with which the jet travels toward the elastic boundary have been measured. These are considerably higher than the well-known jet that is observed for rigid boundaries. The authors explain these high velocities by citing three factors: the strong focusing of the annular or equatorial jet, shock wave–driven flow, and the conical shape that accelerates the jet until it hits the opposite bubble wall [114]. Boundary element simulation (see, e.g., Ohl et al. [115]) accounting for the elasticity and density difference are roughly in agreement with the experimentally observed splitting of the bubbles and acceleration of the liquid jets, although they neglect acoustic emission or propagation of elastic waves in the solid.

For smaller standoffs the interaction becomes more complex, and already during the expansion phase the boundary interacts strongly with the bubble-induced

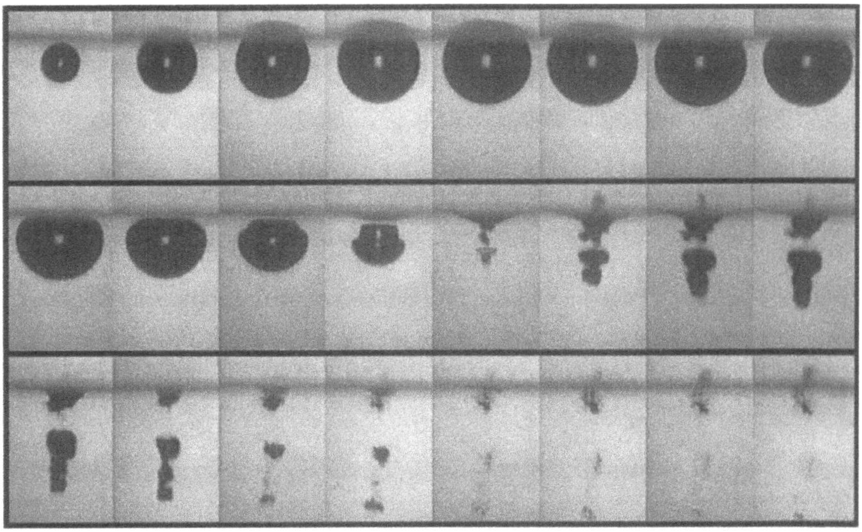

Figure 2.9 Bubble shape dynamics for $\gamma = 0.44$ at an elastic boundary that spans horizontally in the upper part of the image. The bubble shows a characteristic mushroom shape right before its collapse (frame 12). In the expansion phase (first row), the boundary is indented; after the collapse it is elongated, and jetting into the gel material occurs. *Image is reproduced from E.-A. Brujan, K. Nahen, P. Schmidt, A. Vogel, Dynamics of laser-induced cavitation bubbles near an elastic boundary, Journal of Fluid Mechanics 433 (1) (2001) 251–281.*

flow, resulting in the ejection of PAA gel into the liquid. In contrast the jet directed toward the boundary is dampened because the bubble collapses from the boundary-facing side first and interacts with the jet ejecting material from the surface.

Shaw et al. [116] and, in greater detail, Turangan et al. [117] found similar bubble splitting when the solid slab of elastic material was replaced by a thin membrane.

Elastic boundaries return some of the fluid mechanic energy stored during the bubble expansion. Their mechanical properties remain fixed during the bubble dynamics. It is also possible to alter the boundary condition during the bubble dynamics by the presence of the cavitation bubble itself. Consider that the surface traps many small gas bubbles. During the expansion of a cavitation bubble close to that surface, the gas bubbles expand as a result of the reduced pressure in the cavitation bubble. Owing to liquid inertia, the bubbles cannot follow instantaneously this pressure change and remain more expanded during collapse than a simple pressure equilibrium would predict. Then the cavitation bubble sees, owing to the still expanded gas bubbles, a free boundary during its collapse, the consequence of which is jetting away from the boundary. A first implementation using particularly shaped bubble traps has been reported recently by Gonzalez-Avila [118].

2.4.3 Bubble Collapse in Confined Geometries

The bubble dynamics occurring close to more than a single infinitely extended boundary greatly enriches the observed fluid phenomena. A simple approach to understand the complexity may be achieved if we assume that the bubble is a spherical sink-source in a simple potential flow. Then a bubble near a rigid boundary is equivalent to two bubbles located at twice the distance from the boundary (but without the boundary). This approach is called the method of images [1]. Unfortunately, we cannot simply add a third bubble to obtain the geometry of a bubble between two boundaries. Instead, we need to account for the image of source-sink to each boundary that leads to an infinite row of image bubbles to fulfill the boundary condition, that is, to prevent flow through the boundary. Potential flow solvers such as the boundary element method or method of images have been used to correctly predict the gross features of bubble dynamics in gaps and near complex boundaries (see, e.g., Gonzalez-Avila et al. [119]), at least for short enough times that boundary layers do not grow to noticeable heights. The research on the dynamics of cavitation bubbles in confined geometries has been motivated partly by engineering problems, such as loaded ball bearings [120], ultrasound cleaning in thin gaps, or the well-known cavitation in synovial joints. We document first some selected works in thin liquid layers, while distinguishing between rigid and elastic boundaries. Then some more recent works on bubble dynamics in complex environments are highlighted.

2.4.3.1 Narrow Rigid Gaps

Two reviews on the dynamics of cavitation bubbles in narrow liquid gaps [121] and microfluidic confinements are available [122]. Narrow gaps confine one of the three dimensions of the bubble during expansion and collapse. Chen et al. [123] reported on cavitation–gas bubble interaction within 10-μm-thick gaps. Bubbles with a lateral diameter of approximately 0.2–0.4 mm are created through linear absorption of a pulsed laser, and the dynamics is observed with a microscope connected to a single-frame camera. The advantage of the thin gap was the ease of observing the interior of the bubble and thus the complex fragmentation dynamics that is induced by the cavitation bubble into neighboring gas bubbles. In absence of boundaries it was first speculated that bubbles in these narrow gaps can be modeled simply with a two-dimensional Rayleigh equation [124], at least for sufficiently short times [125]. It was later realized that the flow eventually becomes three-dimensional [126] during the collapse of the bubble when boundary layer separation occurs and a recirculating region close to the bubble is formed. Manipulation of objects and cells needs to take these details into account. While in very thin liquid gaps the bubble expands and collapses in a pancake shape, channels with a height H that is similar to the lateral maximum radius R_x offer an unexpected collapse scenario. Here, the introduction of the nondimensional parameter $\eta = H/R_x$ provides some help to order the observations. Additionally, a second parameter can be introduced, namely, similar to the single rigid boundary, the standoff distance of the bubble from the boundary, $\gamma = h/R_{max}$, where h is the distance to the closest boundary between the point of bubble creation and R_{max} is the maximum radius of a sphere having the same volume. Let us start with a bubble that is generated very close to one of the boundaries, that is, $\gamma \approx 0$, and increase the η-parameter steadily from 0. For small η-values, the bubble expands in a pancake shape and collapses at the center of the liquid gap (see Fig. 2.10A). In each row of Fig. 2.10 the bubble is nucleated at the bottom of the liquid gap. For intermediate η-values, the bubble jets

Figure 2.10 Each row shows the bubble dynamics for a different channel height but similar maximum lateral bubble radius of $R_x = 140$ μm. (A) Collapse between the two walls, $\eta = 0.4$. (B) Collapse onto the upper wall, $\eta = 0.8$. (C) Collapse onto the upper wall where the bubble does not reach the upper wall, $\eta = 0.9$. (D) Neutral collapse, $\eta = 1.2$. (E) Nonspherical collapse onto the lower wall, $\eta = 1.6$. (F) Nearly hemispherical collapse onto the lower wall, $\eta = 7.3$. *Image is reproduced from S.R. Gonzalez-Avila, E. Klaseboer, BC Khoo, C.D. Ohl, Cavitation bubble dynamics in a liquid gap of variable height, Journal of Fluid Mechanics, 682 (2011) 241–260.*

and collapses onto the distant side where it was nucleated, and for large η-values it collapses onto the same side, resembling then the dynamics of a single boundary with $\gamma = 0$. This behavior provides a means of inducing strong shear stress away from the side at which the bubble was nucleated, which has been utilized for detachment and sonoporation of cells and cleaning in thin gaps [127]. A more complete picture varying the standoff distance and the length of the gap is provided by Zeng et al. [128]. We want to mention that a bubble created in the center of gap results in a circumferential or radial jet that leads to a dumbbell-shaped bubble prior to collapse.

Studies on interaction of bubbles in thin gaps started with the fragmentation experiments from Chen et al. [123]. The coupling of the radial and translatory dynamics of two cavitation bubbles was demonstrated by Quinto-Su and Ohl [129] for mostly cylindrical bubbles. When the two bubbles are of largely different size [130] or created with time delay [131], high speed jetting can be induced that have been utilized for manipulating of objects or injecting drugs into cells [132].

2.4.3.2 Narrow Elastic Gaps

Microfluidic channels may be composed of a rubber-like transparent material, such as polydimethylsiloxane (PDMS). It has been observed that bubbles oscillating in channels composed of this elastic material [133] may excite surface waves at the interface between the liquid and the solid. These waves lead to a self-organization of multiple bubbles into a crystal-like pattern [134]. Near a single boundary a detailed study revealed Rayleigh waves propagating, caused by the alternating pressure from the bubble being in contact with the boundary [135]. For thin gaps between two elastic boundaries the bubble excites on both boundaries' surface waves. Doinikov et al. [136] explained that in this case the two elastic surfaces couple and give rise to a Lamb-type wave, that is, a symmetric bending of the liquid gap. The dispersion relation provides limits for Rayleigh and Scholte waves and gives of the order of 10–20 m/s for the propagation speed for of the order of 100 kHz excitation of bubbles between channels of PDMS-type material. Less compliant materials with a Young's modulus of several tens of gigapascals, such as glass or acrylic, have a considerably higher surface wave velocity and, because of that, a much higher acceleration of the surface. Rapet et al. [137] recently demonstrated that for sufficiently large displacements as a result of high-frequency surface waves, secondary cavitation can be nucleated (see Fig. 2.11). One method to generate an impulsive high-pressure region on the surface is the optical breakdown of the liquid in contact with the boundary or directly of the boundary. Then multiple waves are launched into the solid and the liquid. One of the waves is a transient Lamb wave nucleating a ring-shaped cavitation pattern with approximately the speed of sound in the water.

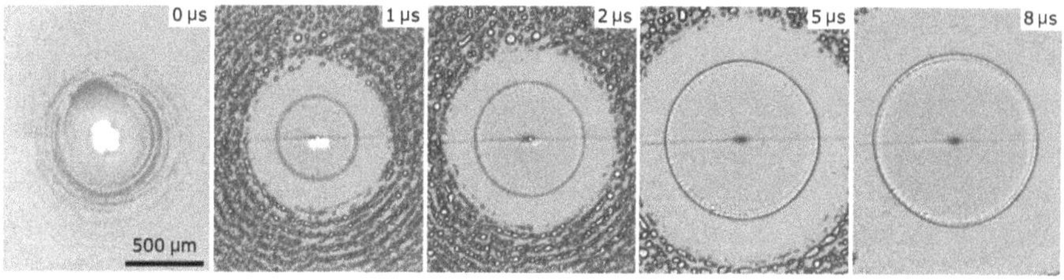

Figure 2.11 Secondary cavitation activity in a 20-μm-thick gap caused by dielectric breakdown at the image center. Centrally, a bubble expands that is surrounded by rings of bubbles nucleated from a surface wave traveling in the liquid. *Image is reproduced from J. Rapet, P.A. Quinto-Su, C.-D. Ohl, Cavitation inception from transverse waves in a thin liquid gap, Physical Review Applied 14 (2) (2020) 1.*

2.4.3.3 Corner Geometries

In recent years the effect on the dynamics of single cavitation bubbles from nonaxisymmetric geometries has received increased attention. An obvious example is a corner that is formed by two plane walls. Brujan et al. [138] studied experimentally the direction of the jet flow of bubbles created at various locations in the corner. They used a right-angled corner, while Molefe and Peters [139] extended it to square and equilateral triangular channels (see also Wang et al. [140]). In potential flow, the corners are modeled with image sources [141]; then the translatory motion of the bubble during the expansion and collapse even in these complex geometries can be roughly explained with the effective secondary Bjerknes force acting on the bubble from all image bubbles. This leads to a motion away from the nearest wall during expansion and toward the wall during collapse. The jet is formed in the direction of the center of mass motion of the bubble during collapse. Placing the bubble between two parallel walls and close to a third wall perpendicular to the two former walls [142] connects the studies of corner geometries with the planar wall geometries discussed previously. As expected, the experimental work [142] revealed that when the bubble is brought closer and closer to the perpendicular wall, the radial jet is changed to an axial jet, that is, the one observed near a single boundary. Again, potential flow simulations in three dimensions are in excellent agreement with the overall bubble dynamics observed experimentally.

2.4.3.4 Rigid and Elastic Tubes

Cavitation bubbles in thin tubes can be used to pump liquids. An everyday example is the drip coffee maker or the putt-putt steamboat (for references, see Dijkink et al. [143]). Consider a cavitation bubble expanding in a viscous narrow tube [144]; that is, the bubble expands to most of the cross-sectional area. When the bubble is positioned midway between the two ends having the same boundary conditions, the bubble due to symmetry expands and collapses at the same location. However, if the bubble is not located at the center, a net flow is generated toward the tube end that is more distant from the point of bubble nucleation [144]. The fluid mechanics behind this seems not to be fully understood. It is partly due to the development of boundary layers and the different mass or inertia of the liquid on both sides of the bubble. But even with symmetric boundary conditions, the volume dynamics during expansion and collapse of the bubble in a narrow tube is nonsymmetric. Sun et al. [145] demonstrated for microscopic channels that thermal effects can lead to such asymmetry. For more macroscopic tubes the pumping of liquid was demonstrated with much larger laser-induced bubbles [146]. Here, the bubble fragments generated a complex residual flow after the first collapse.

In several medical applications, cavitation is unwanted, as it could lead to severe side effects [147]. The relevant geometries are in many cases with bodily liquid–filled elastic tubes. One example is the pulsed laser ablation in laser angioplasty [148], in which a cavitation bubble dilates the blood vessels by growing within. A second example of unwanted cavitation in vessels is shock wave lithotripsy. There, cavitation bubbles may grow in small blood vessels located along the acoustic path [149]. In both modalities, vascular injuries were reported and related to the observed dynamics of the cavitation bubbles. Zhong et al. [149] reported that the rapid, large expansion of intraluminal bubbles causes dilation of the vessel wall that may even lead to rupture of the vessel. Similarly, large bubbles created in angioplasty [148] expand greatly the vessel phantom and may induce cavitation even from outside the vessel. In both experiments [148,149], the tissue and the mechanical coupling of the vessel to it were ignored. The importance of the embedding tissue and elastic properties was reported by Chen et al. [150], who studied the dynamics of individual bubbles in rat abdominal tissue. The limits for medical protocols that use or involve cavitation are only partly defined. The rather complex interaction of the fluid with the elastic vessel structure is still a challenge for modeling, although considerable progress has been achieved in recent years. Ye and Bull [151] coupled the viscous flow with a flexible boundary and revealed marked differences to a rigid tube. Their calculated streamlines reveal single and multiple stagnation points along the elastic tube axis, and the stress on the wall and pressures in the tube are greatly reduced. While the inertial expansion and collapse of a cavitation bubble in an elastic tube are dampened, shock-collapsed gas bubbles may significantly contribute to the injury of blood vessels in shock wave lithotripsy [152].

2.5 Outlook

We have touched on some aspects but have also left out many inspiring works. For this omission we want to apologize to our colleagues. More than 100 years after the pioneering work of Rayleigh [7] revealed the damaging potential of cavitation bubble collapse, the science of nonspherical collapses has greatly progressed. Generally, all boundaries and inhomogeneities in the liquid result in focusing some part of the flow into liquid jets. Their impact, deformation, or even penetration into boundaries and the development of unsteady boundary layers are of current research interest.

Simulations predict for bubbles collapsing very close to a rigid boundary extremely fast jets and, as a result, enormous shear stresses. Both could be important contributors to surface cleaning and erosion. While simulations have advanced, experiments are lagging behind. We expect that in the coming years this knowledge gap will be closed through the use of high-speed microphotography combined with surface measurement of elastic and plastic deformations. On the simulations side of cavitation research on single-bubble dynamics, we expect that more nonaxisymmetric simulations of the flow will explore how instabilities develop and affect the focusing of energy. Finally, experiments on cavitation in soft elastic materials and their simulations have only started in recent years. The importance of being able to predict stress and strain fields for medical therapy will be driving research in the years to come.

References

[1] H. Cole, Underwater Explosions, Princeton University Press, Princeton, 1948.

[2] M.S. Plesset, A. Prosperetti, Bubble dynamics and cavitation, Annual Review of Fluid Mechanics 9 (1) (1977) 145–185.

[3] J.R. Blake, D.C. Gibson, Cavitation bubbles near boundaries, Annual Review of Fluid Mechanics 19 (1) (1987) 99–123.

[4] W. Lauterborn, T. Kurz, Physics of bubble oscillations, Reports on Progress in Physics 73 (10) (2010) 106501.

[5] T. Leighton, The Acoustic Bubble, Academic Press, London, 1994.

[6] K. Yasui, Acoustic Cavitation and Bubble Dynamics, Springer International Publishing, 2018. , for more information see: <https://www.springer.com/gp/book/9783319682365>.

[7] L. Rayleigh, VIII. On the pressure developed in a liquid during the collapse of a spherical cavity, The London, Edinburgh, and Dublin Philosophical Magazine and Journal of Science 34 (200) (1917) 94–98.

[8] N.A. Kudryashov, D.I. Sinelshchikov, Analytical solutions of the rayleigh equation for empty and gas-filled bubble, Journal of Physics A: Mathematical and Theoretical 47 (40) (2014) 405202.

[9] D. Obreschkow, M. Bruderer, M. Farhat, Analytical approximations for the collapse of an empty spherical bubble, Physical Review E 85 (6) (2012) 066303.

[10] M.S. Plesset, The dynamics of cavitation bubbles, Journal of Applied Mechanics 16 (1949) 277–282.

[11] A. Bogoyavlenskiy, Single-bubble sonoluminescence: shape stability analysis of collapse dynamics in a semianalytical approach, Physical Review E 62 (2) (2000) 2158.

[12] R.Gilmore, The Growth or Collapse of a Spherical Bubble in a Viscous Compressible Liquid, 1952.

[13] M.P. Brenner, S. Hilgenfeldt, D. Lohse, Single-bubble sonoluminescence, Reviews of Modern Physics 74 (2) (2002) 425.

[14] D. Fuster, A review of models for bubble clusters in cavitating flows, Flow, Turbulence and Combustion 102 (3) (2019) 497–536.

[15] C.F. Hung, J.J. Hwangfu, Experimental study of the behaviour of mini-charge underwater explosion bubbles near different boundaries, Journal of Fluid Mechanics 651 (2010) 55.

[16] C.F. Naude, A.T. Ellis, On the mechanism of cavitation damage by nonhemispherical cavities collapsing in contact with a solid boundary, Journal of Basic Engineering 83 (4) (1961) 648–656.

[17] N.D. Shutler, R.B. Mesler, A photographic study of the dynamics and damage capabilities of bubbles collapsing near solid boundaries, Journal of Basic Engineering 87 (2) (1965) 511–517.

[18] S.L. Jacques, Role of tissue optics and pulse duration on tissue effects during high-power laser irradiation, Applied Optics 32 (13) (1993) 2447–2454.

[19] A. Vogel, V. Venugopalan, Pulsed laser ablation of soft biological tissues, in: Ashley J. Welch, Martin J.C. van Gemert (Eds.), Optical-Thermal Response of Laser-Irradiated Tissue, Springer, Dordrecht, 2010, pp. 551–615.

[20] J.P. Padilla-Martinez, C. Berrospe-Rodriguez, G. Aguilar, J.C. Ramirez-San-Juan, R. Ramos-Garcia, Optic cavitation with CW lasers: a review, Physics of Fluids 26 (12) (2014) 122007.

[21] D.M. Nguyen, L. Hu, J. Miao, C.-D. Ohl, Oscillate boiling from electrical microheaters, Physical Review Applied 10 (4) (2018).

[22] Y.S. Kannan, B. Karri, K.C. Sahu, Letter: entrapment and interaction of an air bubble with an oscillating cavitation bubble, Physics of Fluids 30 (4) (2018).

[23] A. Vogel, J. Noack, G. Hüttman, G.J.A.P.B. Paltauf, Mechanisms of femtosecond laser nanosurgery of cells and tissues, Applied Physics B 81 (8) (2005) 1015–1047.

[24] A. Vogel, P. Schweiger, A. Frieser, M.N. Asiyo, R. Birngruber, Intraocular Nd:YAG laser surgery: laser-tissue interaction, damage range, and reduction of collateral effects, IEEE Journal of Quantum Electronics 26 (12) (1990) 2240–2260.

[25] A. Vogel, N. Linz, S. Freidank, G. Paltauf, Femtosecond-laser-induced nanocavitation in water: implications for optical breakdown threshold and cell surgery, Physical Review Letters 100 (3) (2008) 038102.

[26] W. Lauterborn, High-speed photography of laser-induced breakdown in liquids, Applied Physics Letters 21 (1) (1972) 27–29.

[27] F. Reuter, Collapsing Bubbles, Induced Flow Phenomena, and Their Application in Ultrasonic Cleaning (Ph.D. thesis), Georg-August-Universität Göttingen, Germany, 2017.

[28] A. Vogel, K. Nahen, D. Theisen, J. Noack, Plasma formation in water by picosecond and nanosecond Nd: YAG laser pulses. I. Optical breakdown at threshold and superthreshold irradiance, IEEE Journal of Selected Topics in Quantum Electronics 2 (4) (1996) 847–860.

[29] Y. Tagawa, S. Yamamoto, K. Hayasaka, M. Kameda, On pressure impulse of a laser-induced underwater shock wave, Journal of Fluid Mechanics 808 (2016) 5–18.

[30] G. Sinibaldi, A. Occhicone, F. Alves Pereira, D. Caprini, L. Marino, F. Michelotti, et al., Laser induced cavitation: plasma generation and breakdown shockwave, Physics of Fluids 31 (10) (2019) 103302.

[31] T.B. Benjamin, A.T. Ellis, The collapse of cavitation bubbles and the pressures thereby produced against solid boundaries, Philosophical Transactions for the Royal Society of London. Series A, Mathematical and Physical Sciences 260 (1966) 221–240.

[32] A.B. Arons, D.R. Yennie, Energy partition in underwater explosion phenomena, Reviews of Modern Physics 20 (3) (1948) 519.

[33] S. Buogo, G.B. Cannelli, Implosion of an underwater spark-generated bubble and acoustic energy evaluation using the rayleigh model, The Journal of the Acoustical Society of America 111 (6) (2002) 2594–2600.

[34] W. Lauterborn, A. Vogel, Shock wave emission by laser generated bubbles, in: C. Delale (Ed.), Bubble Dynamics and Shock Waves, Shock Wave Science and Technology Reference Library, Vol 8, Springer, Berlin, Heidelberg, 2013, pp. 67–103.

[35] T. Sato, M. Tinguely, M. Oizumi, M. Farhat, Evidence for hydrogen generation in laser-or spark-induced cavitation bubbles, Applied Physics Letters 102 (7) (2013) 074105.

[36] Q. Zhang, J. Luo, Y. Zhai, Y. Li, Improved instruments and methods for the photographic study of spark-induced cavitation bubbles, Water 10 (11) (2018) 1683.

[37] M. Tinguely, The Effect of Pressure Gradient on the Collapse of Cavitation Bubbles in Normal and Reduced Gravity, 2013.

[38] S. Fujikawa, T. Yano, M. Watanabe, Vapor-Liquid Interfaces, Bubbles and Droplets: Fundamentals and Applications, Springer Science & Business Media, Berlin, Heidelberg, 2011.

[39] M.S. Plesset, R.B. Chapman, Collapse of an initially spherical vapour cavity in the neighbourhood of a solid boundary, Journal of Fluid Mechanics 47 (2) (1971) 283–290.

[40] W. Lauterborn, H. Bolle, Experimental investigations of cavitation-bubble collapse in the neighbourhood of a solid boundary, Journal of Fluid Mechanics 72 (2) (1975) 391–399.

[41] J.R. Blake, B.B. Taib, G. Doherty, Transient cavities near boundaries. Part 1. Rigid boundary, Journal of Fluid Mechanics 170 (1986) 479–497.

[42] Y. Tomita, A. Shima, Mechanisms of impulsive pressure generation and damage pit formation by bubble collapse, Journal of Fluid Mechanics 169 (1986) 535–564.

[43] M. Rattray, Perturbation Effects in Cavitation Bubble Dynamics (Ph.D. thesis), California Institute of Technology, 1951.

[44] A. Vogel, W. Lauterborn, R. Timm, Optical and acoustic investigations of the dynamics of laser-produced cavitation bubbles near a solid boundary, Journal of Fluid Mechanics 206 (1989) 299–338.

[45] F. Reuter, S.R. Gonzalez-Avila, R. Mettin, C.-D. Ohl, Flow fields and vortex dynamics of bubbles collapsing near a solid boundary, Physical Review Fluids 2 (6) (2017) 064202.

[46] C.-D. Ohl, M. Arora, R. Dijkink, V. Janve, D. Lohse, Surface cleaning from laser-induced cavitation bubbles, Applied Physics Letters 89 (7) (2006) 074102.

[47] M. Koch, J.M. Rosselló, C. Lechner, W. Lauterborn, J. Eisener, R. Mettin. Realistic Prediction and Analysis of Cavitation Bubble Dynamics via Optical Ray-Tracing of the Corresponding Numerical Simulations, arXiv preprint arXiv:2005.10171, 2020.

[48] C. Lechner, M. Koch, W. Lauterborn, R. Mettin, Pressure and tension waves from bubble collapse near a solid boundary: a numerical approach, The Journal of the Acoustical Society of America 142 (6) (2017) 3649–3659.

[49] O. Supponen, D. Obreschkow, P. Kobel, M. Tinguely, N. Dorsaz, M. Farhat, Shock waves from nonspherical cavitation bubbles, Physical Review Fluids 2 (9) (2017) 093601.

[50] O. Lindau, W. Lauterborn, Cinematographic observation of the collapse and rebound of a laser-produced cavitation bubble near a wall, Journal of Fluid Mechanics 479 (2003) 327–348.

[51] W. Lauterborn, A. Vogel, Bubble dynamics and shock waves, Shock Wave Science and Technology Reference Library 8 (2013) 67–103.

[52] M. Kornfeld, L. Suvorov, On the destructive action of cavitation, Journal of Applied Physics 15 (6) (1944) 495–506.

[53] A. Pearson, J.R. Blake, S.R. Otto, Jets in bubbles, Journal of Engineering Mathematics 48 (3–4) (2004) 391–412.

[54] T. Kodama, K. Takayama, Dynamic behavior of bubbles during extracorporeal shock-wave lithotripsy, Ultrasound in Medicine & Biology 24 (5) (1998) 723–738.

[55] J.P. Dear, J.E. Field, A study of the collapse of arrays of cavities, Journal of Fluid Mechanics 190 (1988) 409–425.

[56] E. Lauer, X.Y. Hu, S. Hickel, N.A. Adams, Numerical modelling and investigation of symmetric and asymmetric cavitation bubble dynamics, Computers & Fluids 69 (2012) 1–19.

[57] O. Supponen, D. Obreschkow, M. Tinguely, P. Kobel, N. Dorsaz, M. Farhat. Scaling Laws for Jets of Single Cavitation Bubbles, arXiv preprint arXiv:1703.01088, 2017.

[58] J.R. Blake, The Kelvin impulse: application to cavitation bubble dynamics, The ANZIAM Journal 30 (2) (1988) 127–146.

[59] J.R. Blake, D.M. Leppinen, Q. Wang, Cavitation and bubble dynamics: the Kelvin impulse and its applications, Interface Focus 5 (5) (2015). 20150017.

[60] C. Lechner, W. Lauterborn, M. Koch, R. Mettin, Fast, thin jets from bubbles expanding and collapsing in extreme vicinity to a solid boundary: a numerical study, Physical Review Fluids 4 (2) (2019) 021601.

[61] O.V. Voinov, V.V. Voinov, On the pattern of cavitation bubble collapse near the wall and formation of cumulative jet, Doklady Akademii Nauk SSSR 227 (1) (1976) 85–92.

[62] A.A. Aganin, L.A. Kosolapova, V.G. Malakhov, Dynamics of a spheroidal gas bubble near a rigid surface, Journal of Physics: Conference Series 1328 (2019) 012043.

[63] D. Kröninger, K. Köhler, T. Kurz, W. Lauterborn, Particle tracking velocimetry of the flow field around a collapsing cavitation bubble, Experiments in Fluids 48 (3) (2010) 395–408.

[64] E.A. Brujan, G.S. Keen, A. Vogel, J.R. Blake, The final stage of the collapse of a cavitation bubble close to a rigid boundary, Physics of Fluids 14 (1) (2002) 85–92.

[65] J.P. Best, The formation of toroidal bubbles upon the collapse of transient cavities, Journal of Fluid Mechanics 251 (1993) 79–107.

[66] Q.X. Wang, K.S. Yeo, B.C. Khoo, K.Y. Lam, Vortex ring modelling of toroidal bubbles, Theoretical and Computational Fluid Dynamics 19 (5) (2005) 303–317.

[67] Q.X. Wang, Multi-oscillations of a bubble in a compressible liquid near a rigid boundary, Journal of Fluid Mechanics 745 (4) (2014) 509–536.

[68] A.M. Zhang, S. Li, J. Cui, Study on splitting of a toroidal bubble near a rigid boundary, Physics of Fluids 27 (6) (2015) 062102.

[69] F. Reuter, S.A. Kaiser, High-speed film-thickness measurements between a collapsing cavitation bubble and a solid surface with total internal reflection shadowmetry, Physics of Fluids 31 (9) (2019) 097108.

[70] L.T. Liu, X.L. Yao, N.N. Liu, F.L. Yu, Toroidal bubble dynamics near a solid wall at different reynolds number, International Journal of Multiphase Flow 100 (2018) 104–118.

[71] T. Hupfeld, G. Laurens, S. Merabia, S. Barcikowski, B. Gökce, D. Amans, Dynamics of laser-induced cavitation bubbles at a solid-liquid interface in high viscosity and high capillary number regimes, Journal of Applied Physics 127 (4) (2020) 044306.

[72] W. Lauterborn, C. Lechner, M. Koch, R. Mettin, Bubble models and real bubbles: Rayleigh and energy-deposit cases in a tait-compressible liquid, IMA Journal of Applied Mathematics 83 (4) (2018) 556–589.

[73] A. Philipp, W. Lauterborn, Cavitation erosion by single laser-produced bubbles, Journal of Fluid Mechanics 361 (1998) 75–116.

[74] J.-C. Isselin, A.-P. Alloncle, M. Autric, On laser induced single bubble near a solid boundary: contribution to the understanding of erosion phenomena, Journal of Applied Physics 84 (10) (1998) 5766–5771.

[75] M. Dular, T. Požar, J. Zevnik, et al., High speed observation of damage created by a collapse of a single cavitation bubble, Wear 418 (2019) 13–23.

[76] P. Cui, A.-M. Zhang, S. Wang, B.C. Khoo, Ice breaking by a collapsing bubble, Journal of Fluid Mechanics 841 (2018) 287.

[77] E. Johnsen, T.I.M. Colonius, Numerical simulations of non-spherical bubble collapse, Journal of Fluid Mechanics 629 (2009) 231–262.
[78] G.L. Chahine, C.-T. Hsiao, Modelling cavitation erosion using fluid-material interaction simulations, Interface Focus 5 (5) (2015) 20150016.
[79] G.L. Chahine, A. Kapahi, J.-K. Choi, C.-T. Hsiao, Modeling of surface cleaning by cavitation bubble dynamics and collapse, Ultrasonics Sonochemistry 29 (2016) 528–549.
[80] S.A. Beig, B. Aboulhasanzadeh, E. Johnsen, Temperatures produced by inertially collapsing bubbles near rigid surfaces, Journal of Fluid Mechanics 852 (2018) 105–125.
[81] K.-H. Kim, G. Chahine, J.-P. Franc, A. Karimi, Advanced Experimental and Numerical Techniques for Cavitation Erosion Prediction, 106, Springer, Netherlands, 2014. ; for more information see <https://www.springer.com/de/book/9789401785389>.
[82] R.P. Chhabra, Bubbles, Drops, and Particles in Non-Newtonian Fluids, CRC Press, Boca Raton, 2006.
[83] R. Dijkink, C.-D. Ohl, Measurement of cavitation induced wall shear stress, Applied Physics Letters 93 (25) (2008) 254107.
[84] F. Reuter, R. Mettin, Electrochemical wall shear rate microscopy of collapsing bubbles, Physical Review Fluids 3 (6) (2018) 063601.
[85] F. Reuter, R. Mettin, Mechanisms of single bubble cleaning, Ultrasonics Sonochemistry 29 (2016) 550–562.
[86] Q. Zeng, S.R. Gonzalez-Avila, R. Dijkink, P. Koukouvinis, M. Gavaises, C.-D. Ohl, Wall shear stress from jetting cavitation bubbles, Journal of Fluid Mechanics 846 (2018) 341–355.
[87] F. Denner, F. Evrard, B. van Wachem, Modeling acoustic cavitation using a pressure-based algorithm for polytropic fluids, Fluids 5 (2) (2020) 69.
[88] F. Denner, F. Evrard, F. Reuter, S.R. Gonzalez-Avila, B. van Wachem, C.-D. Ohl, Predicting laser-induced cavitation near a solid substrate, in: Holm Altenbach (Ed.), Proceedings in Applied Mathematics and Mechanics, Wiley-VCH GmbH, Weinheim, 2020, e202000007.
[89] P. Koukouvinis, G. Strotos, Q. Zeng, S.R. Gonzalez-Avila, A. Theodorakakos, M. Gavaises, et al., Parametric investigations of the induced shear stress by a laser-generated bubble, Langmuir: The ACS Journal of Surfaces and Colloids 34 (22) (2018) 6428–6442.
[90] E. Maisonhaute, C. Prado, P.C. White, R.G. Compton, Surface acoustic cavitation understood via nanosecond electrochemistry. Part III: shear stress in ultrasonic cleaning, Ultrasonics Sonochemistry 9 (6) (2002) 297–303.
[91] W. Kim, T.-H. Kim, J. Choi, H.-Y. Kim, Mechanism of particle removal by megasonic waves, Applied Physics Letters 94 (8) (2009) 081908.
[92] H.F. Okorn-Schmidt, F. Holsteyns, A. Lippert, D. Mui, M. Kawaguchi, C. Lechner, et al., Particle cleaning technologies to meet advanced semiconductor device process requirements, ECS Journal of Solid State Science and Technology 3 (1) (2013) N3069.
[93] C.-D. Ohl, M. Arora, R. Ikink, N.D. Jong, M. Versluis, M. Delius, et al., Sonoporation from jetting cavitation bubbles, Biophysical Journal 91 (11) (2006) 4285–4295.
[94] F. Reuter, S. Lauterborn, R. Mettin, W. Lauterborn, Membrane cleaning with ultrasonically driven bubbles, Ultrasonics Sonochemistry 37 (2017) 542–560.
[95] A.N. Hellman, K.R. Rau, H.H. Yoon, V. Venugopalan, Biophysical response to pulsed laser microbeam-induced cell lysis and molecular delivery, Journal of Biophotonics 1 (1) (2008) 24–35.
[96] J.L. Compton, A.N. Hellman, V. Venugopalan, Hydrodynamic determinants of cell necrosis and molecular delivery produced by pulsed laser microbeam irradiation of adherent cells, Biophysical Journal 105 (9) (2013) 2221–2231.
[97] K.R. Rau, A. Guerra III, A. Vogel, V. Venugopalan, Investigation of laser-induced cell lysis using time-resolved imaging, Applied Physics Letters 84 (15) (2004) 2940–2942.

[98] K.R. Rau, P.A. Quinto-Su, A.N. Hellman, V. Venugopalan, Pulsed laser microbeam-induced cell lysis: time-resolved imaging and analysis of hydrodynamic effects, Biophysical Journal 91 (1) (2006) 317–329.

[99] B. Helfield, X. Chen, S.C. Watkins, F.S. Villanueva, Biophysical insight into mechanisms of sonoporation, Proceedings of the National Academy of Sciences 113 (36) (2016) 9983–9988.

[100] T. Kodama, Y. Tomita, Y. Watanabe, K. Koshiyama, T. Yano, S. Fujikawa, Cavitation bubbles mediated molecular delivery during sonoporation, Journal of Biomechanical Science and Engineering 4 (1) (2009) 124–140.

[101] O. Krasulya, V. Bogush, V. Trishina, I. Potoroko, S. Khmelev, P. Sivashanmugam, et al., Impact of acoustic cavitation on food emulsions. Ultrasonics Sonochemistry, 30:98–102, 2016.

[102] P.R. Birkin, H.-M. Hirsimäki, J.G. Frey, T.G. Leighton, Mass transfer enhancement produced by laser induced cavitation, Electrochemistry Communications 8 (10) (2006) 1603–1609.

[103] F. Reuter, C. Cairós, R. Mettin, Vortex dynamics of collapsing bubbles: Impact on the boundary layer measured by chronoamperometry, Ultrasonics Sonochemistry 33 (2016) 170–181.

[104] U. Orthaber, J. Zevnik, M. Dular, et al., Cavitation bubble collapse in a vicinity of a liquid-liquid interface-basic research into emulsification process, Ultrasonics Sonochemistry (2020) 105224.

[105] J.P. Best, J.R. Blake, An estimate of the kelvin impulse of a transient cavity, Journal of Fluid Mechanics 261 (1994) 75–93.

[106] Y. Tomita, P.B. Robinson, R.P. Tong, J.R. Blake, Growth and collapse of cavitation bubbles near a curved rigid boundary, Journal of Fluid Mechanics 466 (2002) 259.

[107] P. Cui, A.-man Zhang, S.-ping Wang, Q.-xi Wang, Experimental investigation of bubble dynamics near the bilge with a circular opening, Applied Ocean Research 41 (2013) 65–75.

[108] N. Saleki-Haselghoubi, A. Dadvand, Simulation of bubble dynamics oscillating near a circular aperture made in a curved rigid plate using boundary element method, Ocean Engineering 151 (2018) 12–22.

[109] D.C. Gibson, J.R. Blake, The growth and collapse of bubbles near deformable surfaces, Applied Scientific Research 38 (1) (1982) 215–224.

[110] J.H. Duncan, S. Zhang, On the interaction of a collapsing cavity and a compliant wall, Journal of Fluid Mechanics 226 (1991) 401–423.

[111] J.H. Duncan, C.D. Milligan, S. Zhang, On the interaction between a bubble and a submerged compliant structure, Journal of Sound and Vibration 197 (1) (1996) 17–44.

[112] A. Shima, Y. Tomita, D.C. Gibson, J.R. Blake, The growth and collapse of cavitation bubbles near composite surfaces, Journal of Fluid Mechanics 203 (1989) 199–214.

[113] Y. Tomita, T. Kodama, Interaction of laser-induced cavitation bubbles with composite surfaces, Journal of Applied Physics 94 (5) (2003) 2809–2816.

[114] E.-A. Brujan, K. Nahen, P. Schmidt, A. Vogel, Dynamics of laser-induced cavitation bubbles near an elastic boundary, Journal of Fluid Mechanics 433 (1) (2001) 251–281.

[115] S.W. Ohl, E. Klaseboer, B.C. Khoo, The dynamics of a non-equilibrium bubble near biomaterials, Physics in Medicine and Biology 54 (20) (2009) 6313–6336.

[116] S.J. Shaw, Y.H. Jin, T.P. Gentry, D.C. Emmony, Experimental observations of the interaction of a laser generated cavitation bubble with a flexible membrane, Physics of Fluids 11 (8) (1999) 2437–2439.

[117] C.K. Turangan, G.P. Ong, E. Klaseboer, B.C. Khoo, Experimental and numerical study of transient bubble-elastic membrane interaction, Journal of Applied Physics 100 (5) (2006).

[118] S.R. Gonzalez-Avila, D.M. Nguyen, S. Arunachalam, E.M. Domingues, H. Mishra, C.D. Ohl, Mitigating cavitation erosion using biomimetic gas-entrapping microtextured surfaces (GEMS), Science Advances 6 (13) (2020) 1–11.

[119] S.R. Gonzalez-Avila, E. Klaseboer, B.C. Khoo, C.D. Ohl, Cavitation bubble dynamics in a liquid gap of variable height, Journal of Fluid Mechanics 682 (2011) 241–260.

[120] D. Dowson, C. Taylor, Cavitation in bearings, Annual Review of Fluid Mechanics 11 (35–65) (2003) 11.

[121] L. Bai, J. Yan, Z. Zeng, Y. Ma, Cavitation in thin liquid layer: a review, Ultrasonics Sonochemistry 66 (2020) 105092.

[122] S.-W. Ohl, C.-D. Ohl, Acoustic Cavitation in a Microchannel, Springer Singapore, Singapore, 2016, pp. 99–135.

[123] Y.-H. Chen, H.-Y. Chu, I. Lin, Interaction and fragmentation of pulsed laser induced microbubbles in a narrow gap, Physical Review Letters 96 (2006) 034505.

[124] V.K. Kedrinskii, V.T. Kuzavov, Dynamics of a cylindrical cavity in a boundless compressible liquid, in: W. Lauterborn (Ed.), Cavitation and Inhomogeneities in Underwater Acoustics, Springer, Berlin, Heidelberg, 1980, pp. 119–124.

[125] Ed Zwaan, S.L. Gac, K. Tsuji, C.-D. Ohl, Controlled cavitation in microfluidic systems, Physical Review Letters, 98, 2007, p. 254501.

[126] M. Mohammadzadeh, F. Li, C.-D. Ohl, Shearing flow from transient bubble oscillations in narrow gaps, Physical Review Fluids 2 (2017) 014301.

[127] S.R. Gonzalez-Avila, A.C.V. Blokland, Q. Zeng, C.D. Ohl, Jetting and shear stress enhancement from cavitation bubbles collapsing in a narrow gap, Journal of Fluid Mechanics 884 (2019) A23.

[128] Q. Zeng, S.R. Gonzalez-Avila, C.D. Ohl, Splitting and jetting of cavitation bubbles in thin gaps, Journal of Fluid Mechanics 896 (2020) A28.

[129] P.A. Quinto-Su, C.D. Ohl, Interaction between two laser-induced cavitation bubbles in a quasi-two-dimensional geometry, Journal of Fluid Mechanics 633 (2009) 425–435.

[130] P.A. Quinto-Su, X.H. Huang, S.R. Gonzalez-Avila, T. Wu, C.D. Ohl, Manipulation and microrheology of carbon nanotubes with laser-induced cavitation bubbles, Physical Review Letters 104 (1) (2010) 014501.

[131] F. Yuan, G. Sankin, P. Zhong, Dynamics of tandem bubble interaction in a microfluidic channel, The Journal of the Acoustical Society of America 130 (5) (2011) 3339–3346.

[132] F. Li, C. Yang, F. Yuan, D. Liao, T. Li, F. Guilak, et al., Dynamics and mechanisms of intracellular calcium waves elicited by tandem bubble-induced jetting flow, Proceedings of the National Academy of Sciences 115 (2017) E353–E362.

[133] F. Mekki-Berrada, P. Thibault, P. Marmottant, Acoustic pulsation of a microbubble confined between elastic walls, Physics of Fluids 28 (3) (2016) 032004.

[134] D. Rabaud, P. Thibault, M. Mathieu, P. Marmottant, Acoustically bound microfluidic bubble crystals, Physical Review Letters 106 (2011) 134501.

[135] M.G. Marc Tinguely, A. Hennessy, O.K. Pommella, Matar, V. Garbin, Surface waves on a soft viscoelastic layer produced by an oscillating microbubble, Soft Matter 12 (2016) 4247–4256.

[136] A.A. Doinikov, F. Mekki-Berrada, P. Thibault, P. Marmottant, Lamb-type waves generated by a cylindrical bubble oscillating between two planar elastic walls, Proceedings of the Royal Society A: Mathematical, Physical and Engineering Sciences 472 (2188) (2016) 20160031.

[137] J. Rapet, P.A. Quinto-Su, C.-D. Ohl, Cavitation inception from transverse waves in a thin liquid gap, Physical Review Applied 14 (2) (2020) 1.

[138] E.A. Brujan, T. Noda, A. Ishigami, T. Ogasawara, H. Takahira, Dynamics of laser-induced cavitation bubbles near two perpendicular rigid walls, Journal of Fluid Mechanics 841 (2018) 28–49.

[139] L. Molefe, I.R. Peters, Jet direction in bubble collapse within rectangular and triangular channels, Physical Review E 100 (6) (2019) 063105.

[140] Q. Wang, M. Mahmud, J. Cui, W.R. Smith, A.D. Walmsley, Numerical investigation of bubble dynamics at a corner, Physics of Fluids 32 (5) (2020) 053306.

[141] Y. Tagawa, I.R. Peters, Bubble collapse and jet formation in corner geometries, Physical Review Fluids 3 (2018) 081601.

[142] E.A. Brujan, H. Takahira, T. Ogasawara, Planar jets in collapsing cavitation bubbles, Experimental Thermal and Fluid Science 101 (2019) 48–61.

[143] R.J. Dijkink, J.P. Van Der Dennen, C.D. Ohl, A. Prosperetti, The "acoustic scallop": a bubble-powered actuator, Journal of Micromechanics and Microengineering 16 (8) (2006) 1653–1659.

[144] E. Ory, H. Yuan, A. Prosperetti, S. Popinet, S. Zaleski, Growth and collapse of a vapor bubble in a narrow tube, Physics of Fluids 12 (6) (2000) 1268–1277.

[145] C. Sun, E. Can, R. Dijkink, D. Lohse, A. Prosperetti, Growth and collapse of a vapour bubble in a microtube: the role of thermal effects, Journal of Fluid Mechanics 632 (2009) 5–16.

[146] H. Li, J. Huang, X. Wu, J. Zhang, J. Wang, Y. Wang, et al., Dynamic behaviors of a laser-induced bubble and transition mechanism of collapse patterns in a tube, AIP Advances 10 (3) (2020) 035210.

[147] E.-A. Brujan, Cardiovascular cavitation, Cavitation in Non-Newtonian Fluids with Biomedical and Bioengineering Applications, Springer, Berlin, Heidelberg, 2011, pp. 175–223.

[148] A. Vogel, R. Engelhardt, U. Behnle, U. Parlitz, Minimization of cavitation effects in pulsed laser ablation illustrated on laser angioplasty, Applied Physics B: Lasers and Optics 62 (2) (1996) 173–182.

[149] P. Zhong, Y. Zhou, S. Zhu, Dynamics of bubble oscillation in constrained media and mechanisms of vessel rupture in SWL, Ultrasound in Medicine and Biology 27 (1) (2001) 119–134.

[150] H. Chen, W. Kreider, A.A. Brayman, M.R. Bailey, T.J. Matula, Blood vessel deformations on microsecond time scales by ultrasonic cavitation, Physical Review Letters 106 (2011) 034301.

[151] T. Ye, J.L. Bull, Microbubble expansion in a flexible tube, Journal of Biomechanical Engineering 128 (4) (2006) 554–563.

[152] V. Coralic, T. Colonius, Shock-induced collapse of a bubble inside a deformable vessel, European Journal of Mechanics-B/Fluids 40 (2013) 64–74.

Chapter 3

Cavitation Nuisance in Ship Propulsion: A Review of Developments

Tom J.C. van Terwisga, Martijn van Rijsbergen, Erik van Wijngaarden, Johan Bosschers, Sören Schenke and Themis Melissaris
MARIN/TU Delft, Faculty 3ME, Delft, The Netherlands

3.1 Introduction

Cavitation in engineering was first noticed by Sir Charles Parsons in his efforts to use a high-powered lightweight steam engine to build the fastest ship in the world at the time, the *Turbinia*, which was designed to speeds up to approximately 34 knots (nautical miles per hour). Steam engines had undergone continuous improvements in the 19th century, and the high-pressure, triple-expansion steam engine led to high power-to-weight ratios, opening up a new design space. Parsons' Marine Steam Engine Company started the development of the *Turbinia* as early as 1894. But at the first trials in 1895 a disappointing maximum speed of approximately 20 knots was attained [1]. By installing a torque-measuring device, Parsons concluded that it was the propeller that was the reason for the disappointing speed. He was strengthened in his conclusion by a paper by Barnaby and Thornycroft from 1895 on the effect of cavitation on propellers [2]. To further study this hitherto unknown effect on marine propellers, Parsons built the world's first cavitation tunnel (see Fig. 3.1).

With the further development from steam to internal combustion engines in the early 20th century and the greater powers that became available, cavitation

Cavitation and Bubble Dynamics. DOI: https://doi.org/10.1016/B978-0-12-823388-7.00011-4

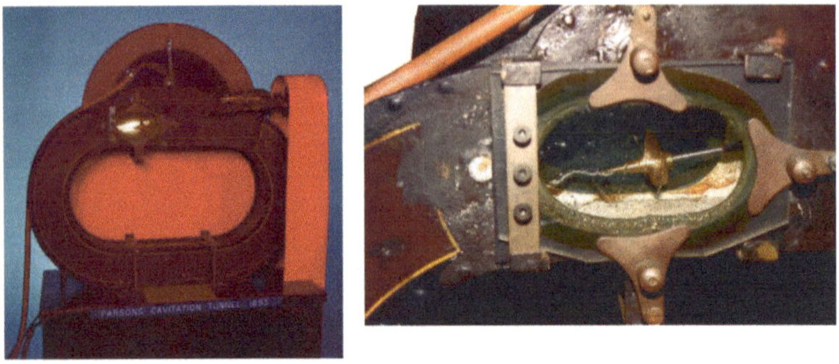

Figure 3.1 The world's first cavitation tunnel for propeller studies, built by Sir Charles Parsons (https://ocw.tudelft.nl/wp-content/uploads/Chapter-2_01.pdf) [3].

problems became an inherent part of ship propulsion. After World War II the larger ships started to suffer from cavitation-induced vibrations in the aftbody of the ship, cavitation erosion on the propeller blades and possibly the rudders, thrust breakdown (similar to the thrust problem on *Turbinia*), and inboard noise. In particular, the increasing powers installed in tankers and bulk carriers led to serious vibration and erosion problems. In the 1990s, the cavitation problems were caused by increasing power densities on propellers and increasing ship speeds, and it seemed that propeller designers were facing the cavitation limits of their propeller designs. Since 2008, when a serious global financial crisis led to a reduction in ship speeds and the increasing awareness of global warming led to a drive for low fuel consumption and high propulsive efficiency, propeller designers have been more and more searching for the balance between efficiency and acceptable cavitation nuisance. Since the early 21st century a growing awareness of the impact of underwater radiated noise (URN) by shipping has added further reasons to reduce noise emission.

The importance of cavitation for the marine propeller can readily be observed when comparing the blade area ratios of airplane propellers with those of marine propellers. The typical difference is that the blade area relative to the propeller disk for the marine propeller is larger. The larger blade area is needed to distribute the pressure difference needed for a given thrust requirement so that the absolute minimum pressures remain within the bounds of acceptable cavitation extent.

What is the effect of cavitation hindrance constraints on ship propulsive efficiency? A systematic study to answer this question has hitherto not been conducted to the knowledge of the author. But collected experience [4] and systematic studies of a limited extent indicate that the losses could be of the order of 10%–15% if one includes the limitations imposed by vibration considerations and the consequent limitations to the propeller diameter. To illustrate the battle between efficiency and cavitation nuisance, Fig. 3.2 gives a nice impression of

Figure 3.2 Example of cavitation erosion near the trailing edge of a propeller blade as caused by the tip vortex coming off the tip of the preswirl stator. The rubber casting gives an impression of the surface structure of the damaged propeller blade (Courtesy Lloyd's Register) [5].

the damage done to a propeller blade by the tip vortex coming off a preswirl stator.

The objective of this chapter is to describe the problems caused by cavitation, most often occurring on propellers but sometimes on rudders and energy-saving devices as well. This is followed by a brief description of the essential physics that govern the dynamics of the multiphase flow. We will seek answers to questions such as "What is the role of viscous stresses?" and "Can we apply Reynolds averaging to the Reynolds stresses without affecting the dynamics too much? And if we do affect the dynamics, in which frequency range and to what extent? And how does this feed back to the power spectrum of pressure pulses radiated on the hull?" Many answers to these questions have been collected in the course of time by several researchers. Some questions are still open. We should appreciate that the revelation of the essential physics depends on the suitability and shortcomings of experiments and numerical models used in research. This is why we give a characterization of salient experiments and numerical models at the close of each section.

3.2 Cavitation Inception

This section describes the problem of cavitation inception and its consequences. It starts with a historical experiment showing a complete lack of agreement in results, followed by a summary of our understanding of the physics and a notional experiment as well as the state of art in numerical modeling of the inception process. Although from a physical point of view, it seems logical to start this chapter with cavitation inception, the physics that control cavitation inception do so at a micro scale and appear more complicated than those for developed cavitation. This is the reason that direct observations give us insight into the consequences of parametric variations but not directly into the mechanisms. These have to be derived from hypotheses and their falsification.

The Report of the ITTC Cavitation Committee [6] presents the results of cavitation inception tests for sheet cavitation that were conducted on a cylindrical half body in various facilities around the world. Each water tunnel was to use its own procedure. The tests were to be performed at constant water velocity and air content ratio, and the pressure was to be lowered to find the point of cavitation inception. Fig. 3.3 gives a summary of results of the cavitation number at inception. The wide scatter in results is remarkable and is characteristic of the complex mechanisms that influence cavitation inception.

Cavitation on a propeller may occur in different forms as indicated in Fig. 3.4. We will focus here on the most annoying types of cavitation causing serious restrictions for propeller designers: sheet cavitation and vortex cavitation. Sheet cavitation typically occurs near the leading edge of the propeller blade or any other foil, whereas vortex cavitation may occur in different locations. The most notorious are the tip vortex and sometimes leading edge vortices that cavitate. Most often hub vortex cavitation also occurs. However, the latter type of vortex cavitation appears to produce less noise and contributes little to pressure-induced vibrations on the hull.

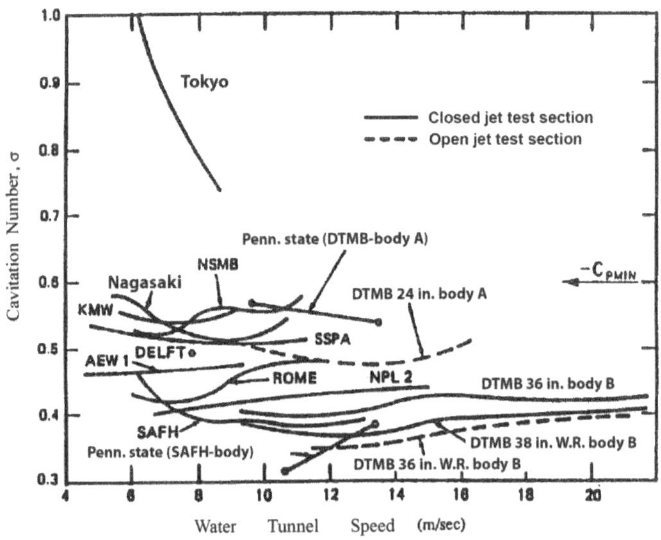

Figure 3.3 Comparison of cavitation inception numbers from tests on the same half body in different facilities. Air content ratio: $\alpha/\alpha_s = 0.2$. Values have been corrected for blockage. *Taken from J.D. van Manen, S. Bindel, P. Eisenberg, H. Lindgren, L. Mazarredo, W.B. Morgan, et al., 11th international towing tank conference, in: Report of Cavitation Committee, 167–269. ITTC, 1966.*

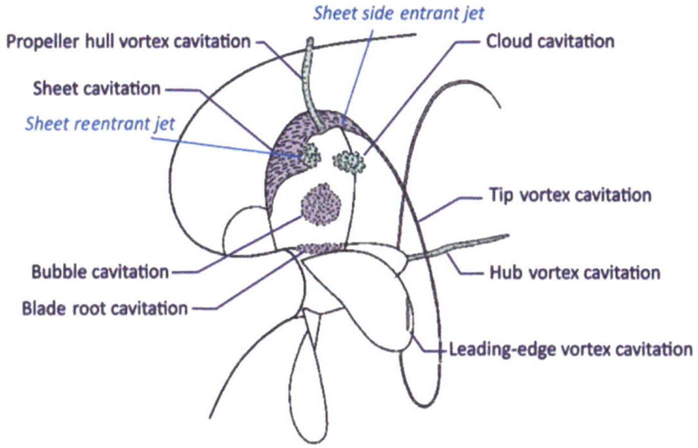

Figure 3.4 Types of cavitation and nomenclature typically occurring on a propeller [8], adapted from ITTC procedure 7.5-02-03-03.2.

The dynamics of unsteady cavitation are responsible for the cavitation-induced pressures, with the first and second time derivatives of the cavity volume determining the radiated pressures [7]. If we are to measure these cavitation-induced pressures on a model scale, which is common practice for ships that are expected to come close to the acceptable limits, we need to make sure that the cavity volume dynamics on the model scale are identical to those on the full scale. However, experimental observations by van Rijsbergen and van Terwisga [9] showed that even for developed sheet cavitation, cavitation inception may not occur over the entire span of the sheet cavity, which would be expected on the full scale (Fig. 3.5). This example illustrates that cavitation inception may not only be relevant for incipient cavitation, but could also affect cavitation extent and therefore pressure fluctuations radiated by developed sheet cavity dynamics.

It is therefore important to understand the physical mechanisms at work in cavitation inception. A review of hitherto known mechanisms for sheet cavitation inception is given by van Rijsbergen [10]. He distinguishes different mechanisms that involve either the presence of free stream nuclei or the presence of surface (wall) bound nuclei. The first mechanism makes use of particles in the flow, which could be either a gaseous bubble or a solid particle, as summarized in Fig. 3.6. It is hypothesized that the gas bubble (gas-filled nucleus) or the gas-filled wake of a solid particle touches the separation streamline of a roughness element on the surface of the foil or propeller blade. Such an inception process may also occur on the separation streamline of a laminar separation bubble where no roughness element on the surface is present.

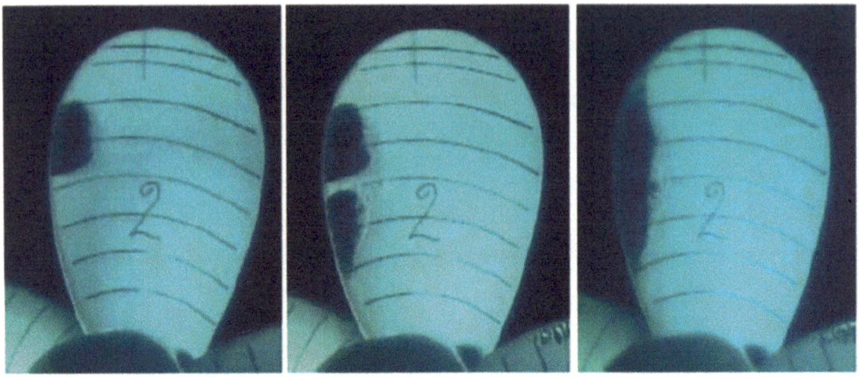

Figure 3.5 Typical observed sheet cavities at $\sigma_n = 1.7$, $K_T = 0.15$, and $R_n = 1.2 \times 10^6$ with electrolysis grid upstream of the propeller. Left: current $I = 0.0$ A. Middle: $I = 0.0$ (30 s later in the same run as left photo). Right: $I = 3.6$ A. *From M.X. van Rijsbergen, T.J.C. van Terwisga, Water quality effects on sheet cavitation inception on a ship propeller model, in: 7th International Conference on Multiphase Flow. Tampa, Florida USA, 2010. file:///D:/Literature/Rijsbergen/Fluid_dynamics/Cavitation/Inception/Rijsbergen_2010_ICMF.pdf.*

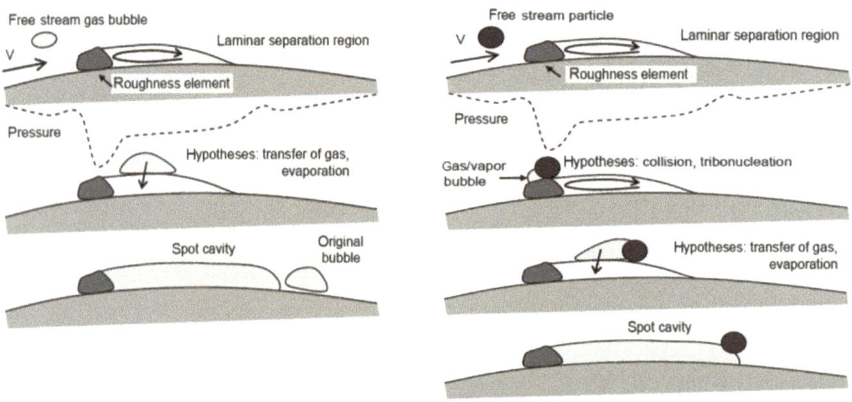

Figure 3.6 Cavitation inception process initiated by free stream nuclei. Left: Inception by a gas bubble (nucleus) and a wall bound roughness element. Right: Inception by a solid particle [10]. *Adapted From M.X. van Rijsbergen, A review of sheet cavitation inception mechanisms, in: International Symposium on Transport Phenomena and Dynamics of Rotating Machinery. Honolulu, Hawaii, 2016. file:///D:/Literature/Rijsbergen/Fluid_dynamics/Cavitation/Inception/Rijsbergen_2016_ISROMAC.pdf.*

The second mechanism is based on surface-bound gas nucleation, which can be produced by microscopic crevices. It has been demonstrated by Groß [11] that such crevices may act as sites where dissolved gas diffuses into gas bubbles, which subsequently act as nucleation sites for cavitation streaks. This mechanism is sketched in Fig. 3.7. The appearance of the gas-filled streak seems to depend on the pressure gradient over the surface, whether it is an accelerating or decelerating flow.

In all inception mechanisms the gas and particle content of the flow plays an important role in the point of cavitation inception. The mechanisms described here give only qualitative relationships between water quality and cavitation inception. The importance of water quality effects has been known since researchers started to investigate the large variation of cavitation inception as demonstrated in Fig. 3.3. However, it is cumbersome to measure the nuclei spectrum in a flow. Techniques to do so are being developed and have been applied since the 1990s. A nice example of early water quality measurements in relation to the various forms of cavitation on a propeller is given by Gindroz and Billet [12]. A more recent example is given by Russell et al. [13], who also present an experimentally determined relationship between the nuclei spectra and the character of sheet cavitation (Fig. 3.8).

Further research on a microscopic level is needed to derive quantitative relationships. Experiments need to be performed that allow for accurate measurement of the gas and particle content in the flow and that allow for detailed cavitation inception observations on a micro-scale.

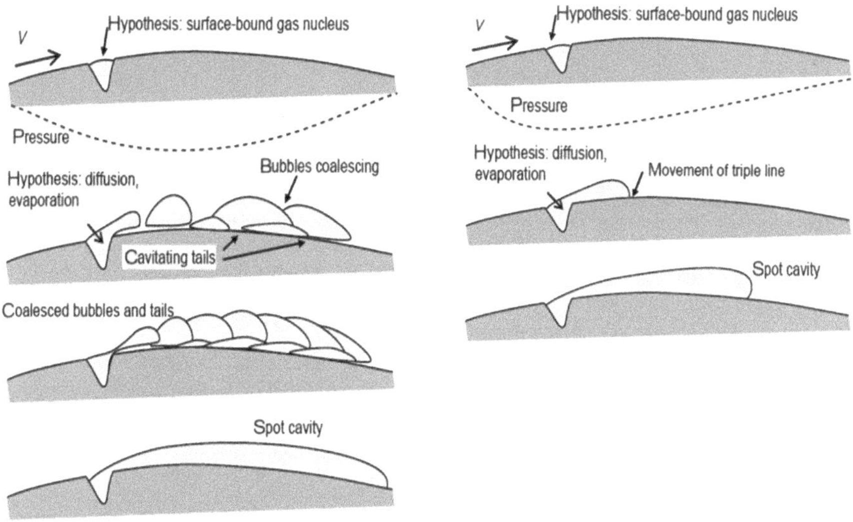

Figure 3.7 Inception process for a surface-bound gas nucleus from a crevice in an accelerating flow (left) and under a decelerating flow (right). *Adapted From M.X. van Rijsbergen, A review of sheet cavitation inception mechanisms, in: International Symposium on Transport Phenomena and Dynamics of Rotating Machinery. Honolulu, Hawaii, 2016. file:///D:/Literature/Rijsbergen/Fluid_dynamics/Cavitation/Inception/Rijsbergen_2016_ISROMAC.pdf.*

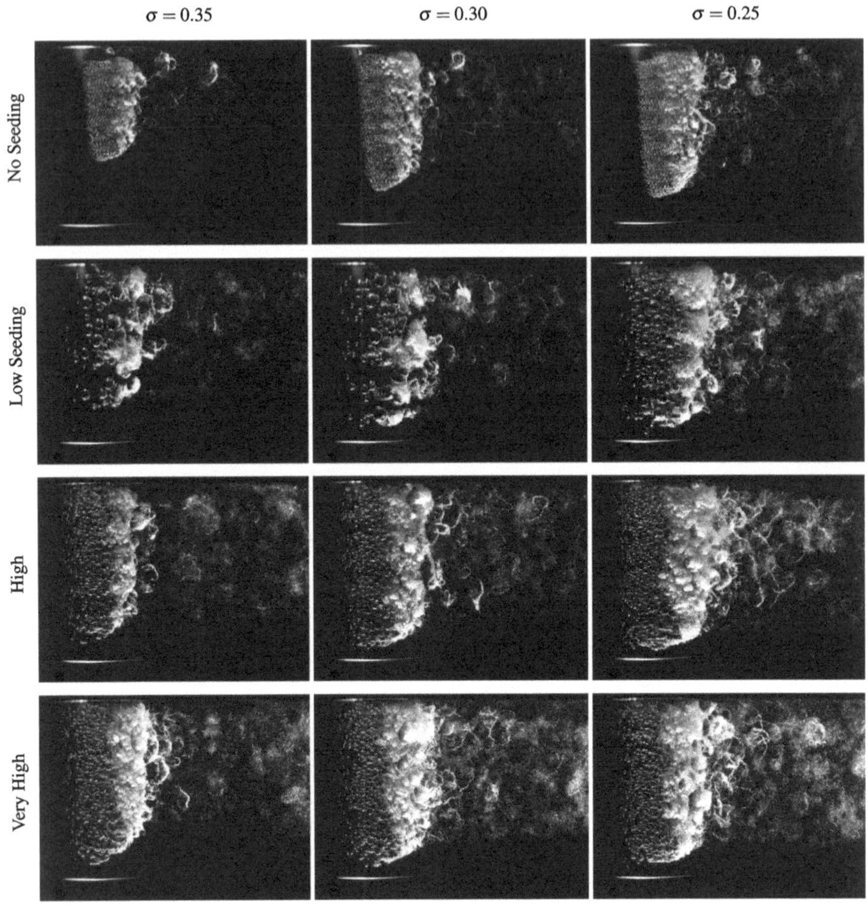

Figure 3.8 Effect of seed concentration on the inception structure and collapse of the sheet/bubble cavities. The number of large bubbles in the near wake increase with the concentration of seeding and decrease with cavitation number. *Copied from P.S.Russell, J.A. Venning, P.A. Brandner, B.W. Pearce, D.R. Giosio, S.L. Ceccio, Microbubble disperse flow about a lifting surface, in: 32nd Symposium on Naval Hydrodynamics, no. August: 5–10, 2018. https://jamesvenning.net/wp-content/uploads/2020/04/2018-SNH-32nd-Russell_et-al.pdf%0A, https://jamesvenning.net/wp-content/uploads/2020/04/SNH2018_seeding.png.*

With our current understanding, it seems that in order to model the mechanisms numerically, we need to distinguish between the individual bubbles and particles in the flow. We also need a gas diffusion model to model the correct gas content in the bubble or cavity and a wall nucleation model. Pioneers in

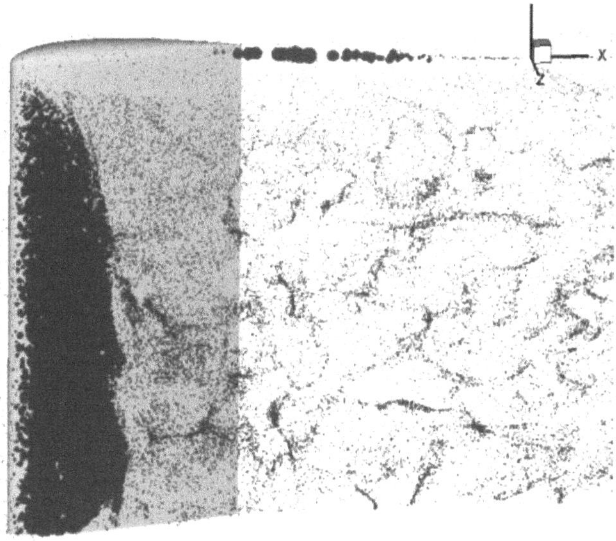

Figure 3.9 Bubble distribution over a finite-span hydrofoil shown at a selected time with both free nuclei and a boundary nucleation model ($f_n = 0.8$ kHz and $N_s = 10\ \text{cm}^{-2}$). Copied from [13].

predicting cavity sheet and vortex dynamics from such an Euler-Lagrangian perspective are found in [14]. An example of a numerical simulation of sheet cavitation over a foil is presented in Fig. 3.9.

3.3 Cavitation-Induced Vibrations

Perhaps the most important problem caused by cavitation is the fluctuating pressures, leading to forces on the nearby hull, contributing to vibrations and onboard noise. The hull thereby amplifies those frequencies that coincide with the natural frequencies of the hull structure.

The comfort level of a ship is not expressed in terms of pressure fluctuations but by its vibration and noise levels. Limits on onboard noise levels are usually specified as frequency weighted rms noise levels in dB(A) between 10 Hz and 30 kHz, e.g. IMO MSC.337 (2000). Most class rules on vibration levels refer to ISO Standard 6954:2000. This standard, released in 2000, poses limits to the frequency weighted rms vibration values between 1 Hz and 80 Hz. Maximum vibration levels are usually found at frequencies that correspond to multiples of the blade-passing frequency (BPF).

URN has the same source and obeys largely the same physical laws as the propeller-induced vibrations: Cavitation dynamics on or in the vicinity of the propeller. Interest in URN has grown over the past years, as there is accumulating evidence that URN often interferes with marine life [15]. For reasons of available space, this topic is not discussed here.

3.3.1 On the Mechanisms of Propeller-Induced Hull Vibrations

The simplest case of bubble dynamics is a vapor bubble in an infinite domain that is excited by a pressure disturbance. Using the conservation of momentum and mass for a control volume that contains (part of) the vapor-liquid interface, one can derive an expression for the pressure $p(r,t)$ at some radius r in the domain and a time instant t (see, e.g., Franc et al. [16], eq. 3.9)

$$\frac{p(r,t) - p_{\infty}(t)}{\rho} = \ddot{R}\frac{R^2}{r} + 2\dot{R}^2\left(\frac{R}{r} - \frac{R^4}{4r^4}\right) \tag{3.1}$$

where ρ is the specific mass of the fluid outside the vapor bubble and R is the instantaneous bubble radius at time t. This relation can be rewritten as a function of the bubble volume, which leads to

$$\frac{p(r,t) - p_{\infty}(t)}{\rho} = \frac{\ddot{V}}{4\pi r} - \frac{\dot{V}^2}{32\pi r^4} \tag{3.2}$$

where $\dot{V}$ and $\ddot{V}$are the first and the second time derivatives of the bubble volume. This relation shows that both the volume acceleration and velocity affect the local pressure but that the effect of the volume velocity decays much faster with r than the acceleration term.

The cavity volume and its dynamics on a marine propeller deviate significantly from the ideal vapor bubble in an infinite environment. First, the cavity morphology on a propeller is different; second, the excitation of these cavities is more complicated than that of the sudden and often stylized pressure change on a cavity bubble. Nevertheless, Eq. (3.2) remains a useful approximation for the cavitation induced pressure in the far field.

Research over recent years has given us more insight into the relationship between the cavity morphology and the pressure spectrum. A representative pressure spectrum as measured on a pressure sensor in the ship's hull above the propeller showing predominantly tip-vortex cavitation is presented in Fig. 3.10. Clearly noticeable in this spectrum are the tonal components at the blade-passing frequency (BPF) and multiples thereof. Also characteristic is a broadband energy hump that stretches roughly from approximately 40–60 Hz. In the presence of sheet cavitation, the hull pressure spectrum is dominated by the tonal harmonics of the blade passage frequency.

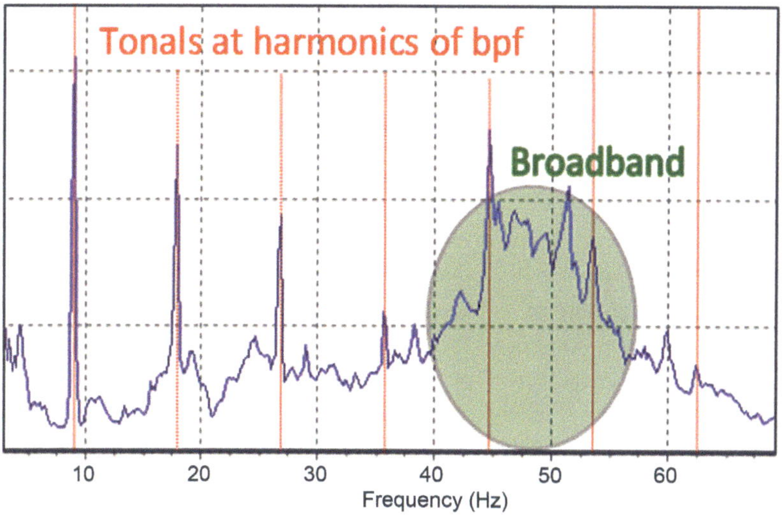

Figure 3.10 Typical pressure amplitude spectrum induced by a ship's propeller with predominant tip-vortex cavitation showing broadband excitation in between spikes at multiples of the blade-passing frequency (BPF) [17]. *From E. van Wijngaarden, J. Bosschers, G. Kuiper, Aspects of the cavitating propeller tip vortex as a source of inboard noise and vibration, in: 2005 ASME Fluids Engineering Division Summer Meeting and Exhibition, 1–6, 2005. https://doi.org/10.1115/FEDSM2005-77271.*

What is governing the dynamics of the cavity? Is it the pressure variation on the blade due to the ship wakefield or is it the pressure varations by turbulence or is it affected by the natural vibrations of the cavity? As for hull vibrations, the wakefield appears to be the main excitation source for the forced cavity dynamics. Natural shedding frequencies of a sheet cavity are typically of a higher frequency than the BPF and its multiples and can typically not be discerned in pressure amplitude spectra.

If the cavity pattern is dominated by tip-vortex cavitation, the tonals at higher order harmonics of the blade passage frequency are small and the broadband part becomes more prominent. The repetitive passage of the propeller blade through the ship wake field leads to a repetitive cavity collapse and pressure pulses. If the cavity collapse would be perfectly repetitive, the pressure spectrum would consist of tonals at harmonics of the blade passage frequency only. However, in reality the cavity collapse varies somewhat between blade passages which leads to a combination of tonals and broadband content of the spectrum. The variations in cavity collapse can be due to among others variations in ship wake field, cavitation inception, ship motions, and propeller operating condition. The collapse and rebound of the vortex cavity suggests the presence of a resonance frequency of the deformation mode, but further research is required to understand the mechanisms involved. A detailed discussion of the mechanisms is given by

Bosschers [18]. The lowest natural frequencies are associated with the lowest-order deformation modes of a line vortex. The three lowest deformation modes are visualized in Fig. 3.11.

3.3.2 Experimental Techniques

When conducting experiments on cavitating propellers, one should realize that both the noncavitating propeller blade and the cavity produced by the propeller blade cause a pressure signature of a different character (see Fig. 3.12). The pressure signature of the cavity is dominated by a monopole character, whereas the noncavitating blade is typically represented by a dipole character. The latter contribution does not propagate as effective as the cavitating monopole contribution. However in the near field, such as on the hull above the propeller, the non-cavitating blade has a significant contribution, especially at the blade rate frequency.

Because the pressure field on the hull typically shows a strong spatial and temporal variation, it is important to sample pressure fluctuations on the hull at sufficient resolution. This is especially important because it is the integrated pressure force acting on the hull that sets the hull into vibration rather than the pressures at separate sensor locations. This is why a relatively large number of pressure sensors needs to be mounted in the hull of a ship model when performing experiments (Fig. 3.13).

Cavitation is a highly nonlinear phenomenon in the sense that small disturbances in the flow can cause large effects in cavitation dynamics and hence in

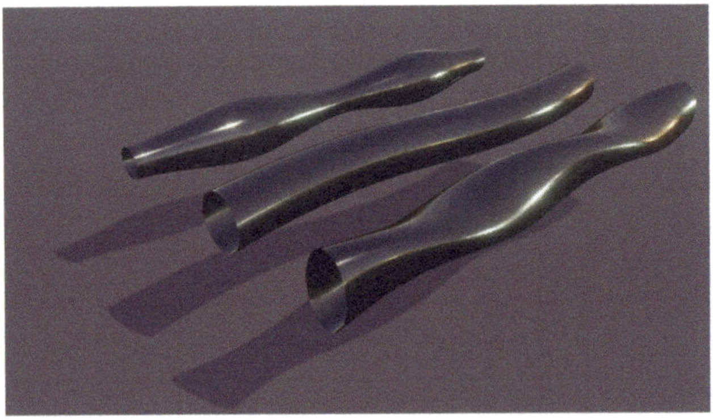

Figure 3.11 Deformation modes of the vortex cavity. From top to bottom: $n=0$ (volume variation), $n=1$ (centerline oscillation), and $n=2$ mode (elliptical axis oscillation or twisted ribbon). Visualization done in Blender [8,19]. *Based on J. Bosschers, Cavitating Vortices on Ship Propellers and the Related Hull Pressure Fluctuations and Radiated Noise Proefschrift. University of Twente, 2018; M. Klapwijk, Elliptical Foil with Turbulence Generator, T.b.D., 2020.*

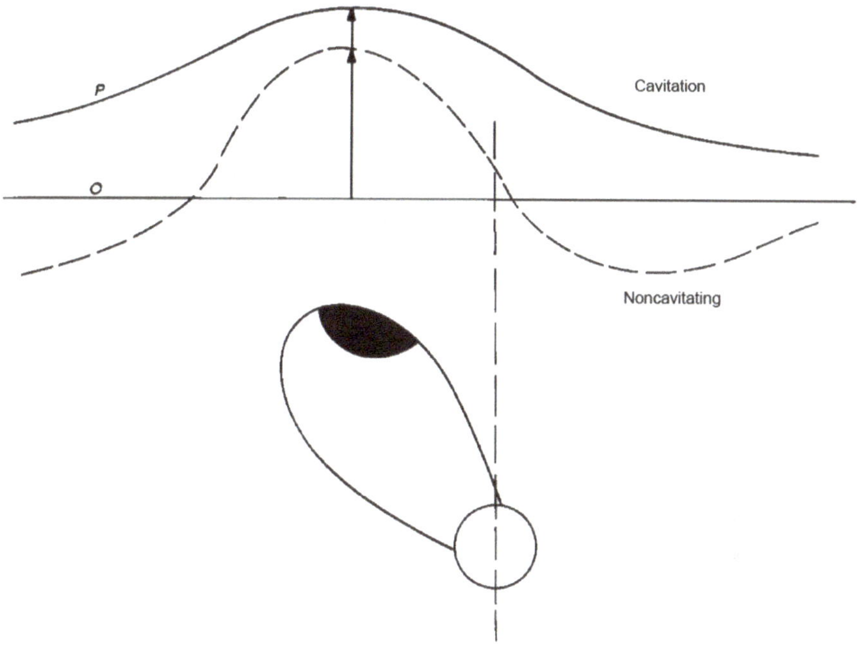

Figure 3.12 Schematic of the difference in phase angles between the pressures radiated on the hull from a noncavitating propeller (due to displacement and loading) and the cavitation dynamics.

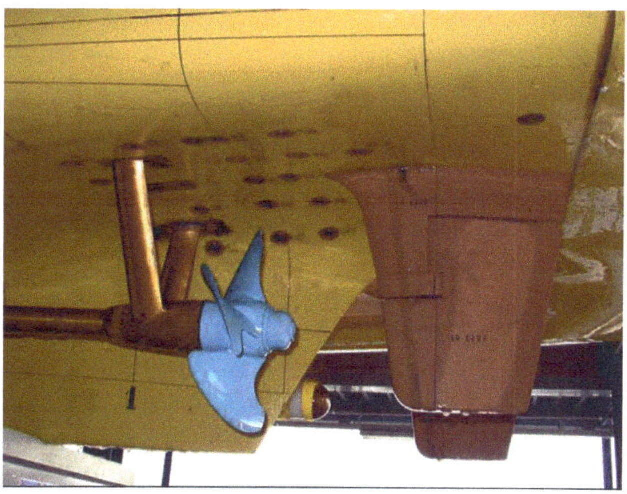

Figure 3.13 Outfitting of a ship model with a distribution of pressure transducers on the hull to sample the propeller-induced pressures.

pressure fluctuations. As a consequence, much of our understanding of the physics of cavitation derives from experiments.

Experiments are also subject to the sensitivities of cavitation dynamics. This appears, for instance, through scale effects (notably viscous scale effects and water quality effects) and laboratory effects, such as wall effects in laboratories that are constrained in size.

3.3.2.1 Sheet Cavitation and Tonal Pressure Pulses

Cavitation-induced pressure fluctuations are traditionally most reliably determined from model-scale experiments with relatively large ship hulls and propellers. The large scale is required to limit viscous-scale effects to a minimum. A large facility is required to limit blockage effects on the flow by the model. Large facilities could be realized through a cavitation tunnel concept with typical cross sections of the order of 3 m in width or diameter or by a depressurized towing basin.

Despite these large-scale experiments, it appeared in the course of time that significant deviations could occur between a full-scale measurement of propeller-induced pressure fluctuations and similar measurements on large-scale models. A typical example of such deviations is shown in Fig. 3.14. This graph shows the propeller-induced pressures on the hull of a container ship. This class of ships showed notorious deviations in pressure amplitudes between model scale and full scale.

Other ship types, notably twin-screw vessels such as cruise ships and ferries but also full block ships such as single-screw tankers and bulk carriers, suffered much less from such discrepancies [23].

van Wijngaarden [20] set out to investigate factors that would contribute to the uncertainty of pressure pulse predictions and the discrepancies found with full

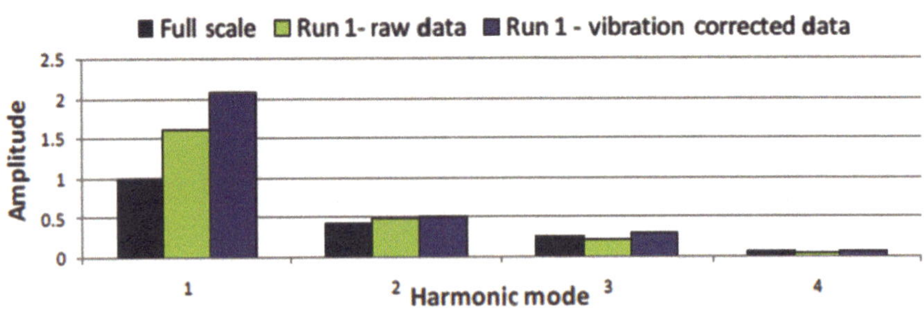

Figure 3.14 Comparison of propeller-induced pressure amplitudes on a hull-mounted transducer for the first four BPFs [20]. *From H.C.J. van Wijngaarden, Prediction of Propeller-Induced Hull-Pressure Fluctuations Proefschrift, 2011.*

scale. He acknowledged three important factors that would complement the existing measurement uncertainty:

- Viscous-scale effects on the effective wake and its influence on sheet cavity dynamics
- Vibration-induced contributions to the pressure signals
- Differences in the variability of measured pressure fluctuations from one blade passage to another, for example due to water quality, scale and/or laboratory effects. These may significantly affect the higher order harmonics in the pressure fluctuations.

A fourth factor was not investigated by him but was later also demonstrated to be responsible for the discrepancies between full-scale and model-scale measurements and for the sometimes disappointing reproducibility of repeat measurements at different times. This factor is the water quality (see e.g. [22]). Water quality in this respect is defined both in terms of micro bubble (nuclei) spectrum in the incoming flow (distribution of size over number of bubbles) and dissolved gas. Recent experimental observations suggest that water quality can have a significant effect on the appearance and dynamics of sheet cavitation and possibly on those of vortex cavitation. One of the most convincing examples is reported in [21] and shown here in Fig. 3.15.

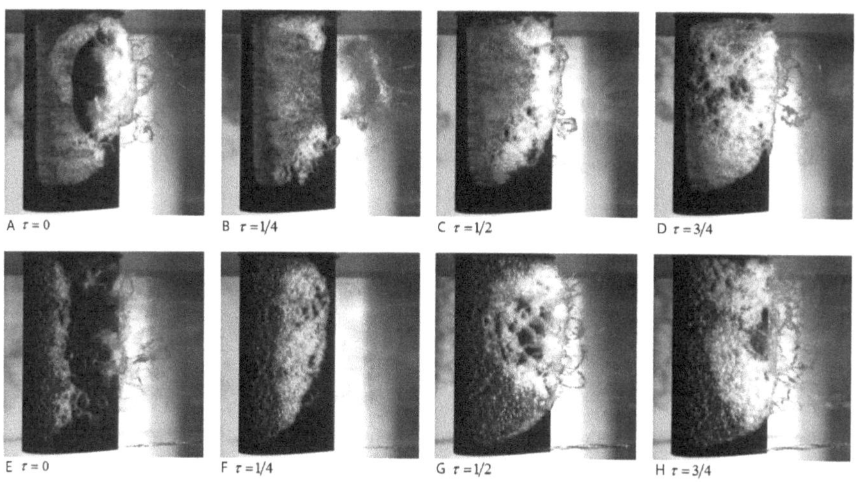

Figure 3.15 Photographs at one-quarter increments of the shedding cycle for the nuclei-depleted (top) and abundant (bottom) conditions. The shedding cycle for the depleted case is 1.8 times faster than that for abundant nuclei [23].

3.3.2.2 Tip Vortex Cavitation and Broadband Pressure Pulses

Which energy content in a typical hull pressure fluctuation spectrum (Fig. 3.10) is caused by vortex cavitation? And which part can be ascribed to sheet cavitation and which part to vortex cavitation? And how big is the role of sheet-vortex interaction? These questions have driven cavitating vortex research for the past few decades, and the contours of an answer have recently emerged. Let us focus here on the contribution of the cavitating vortex.

A classical study of the dynamics of a cavitating vortex has been carried out on the so-called singing vortex, which was reported by Maines and Arndt [24]. To better understand its dynamic behavior, these experiments were repeated by Pennings et al. [25]. The dynamics of the cavitating vortex were observed with two high-speed cameras (see Fig. 3.16) and were subsequently analyzed with Fourier analysis. The deformation mode shapes and corresponding frequencies could well be matched to the modes and frequencies from Thomson [27], who derived them from the model of a columnar cylinder. It appeared possible to match these dynamics in terms of the relationship between wavelength and frequency in a so-called dispersion diagram (wave number versus frequency) for the three deformation modes sketched in Fig. 3.11.

If we now have an understanding of the dynamics of a cavitating tip vortex from a stationary foil in a uniform and steady inflow, how can we use this knowledge for cavitating vortices coming off a propeller blade that operates in a nonuniform flow, inherently producing an unsteady flow over the propeller blade?

It is difficult to convert our knowledge of deforming two-dimensional cylinder models in a uniform flow to the reality of an unsteady cavitating tip vortex from a propeller in a nonuniform wakefield. Let us consider the still pictures in Fig. 3.17 of cavitating vortices from the propeller blade of two different ships: a full block bulk carrier and a container ship. The tip vortex coming off the propeller blade of the bulk

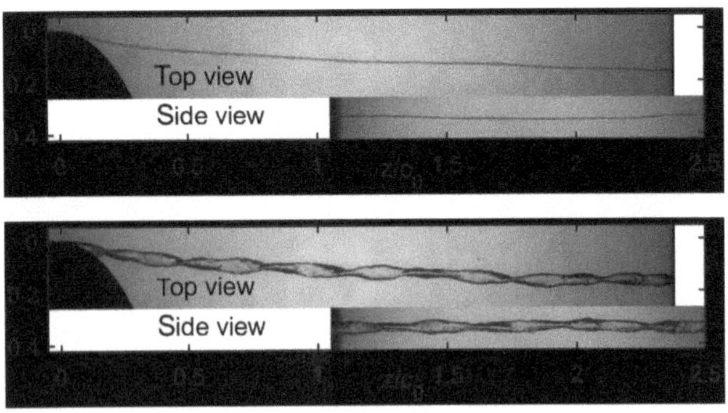

Figure 3.16 Top and side views of a vortex cavity from an elliptical foil. Flow is from left to right [26]. *Reproduced from P. Pennings, Dynamics of Vortex Cavitation. TU Delft, 2016.*

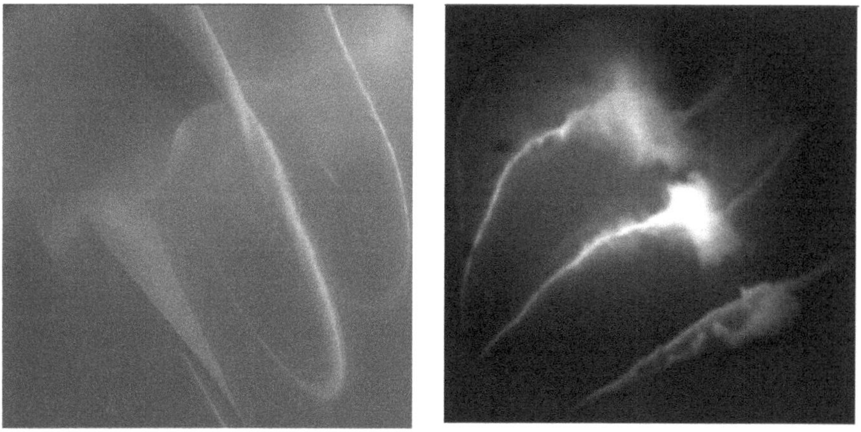

Figure 3.17 Stills from high-speed video recordings of full-scale cavitating tip vortices. Left: Observation from a single-screw bulk carrier (EU FP7 GRIP project). Right: Observation from a single-screw container vessel (EU FP5 EROCAV project). *From J. Bosschers, Cavitating Vortices on Ship Propellers and the Related Hull Pressure Fluctuations and Radiated Noise Proefschrift. University of Twente, 2018.*

carrier appears as a twisted ribbon developing smoothly along a so-called helicoidal path without apparent major distortions for the upcoming blade. However, this is not the case for the tip vortex coming off the blade of the container ship in the right picture. The latter seems to be composed of at least two vortex strings, which cause a major disruption shortly after leaving the blade as shown in the right picture. One vortex string is likely produced by the closure region of the sheet cavity, the other is expected to be the tip vortex.

Bosschers [7] argued that the collapse of the cavity on the blade, which can be either an attached vortex cavity or a sheet cavity, excites the natural vibration modes of the vortex, where the volume variation gives a dominant contribution to the radiated noise in the far field. This volume variation is linked to the first deformation mode of the vortex cavity (Fig. 3.11).

Although the dispersion relationship for this first deformation model allows for a wide range of natural frequencies, only the vibration model in which the wave group velocity over the interface is zero is effective in radiating sound. This condition then yields one dominant natural frequency in the radiated noise spectrum, which depends on the circulation strength of the vortex, the incoming velocity, and the diameter of the cavity [25].

This analysis has led Bosschers [7] to derive the so-called ETV model (Empirical cavitating Tip Vortex model), in which an estimate of this natural frequency is used as the center frequency of the broadband hump which is typically situated between the fourth and seventh BPF for propellers operating in a ship's wake.

It was demonstrated by Pennings [35] that this natural deformation mode generates a significant noise level only after the corresponding deformation mode is excited in mode-dependent frequencies. We assume, however, that for propellers operating behind the ship there is always an excitation source available in the nearby flow.

The smearing of the energy in the spectrum around this vortex-induced broadband hump is argued to be caused by variations in cavity collapse between blade revolutions [36].

3.3.3 Computational Techniques

Having discussed our current understanding of the mechanisms of the tonal and broadband pressure fluctuations, we may infer that we need the cavity volume accelerations and velocities in order to compute the radiated pressures at any one point in the flow field. So the question is which computational model is efficient and sufficiently reliable to give us the requested information.

Boundary element methods solving the potential flow problem, supplemented with empirical corrections for viscosity, have shown to be a powerful tool in unsteady computations of both noncavitating and cavitating propellers (see, e.g., [30] and [29]).

As can be inferred from Eq. 3.2, the cavity volume acceleration dominates the radiated pressure field, so this quantity needs to be predicted reliably and correctly. Although the volume acceleration can be derived from the computations with relative ease, this is not the case for cavitation experiments on propellers. An attempt to make this comparison between computations and experiments was made by Bosschers et al. [35]. The authors compared the computed and experimentally observed cavity-covered area on the blade (see Fig. 3.18). Computations and experiments were made for INSEAN propeller E779A, which has been tested behind a wake generator to simulate a representative nonuniform wakefield.

A validation study of radiated propeller induced pressures was conducted for a more realistic case, in which full-scale computations were made for the propeller behind a single-screw container ship (P&O *Neddlloyd Shackleton*). The effective wakefield that was necessary as input for the unsteady boundary element method (BEM) computations was obtained from a coupled RANS-BEM code, where the viscous flow around the ship hull has been computed by the RANS solver, in which the propeller is represented by a stationary force field, and the propeller forces have been computed by a BEM. The effective wakefield is then computed from the total wakefield computed by RANS

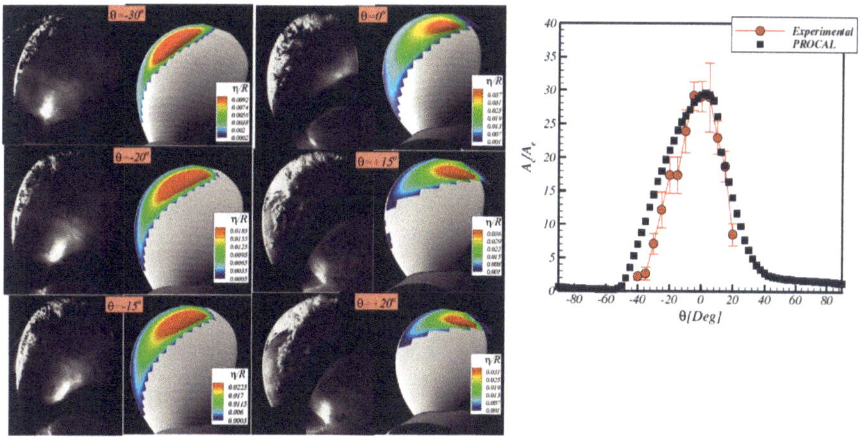

Figure 3.18 Boundary element method results for sheet cavitation dynamics on INSEAN propeller E779A propeller in a wakefield. Cavity thickness plots for six blade positions are shown in the left-hand graphs [36]. The normalized computed cavity area is compared to an estimate of the cavity area from experiments in the right-hand graph [35] and [33].

minus the propeller-induced velocity computed by BEM, thus correctly accounting for the interaction between the wakefield of the ship and the propeller forces [35]. An example of the attainable correspondence in pressure amplitude between ship trial measurements and computations is given in Fig. 3.19.

From a number of validation and design studies, it has been concluded that only the first two tonal components in the pressure spectrum can be computed by BEM. The higher order pressure fluctuations are not captured due to limitations of the numerical model and the absence of tip-vortex cavitation.

Engineering methods have been proposed that use the computed strength of the tip vortex to compute the tip-vortex cavity induced pressure fluctuations and noise. Such a method was initially proposed by Raestad [34] and later further developed by Bosschers [35]. A typical result of the latter method, which is referred to as the ETV model, is presented in Fig. 3.20.

To capture the radiated pressure fluctuations due to the dynamics of the natural vibration modes of the vortex, both the vortex strength and the interface dynamics need to be captured by the numerical model. Asnaghi et al. [36] demonstrate that RANS methods do not calculate the flow in the vortex core correctly. In their study, they show that implicit large eddy simulation (ILES) comes significantly closer to the predicted in-plane velocities near

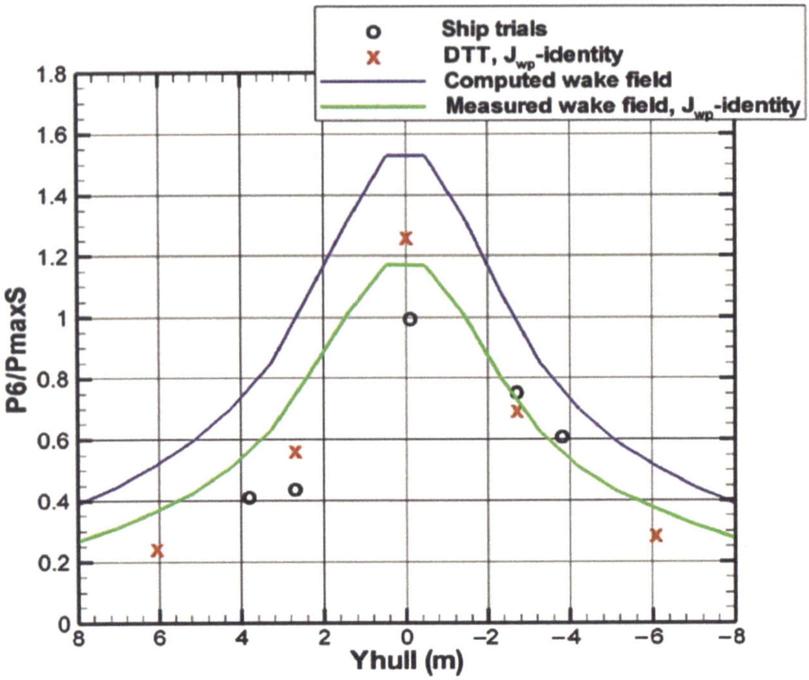

Figure 3.19 Comparison of the spatial distribution between nondimensional hull pressure amplitudes at first blade passage frequency (P6) above the propeller disk at full scale. The left part of the graph is above the upcoming blade, and the right part is above the downgoing blade. Pressure amplitudes have been normalized with the maximum pressure amplitude measured during ship trials. The DTT symbols refer to model scale measurements in the Depressurized Towing Tank of MARIN, the two lines refer to computations with different wakefields.

the vortex core than RANS (Fig. 3.21). Due to the time averaging approach in turbulence modelling in RANS methods, they are not able to capture the variability in the flow with each blade passage. Scale Resolving Simulations will be needed to capture these.

Research on the capabilities of advanced flow modeling by scale-resolving simulations such as ILES and partially averaged Navier–Stokes (PANS) is ongoing. Recently, Klapwijk et al. [19] showed that PANS methods in combination with a turbulence generation model for the incoming flow give dynamics on the vortex interface that show some correspondence to experimental observations. However, this type of flow modelling comes at huge computational costs, rendering these simulations unsuited for multiple design evaluations (Fig. 3.22).

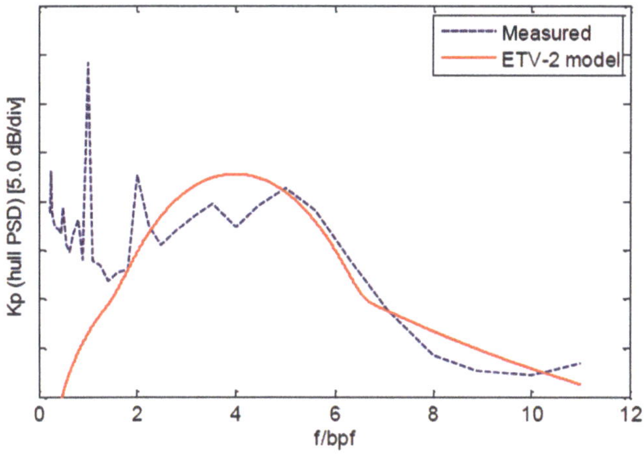

Figure 3.20 Example of a measured and predicted nondimensional hull pressure spectrum for a ship propeller tested at full scale [35].

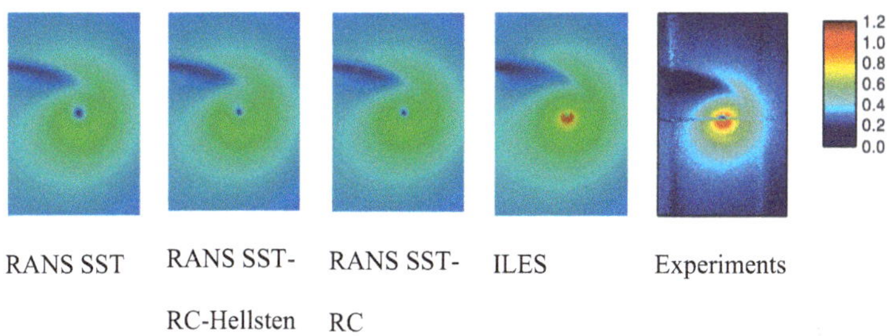

Figure 3.21 Comparison of in-plane velocities in a cross-sectional plane at 0.5 chord length behind the tip. Results are obtained by RANS (three leftmost graphs), ILES, and Particle Image Velocimetry (PIV) experiments. *ILES*, Implicit large eddy simulation; *RANS*, Reynolds-averaged Navier–Stokes. *From A. Asnaghi, R.E. Bensow, U. Svennberg, Comparative analysis of tip vortex flow using rans and les marine 2017, in: 7th International Conference on Computational Methods in Marine Engineering, MARINE 2017 2017-May (May), 2017, 836–47.*

3.4 Cavitation Erosion

3.4.1 On the Physics of Cavitation Erosion

Cavitation erosion on marine propellers, rudders, and other appendages is perhaps the most difficult type of cavitation nuisance to predict from both experimental and computational points of view. Let us start by exploring the physical

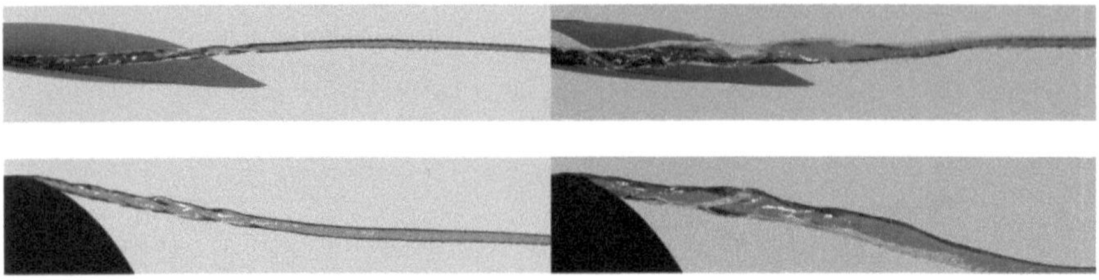

Figure 3.22 Instantaneous top and side views of the vortex cavity as computed on an elliptical foil with a PANS turbulence model by REFRESCO for two different cavitation numbers: $\sigma = 1.60$ (left) and $\sigma = 0.87$ (right). The side views show the wing pressure side. The flow is from left to right. Renders were made using Blender [18].

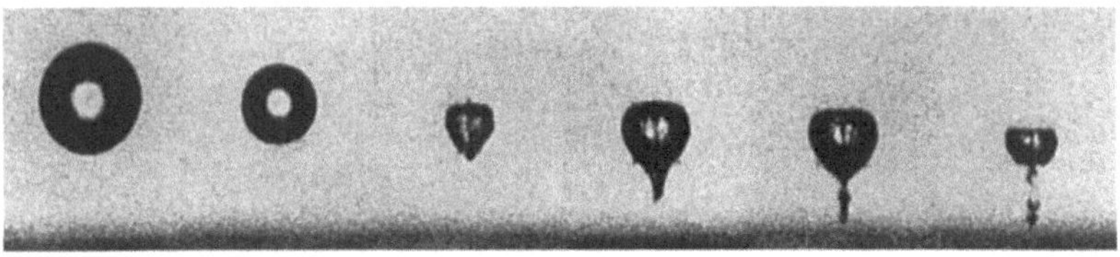

Figure 3.23 Photographs of a collapsing vapor bubble near a wall, clearly illustrating the asymmetry of the bubble collapse and the microjet impinging on the wall by Vogel et al. [37]. The maximum radius $R_0 = 2$ mm, and the wall distance is 2.3 R_0.

mechanisms of cavitation erosion, starting with studies on imploding cavity bubbles near a wall. With a better understanding of the physics, we can then improve our understanding of the problem for marine propellers and rudders.

High-quality photographs of an imploding vapor bubble near a wall have been presented by Vogel et al. [37] (see Fig. 3.23). We clearly observe that the initial spherical bubble develops into an asymmetric cavity with a microjet impinging on the wall below the bubble. The asymmetry of the implosion is caused by the nearby wall. Observations of imploding vapor bubbles in an infinite domain showed that in such cases a symmetric collapse may occur.

Several groups have reported numerical simulations of this process with a variety of fluid dynamics codes. To better understand the implications of the bubble collapse for the material response, an interesting analysis with smooth particle hydrodynamics (SPH) was made by Joshi et al. [38]. By modeling both the fluid dynamics and the solid dynamics, the authors were able to show the effect of the impinging microjet and the shock wave that was produced upon collapse. The authors simulated collapsing bubbles for three distances from the bubble to the wall: 1.2, 1.33, and 1.5 times the initial bubble radius R_0. Fig. 3.24 presents the resulting pressure field in the fluid and the plastic

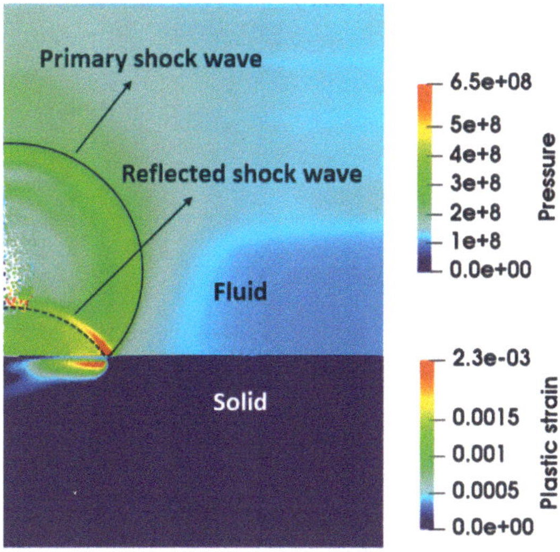

Figure 3.24 Pressure field in the fluid resulting from a vapor bubble collapse as simulated by an SPH method. Indicated in the fluid are the primary shock wave front and the reflecting shock wave. Colors in the solid indicate the plastic strain [38].

strain field in the solid material that result from an imploding vapor bubble at an original bubble distance of 1.33.

These coupled fluid-solid simulations of a collapsing vapor bubble also give input to a long-lasting debate about the aggressiveness of microjets versus shockwaves from bubble collapses.

The results show that the peak pressures excited on the solid-liquid interface by the microjet resulting from the imploding bubble could potentially cause a higher maximum strain rate in the material. However, the affected volume where plastic deformation occurs is minuscule relative to the volume affected by the radially spreading shock wave. This is why there is potentially more damage caused by shock waves produced by imploding bubbles near liquid-solid interfaces.

A nice example of this phenomenon is presented in Fig. 3.25, where we see the collapse of a bubble that is initially positioned at 1.33 times the bubble radius from the fluid-solid interface. We see the asymmetric bubble collapse in the axial velocity plot in Fig. 3.25B, where a moment in time later, the moment of arrival of the pressure shock front at the fluid-solid interface is plotted in Fig. 3.25C. The resulting plastic strain in the material at the end of the simulation is given in Fig. 3.25D, where we clearly see that the biggest material strain has occurred just outside the projected bubble radius just at and below the material surface.

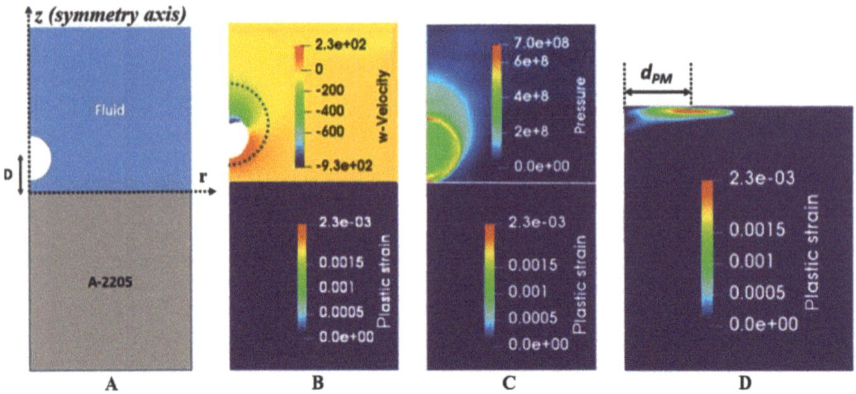

Figure 3.25 (A) Sketch of the initial simulation domain for axisymmetric SPH FSI simulation of a detached cavity of radius $R = 0.15$ mm and standoff ratio SR = 1.33. (B) Contours of axial velocity in the fluid and plastic strain in the solid. A microjet formation can be observed (the dotted semicircle represents the initial bubble shape) at $\tau = 0.92$. (C) Contours of pressure in the fluid and plastic strain in the solid. A shock wave generated as a result of the bubble collapse can be observed as it just reaches the interface at $\tau = 1.053$. (D) Plastic strain contours in the solid at the end of the simulation. The maximum plastic strain occurs at a radial offset d_{PM} from the symmetry axis. *From S. Joshi, J.P. Franc, G. Ghigliotti, M. Fivel, SPH modelling of a cavitation bubble collapse near an elasto-visco-plastic material, Journal of the Mechanics and Physics of Solids 125 (2019) 420–439. https://doi.org/10.1016/j.jmps.2018.12.016.*

3.4.2 Numerical Techniques

The SPH computations by Joshi et al. [38] discussed earlier give a good physical insight into the collapse of a vapor bubble but are not practical for the assessment of cavitation erosion on a ship propeller or rudder. In the former simulations a single imploding bubble with a diameter of 0.15 mm and time steps of the order of 10^{-9} were used. This should be compared to typical length scales of a propeller, which are of the order of meters (10^4 times larger than the vapor bubble) and blade-passing times of the order of 10^{-1} seconds (a factor 10^8 larger). This wide separation of microbubble scales and large cavity dynamics on ship propellers characterizes the multiscale problem of cavitation erosion on ship propellers.

The scale gap between the dynamics of cavity structures and the collapse of bubbles can be solved only by additional modeling of the cavitation erosion process. Following [5], [39] concluded that additional models could best be based on considerations of energy conversion. In its simplest form, Hammitt [39] postulated that material damage would occur once the potential energy in a shed cavity exceeded a certain damage threshold. This threshold would essentially be a function of the material properties only.

Flageul et al. [40] developed this energy model further, explicitly mentioning the energy conversion steps and coupling the acoustic energy flux on the material to a material damage, as depicted schematically in Fig. 3.26. They define a potential power P_{pot} that derives from the time derivative of the potential energy E_{pot} captured by a cavity volume:

$$E_{pot} = (p_{\infty} - p_v)V_v \tag{3.3}$$

They then derive a potential power that can be related to the material damage by introducing an efficiency η^{**}, which depends on the distance of the cavity to the solid surface, the cavity's unsteadiness, and its geometry, which is said to depend on the angle of attack and the shape of the leading edge of the hydrofoil.

The acoustic power that is associated with the pressure waves traveling over the solid surface is referred to as P^{mat}_{waves}. The conversion efficiency η^{*} from the potential power P^{mat}_{pot} to acoustic wave power P^{mat}_{waves} is supposed to depend on the air content in the fluid and is reported to have values between 0.4 and 0.8.

To then arrive at the pit volume caused by plastic deformation in the material, the authors introduce another transfer function, β, which relates the acoustic wave power to the power needed for the pitting damage. This model has been worked out for use with a multiphase RANS code at a later stage in Leclercq et al. [41].

The concept of energy conversion was elaborated by Tinguely et al. [42], who state that the initial potential energy is eventually converted into shock wave energy, dissipative thermal energy, and rebound energy, where the latter is stated

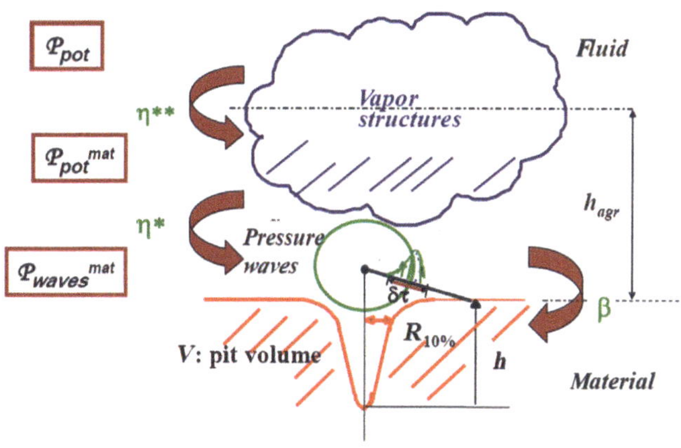

Figure 3.26 Scheme of the physical scenario of cavitation erosion based on energy balance considerations [40].

to be relevant only for low pressures (below 1 bar). Thermal dissipation was argued to be negligible.

Schenke et al. [43] worked further on the concepts of energy conversion of cavities in the flow over a foil in order to apply such a method to ship propellers. The objective of their study was to investigate the extent to which a collapsing cavity can be simulated with an incompressible RANS code for the assessment of erosion risk. It is important that both the amount and the distribution of radiated energy over the surface and over time be predicted reliably. A good test case to verify this energy distribution is given by Schenke [44] based on a numerical test case by Schmidt et al. [45], in which a bubble cluster is exposed to a high pressure of 40 bar and the pressure inside the cavitation bubbles is equal to the vapor pressure, $p_v = 2,340$ bar. Fig. 3.27 presents the bubble cloud in its initial stage [44].

An additional contribution to the energy conversion model is proposed by these authors in the conversion of potential energy into acoustic energy. They propose adding a scalar field for the storage of kinetic energy so that the potential energy is not instantaneously emitted on the nearby surfaces. The kinetic energy that accumulates during the collapse phase is stored in the vapor-liquid interface of the cavity. This accumulated kinetic energy is released only upon the final collapse of the bubble. This is also referred to as the energy-focusing approach. The consequence of this energy-focusing approach in relation to the nonfocusing approach for the bubble cloud collapse is illustrated in Fig. 3.28 in terms of time-dependent radiated surface energy.

The consequences of the different approaches in time-dependent pressure are illustrated in the time evolution of the pressure signal from the sensor underneath the bubble cloud (Fig. 3.27). It is observed that the peak pressure

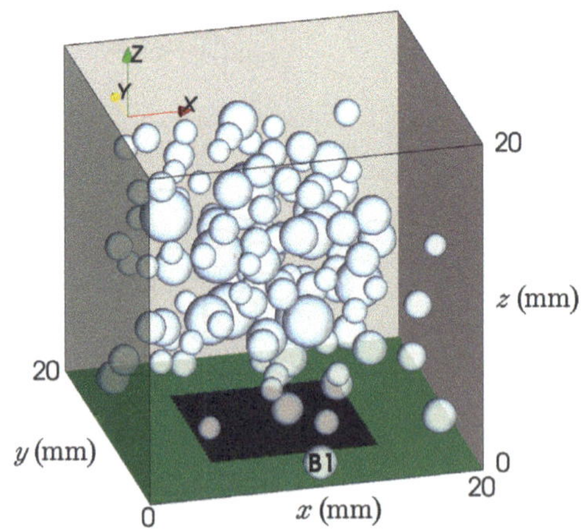

Figure 3.27 Bubble cluster and a 1 × 1 cm numerical pressure sensor indicated in black on the bottom surface. Test case borrowed from [41].

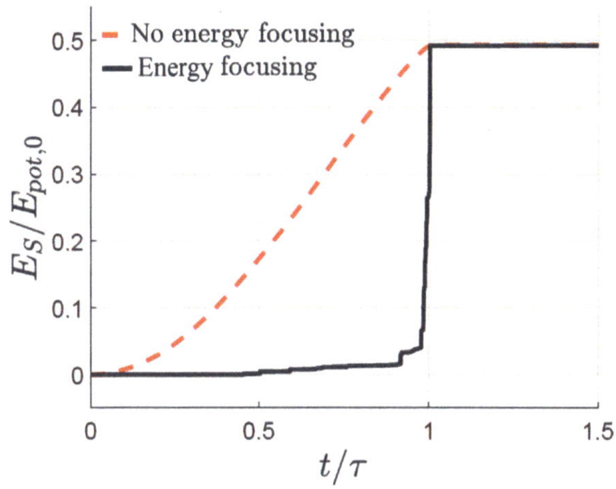

Figure 3.28 Evolution of the nondimensional energy accumulated on the entire bottom underneath the imploding bubble cloud. It shows the normalized energy radiated on the bottom surface for a non-energy-focusing approach and an energy-focusing approach.

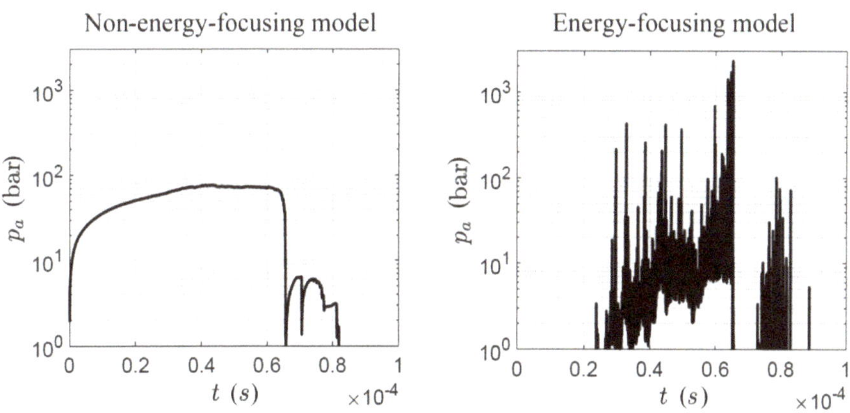

Figure 3.29 Temporal evolution of pressure signal of the bottom pressure sensor for the non-energy-focusing model (left) and the energy-focusing model (right).

for the energy-focusing model is an order of magnitude larger than that for the non-energy-focusing model (Fig. 3.29).

To assess the quality of the pressure signal in the energy-focusing model, the time development of the pressure measured by the sensor is compared with a similar pressure obtained by a fully compressible simulation by Schmidt [45] in

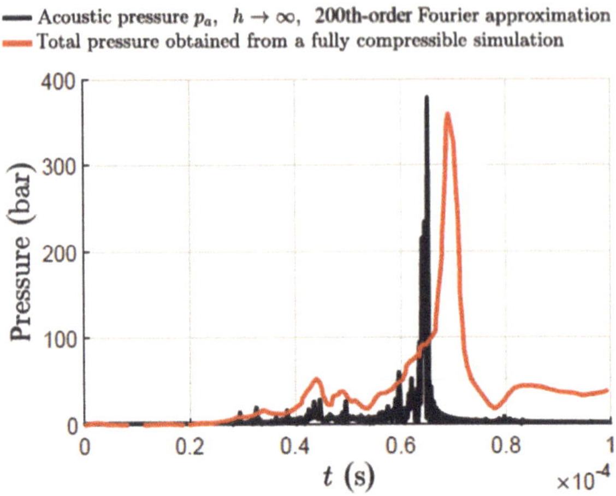

Figure 3.30 Comparison of the acoustic pressure signal obtained from the energy focusing approach of an incompressible inviscid model with the total pressure signal computed by Schmidt [45] with a fully compressible inviscid model.

Fig. 3.30. It should be noted here that the black line represents the acoustic pressure, so it is a pressure perturbation as computed with the incompressible flow solver, whereas the red line represents the total pressure as computed with the fully compressible flow solver. As a consequence, the total pressure approaches the ambient pressure of 40 bar at the end of the collapse, whereas the acoustic pressure disturbance returns to zero. Another difference is the time at which the peak pressure develops. We see that the peak pressure from the incompressible model arrives earlier at the pressure sensor than the peak from the acoustic model. This difference can be explained by the difference in speed of sound for both models. Whereas the speed of sound is infinite in the incompressible model, it has a finite value in the compressible model, explaining the slightly later time of arrival at the pressure sensor on the bottom wall.

For a better understanding of the energy conversion process, let us take a look at the case of the single bubble, for which the energy conversion process is depicted in Fig. 3.31.

A careful derivation of relations is essential in this energy approach, as the initial potential energy of a vapor bubble in an infinite undisturbed domain is perhaps trivial [46]:

$$E^b_{pot,0} = \frac{4}{3}\pi R_0^3(p_\infty - p_v) \tag{3.4}$$

where R_0 is the initial bubble radius and superscript b indicates that this relation is valid for a spherical undisturbed bubble. During the collapse of the bubble, the

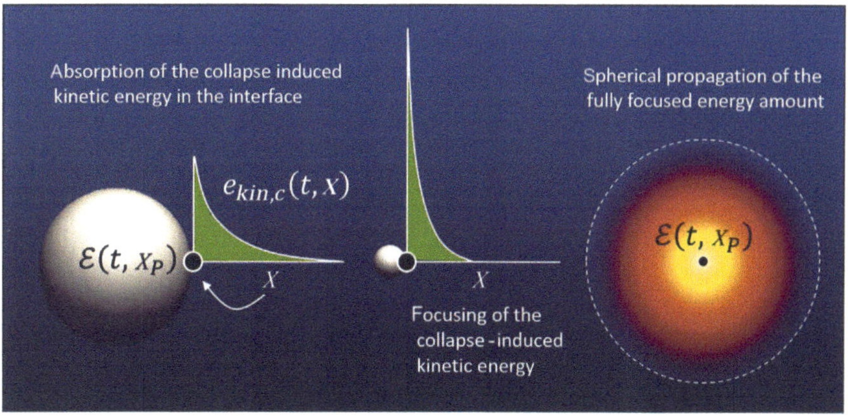

Figure 3.31 Illustration of the conversion of kinetic energy into acoustic collapse energy [44].

surrounding flow gets accelerated and hence starts to contain kinetic energy. Using the mass continuity relation, the instantaneous radial kinetic energy density at some distance r from the bubble center can be derived and can be integrated for the entire body of liquid:

$$E^b_{kin} = \int_R^\infty \left(\frac{\mathrm{d}E_{kin}}{\mathrm{d}r}\right)\mathrm{d}r = 2\pi\rho_l \dot{R}^2 R^3 \tag{3.5}$$

where $R = R(t)$ is the instantaneous bubble radius.

Upon collapse, the initial potential energy of the bubble is fully converted into kinetic energy. During this process, the kinetic energy in the body of liquid also focuses toward the bubble interface (see Fig. 3.31).

The shock wave energy E_{SW} radiated from a point p in the fluid at some point s on the surface can be written as a function of the time dependent acoustic pressure p_a at a distance $r = \left(\overline{x}_p - \overline{x}_s\right)$

$$E_{\mathrm{SW}} = \frac{4\pi r^2}{\rho_l c_l}\int_{T_{\mathrm{SW}}} p_a^2(t,r)\mathrm{d}t \tag{3.6}$$

If we consider a cloud of bubbles instead of a single bubble in an undisturbed environment, we need to deal with their mutual interaction. A similar reasoning can be followed for this cloud of bubbles, as explained in Fig. 3.32. Following [47], the entire potential energy of a bubbly cloud is focusing toward the cloud collapse center.

To allow for the storage of kinetic energy during the collapse, an accumulated kinetic energy field $\varepsilon\left(t,\overline{x}_p\right)$ at a time instant t and a location $\overline{x}_p$ is introduced. The

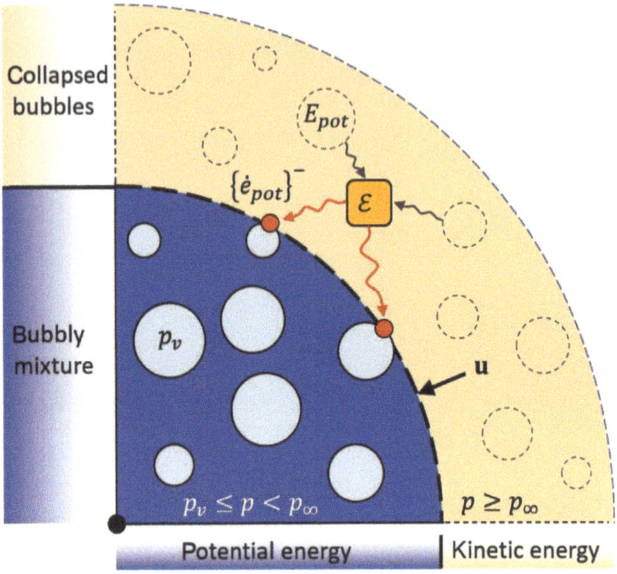

Figure 3.32 Schematic of energy cascade during the collapse of a cloud of vapor bubbles.

energy accumulation and redistribution of kinetic energy can be described by the energy continuity equation:

$$\frac{\partial \varepsilon}{\partial t} + u.\nabla\varepsilon = \underbrace{-\varepsilon\{\nabla.u\}^-}_{\text{kin. energy source}} - \underbrace{\dot{e}_{\text{rad}}}_{\text{acoustic rad. source}} \tag{3.7}$$

The right-hand side of Eq. (3.7) serves as the source term in the kinetic energy balance. Regions of negative flow divergence, where $\nabla.u < 0$, represent the sources where potential energy is converted into kinetic energy. The second term represents the rate at which the collapse-induced kinetic energy is converted into radiated acoustic energy. On the left-hand side the convective term $u.\nabla\varepsilon$ represents the advective transport of kinetic energy, which is responsible for the storage and spatial focusing of the kinetic energy. However, the exact spatial distribution of the kinetic energy is unknown. This will require an additional modeling assumption. As we did for the single bubble, we assume in this case of a bubbly cloud that the locations of potential energy reduction act as attractors for the collapse-induced kinetic energy. We further assume that kinetic energy is stored at the interface of the shrinking cavities until the criterion for the conversion of kinetic energy into radiated acoustic energy is met. As a consequence, the model cannot reflect the spatial kinetic energy distribution around the cavity before the final collapse stage.

The simulated collapse of the bubble cloud at five distinct time instants is presented in Fig. 3.33. It is interesting to note that not all bubbles contribute energy to a synchronized collapse of the bubble cloud. There are two bubbles that do not participate but instead collapse individually. This finding corresponds to the results obtained from a simulation with a compressible Euler code by Schmidt [45], in which it should be noted that the simulations were conducted with an incompressible RANS code. The spatial and temporal distribution of the radiated surface energy on the bottom plate illustrates the added value of the focusing model.

To check the usefulness of the cavitation erosion model in predicting the risk and location of cavitation erosion, a test case was selected for validation. Computations were made on the King's College D (KCD)-193 propeller (see Melissaris et al. [48]). Experiments on this propeller were made in the Emerson Cavitation Tunnel of Newcastle University, where the propeller was mounted behind a narrow wake screen to create a nonuniform wake with a narrow wake peak [49,50]. This setup was modeled in the computational fluid dynamics computations carried out with an incompressible unsteady Reynolds-averaged Navier–Stokes (URANS) solver.

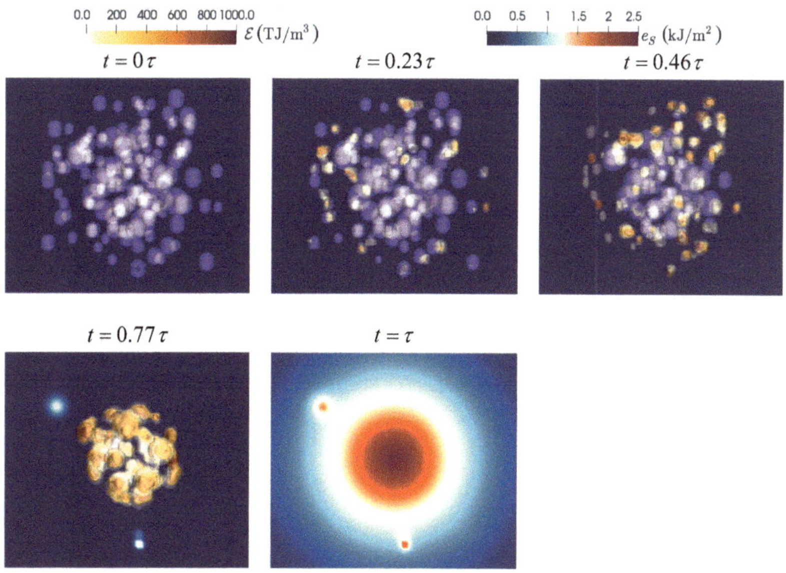

Figure 3.33 Top view of the collapsing vapor bubble cloud and isosurfaces of the accumulated collapse-induced kinetic energy per unit volume ε absorbed in the bubble interfaces at different instants in time of the bubble cloud collapse. Indicated on the blue bottom plate is the distribution of collapse-induced surface energy e_s, which is obtained from time integration of the surface impact power per unit area.

Fig. 3.34 shows the results of the wall resolved computations for two different grids. Table 3.1 presents the features of two of the grids that were used in the simulations.

A few observations on the differences between the damage from the experiments and the distribution of accumulated surface specific energy from the simulations can be made. First, we observe that there is a fair agreement in the location of damage, where it is noted that the damaged area from the experiments is larger than the area covered with peaks of radiated surface energy. The most likely explanation here is that the tests were conducted during half an hour, whereas the simulations covered only eight revolutions (corresponding to 0.32 s). The scatter in the simulations is caused by the intrinsic instabilities of the partial cavity and the modeling/numerical instabilities that occur in the simulations. As a consequence, more time is needed to obtain a converged damage area in the simulations, especially for finer grids where smaller structures are resolved.

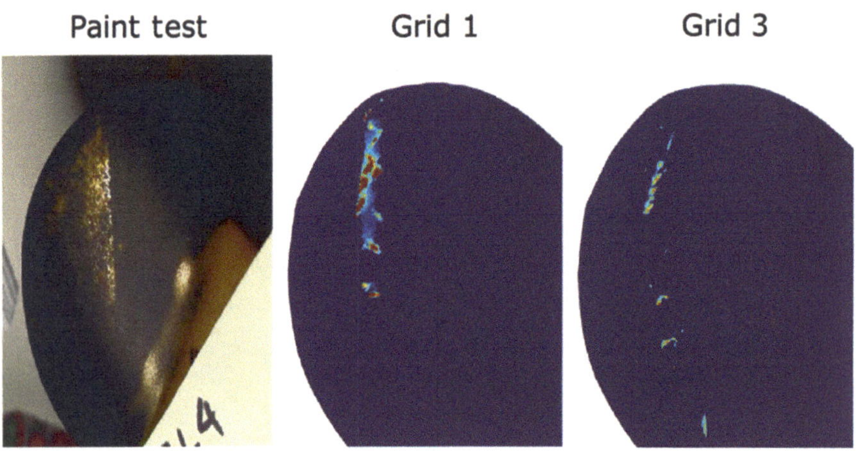

Figure 3.34 Comparison of the erosion damage from an experimental paint test and the surface-specific accumulated energy on the propeller blade after eight propeller revolutions. Grid 1 is a coarse grid. Grid 3 is a refined grid.

TABLE 3.1 Features of grids used in the simulations for the King's College D (KCD) propeller.

Grid	No. of cells total (M)	No. of cells on refined blade (k)	No. of prism layers	Average y^+ on refined blade
Grid 1	3.64	43.3	10	0.24
Grid 3	10.1	112.6	20	0.11

It is good to mention here that the accumulated radiated energy as indicated in the simulation results in Fig. 3.34 is not necessarily the best indicator for damage, as discussed in Schenke et al. [43]. The aggressiveness is likely to be more dependent on radiated acoustic power than on radiated acoustic energy. This might be indicated by using aggressiveness indicators. However, this is considered to be outside the scope of the present section.

3.4.3 Experimental Techniques

A review of experimental techniques for cavitation erosion studies is given by the ITTC [51]. It can be inferred from this report that the propeller and ship design community was still struggling with the assessment, including the experimental assessment, of cavitation erosion risks. The most applied test methods were cavitation tests with video or preferably high-speed video observations of propellers and/or rudders or soft paint tests. In the latter, a thin film of soft paint is applied to the propeller or rudder, which is then tested for a limited duration under the critical condition. The results from a questionnaire indicated that the typical duration of such a test varies between 0.5 and 3 h. Critical to the reliability of such tests is that the proper condition for erosion damage is selected and that scale and laboratory effects in the test setup are controlled. Both these conditions typically cause a large uncertainty in the erosion risk predictions from experiments.

This specialist committee concluded that "the recommended guidelines to reduce cavitation erosion are only qualitative and more research into the physics of cavitation structures/material interactions is required before damage rates at full scale can be quantified" [47]. Since this report, our understanding of the physical mechanisms has been much enhanced by numerical simulations, as described in the preceding section, and numerical methods to indicate the erosion risks and location seem to have become effective. However, the step toward assessing the damage rate has yet to be taken.

3.5 Summary and Outlook

Since the early 20th century, when the consequences of cavitation on the marine propeller became relevant through the increased power delivery to the propeller by modern steam engines, cavitation started to dominate the limiting constraints in propeller design. These limiting conditions are imposed by the cavitation nuisance and can be distinguished as thrust breakdown, cavitation-induced pressure fluctuations, vibrations, noise, and cavitation erosion.

The sensitivity of cavitation inception has been clearly illustrated by the comparative inception study on a headform [6]. At present, we are working on physical models that will explain our experimental observations. The last results suggest that Eulerian-Lagrangian descriptions of the nuclei-laden flow are needed to get a fair agreement with experimental observations.

The physical mechanisms leading to the typical pressure spectra measured on board ships and ship models are still a subject for research, although much has been learned about the sources leading to the tonal signals and the broadband signals. Tonal signals are typically dominated by sheet cavity volume variations, whereas the low-frequency broadband spectrum (typically between the fourth and seventh BPFs) is dominated by volume variations of the cavitating tip vortex (or a structure of interacting tip vortices). The effects of interaction between sheet and vortex cavity dynamics are not yet understood. The bubbly flow in the wake of a cavitating propeller is relevant for radiated noise at higher frequencies.

Cavitation erosion is the phenomenon in which material damage is caused by cavities imploding in the vicinity of the material surface. The mechanism of cavitation erosion has long been uncertain. Theories blaming the material impinging microjet and theories blaming the shockwave propagating over the material surface afterwards were objects of debate over a long period between scientists. The simulation work for a single bubble collapse by Joshi et al. [38] has demonstrated that at least for a single bubble, the shock wave is more devastating than the impinging microjet. Schenke et al. [43] have shown that an estimate of the erosiveness can be given by a model that is based on energy conservation, following the cavity from its maximum size until collapse, thereby largely converting its maximum potential energy into radiated acoustic energy. This model was demonstrated to work in an incompressible RANS model, which is widely used by the industry nowadays.

The need for high-efficiency ship propellers is perhaps greater now than ever before in the development toward zero-emission shipping. An important aspect of optimizing for efficiency is that efficiency and cavitation hindrance give different directions to the propeller design. This increases the need for efficient and reliable cavitation hindrance computations and leads to an increase in the use of incompressible RANS methods for cavitation constraints in optimization studies. However, BEM will remain a workhorse for ship designers, especially if many propeller evaluations are needed, as in optimization studies or stochastic problems such as a ship sailing in waves. Scale- resolving simulations teach us more about the physical mechanisms and their relevance. This knowledge is essential for the development of models for cavitation nuisance that can be used in combination with BEM or RANS models.

References

[1] L.C. Burrill, Sir Charles Parsons and cavitation, The Institute of Marine Engineers, Transactions LXIII (8) (1963) 149–167. Available from: https://scholar.google.co.uk/scholar?q = sir + charles + parsons + and + cavitation&btnG = &hl = en&as_sdt = 0%2C5#0#0.

[2] S. Li, C.E. Brennen, Y. Matsumoto, Cavitation in medicine, Interface Focus 5 (5) (2015) 1–12. Available from: https://doi.org/10.1098/rsfs.2015.0022.

[3] B. Aktas, M. Atlar, S. Turkmen, W. Shi, R. Sampson, E. Korkut, et al., Propeller cavitation noise investigations of a research vessel using medium size cavitation tunnel tests and full-scale trials, Ocean Engineering 120 (2016). Available from: https://doi.org/10.1016/j.oceaneng.2015.12.040.

[4] D. Ligtelijn, The pay-off between cavitation and efficiency, in: IMAREST Ship Propulsion Systems Symposium, (2010) 1–15.

[5] T.J.C. Van Terwisga, P.A. Fitzsimmons, L. Ziru, and E.J. Foeth, Cavitation erosion – a review of physical mechanisms and erosion risk models, in: Proceedings of the 7th International Symposium on Cavitation, (2009) 1–13.

[6] J.D. van Manen, S. Bindel, P. Eisenberg, H. Lindgren, L. Mazarredo, W.B. Morgan, et al., 11th international towing tank conference, in: Report of Cavitation Committee, (1966) 167–269. ITTC.

[7] M.C. Junger, D. Feit, Sound, Structures and Their Interaction, second, MIT Press, Cambridge, MA, USA, 1986.

[8] J. Bosschers, Cavitating Vortices on Ship Propellers and the Related Hull Pressure Fluctuations and Radiated Noise Proefschrift. University of Twente, (2018).

[9] M.X. van Rijsbergen, and T.J.C. van Terwisga, Water quality effects on sheet cavitation inception on a ship propeller model, in: 7th International Conference on Multiphase Flow. Tampa, Florida USA, (2010). <file:///D:/Literature/Rijsbergen/Fluid_dynamics/Cavitation/Inception/Rijsbergen_2010_ICMF.pdf>.

[10] M.X. van Rijsbergen, A review of sheet cavitation inception mechanisms, in: International Symposium on Transport Phenomena and Dynamics of Rotating Machinery. Honolulu, Hawaii. (2016) <file:///D:/Literature/Rijsbergen/Fluid_dynamics/Cavitation/Inception/Rijsbergen_2016_ISROMAC.pdf>.

[11] T.F. Groß, G. Ludwig, and P.F. Pelz, Experimental and theoretical investigation of nucleation from wall-bounded nuclei in a laminar flow, in: International Symposium on Transport Phenomena and Dynamics of Rotating Machinery. Honolulu, Hawaii. (2016) <file:///D:/Literature/Rijsbergen/Fluid_dynamics/Nucleation/Gross_2016_ISROMAC.pdf>.

[12] B. Gindroz, M.L. Billet, Influence of the nuclei on the cavitation inception for different types of cavitation on ship propellers, Journal of Fluids Engineering 120 (1) (1998) 171–178. Available from: https://doi.org/10.1115/1.2819643.

[13] P.S. Russell, J.A. Venning, P.A. Brandner, B.W. Pearce, D.R. Giosio, and S.L. Ceccio, Microbubble disperse flow about a lifting surface, in: 32nd Symposium on Naval Hydrodynamics, no. August: (2018) 5–10. <https://jamesvenning.net/wp-content/uploads/2020/04/2018-SNH-32nd-Russell_et-al.pdf%0A, https://jamesvenning.net/wp-content/uploads/2020/04/SNH2018_seeding.png>.

[14] C.-T. Hsiao, J. Ma, G.L. Chahine, Multiscale two-phase flow modeling of sheet and cloud cavitation, International Journal of Multiphase Flow 145 (November) (2017) 68–84.

[15] C.M. Duarte, L. Chapuis, S.P. Collin, D.P. Costa, R.P. Devassy, V.M. Eguiluz, et al., The soundscape of the anthropocene ocean, Science (New York, N.Y.) 371 (6529) (2021) eaba4658. Available from: https://doi.org/10.1126/science.aba4658.

[16] J.-P. Franc, J.-P. Franc, J.-M. Michel, and J.-M. Michel, Fundamentals of Cavitation. Fluid Mechanics and Its Applications. (2004) <https://doi.org/10.1007/1-4020-2233-6>.

[17] E. Van Wijngaarden, J. Bosschers, and G. Kuiper, Aspects of the cavitating propeller tip vortex as a source of inboard noise and vibration, in: 2005 ASME Fluids Engineering Division Summer Meeting and Exhibition, (2005) 1–6. <https://doi.org/10.1115/FEDSM2005-77271>.

[18] J. Bosschers, Mechanisms of low-frequency broadband noise by cavitating tip- vortices on marine propellers, Sixth International Symposium on Marine Propulsors SMP'19 (2019).
[19] M. Klapwijk, Elliptical foil with turbulence generator. T.b.D. (2020).
[20] H.C.J. van Wijngaarden, Prediction of Propeller-Induced Hull-Pressure Fluctuations Proefschrift. (2011).
[21] J.T. Ligtelijn, H.C.J. van Wijngaarden, J.B. Verkuyl, Correlation of cavitation; comparison of full-scale data with results of model tests and computations, SNAME Annual Meeting 112 (2004) 80–110.
[22] J. Bosschers, E. Van Wijngaarden, Scale effects on hull pressure fluctuations due to cavitating propellers, ICHD Proceedings (2012).
[23] J. Venning, S. Smith, P. Brandner, D. Giosio, B. Pearce, The influence of nucleation on cloud cavitation about a sphere, Irsomac 2017 2 (2017) 1–6. Available from: https://doi.org/10.1017/jfm.2016.819.
[24] B. Maines, R.E.A. Arndt, The case of the singing vortex, Journal of Fluids Engineering, Transactions of the ASME 119 (2) (1997) 271–276. Available from: https://doi.org/10.1115/1.2819130.
[25] P.C. Pennings, J. Bosschers, J. Westerweel, T.J.C. van Terwisga, Dynamics of isolated vortex cavitation, Journal of Fluid Mechanics 778 (2015) 288–313. Available from: https://doi.org/10.1017/jfm.2015.379.
[26] P. Pennings, Dynamics of Vortex Cavitation, TU Delft, 2016.
[27] S.W. Thomson, XXIV. Vibrations of a columnar vortex, The London, Edinburgh, and Dublin Philosophical Magazine and Journal of Science 10 (61) (1880) 155–168. Available from: https://doi.org/10.1080/14786448008626912.
[28] P. Pennings, J. Westerweel, T. van Terwisga, Cavitation tunnel analysis of radiated sound from the resonance of a propeller tip vortex cavity, International Journal of Multiphase Flow 83 (2016) 1–11.
[29] J. Bosschers, On the relation between tonal and broadband content of hull pressure spectra due to cavitating ship propellers, Journal of Physics: Conference Series 656 (1) (2015) 1–4. Available from: https://doi.org/10.1088/1742-6596/656/1/012101.
[30] S.A. Kinnas, N.E. Fine, A nonlinear boundary element method for the analysis of unsteady propeller sheet cavitation, Proceedings of the 19th Symposium on Naval Hydrodynamics (1992) 717–737.
[31] J. Bosschers, G. Vaz, A.R. Starke, and E. van Wijngaarden. 2008. Computational analysis of propeller sheet cavitation and propeller-ship interaction, in: RINA - International Conference - Marine CFD 2008 - Papers, no. March: 147–60. <http://www.marin.nl/upload_mm/1/e/a/1807585792_1999999096_Computational_analysis_of_sheet_cavitation_and_propeller-ship_interaction_-_RINA_CFD2008.pdf>.
[32] G. Vaz, and J. Bosschers, Modelling three dimensional sheet cavitation on marine propellers using a boundary element method, in: Cav2006, no. September. (2006).
[33] F. Pereira, F. Salvatore, F. Felice, M. Soave, Experimental investigation of a cavitating propeller in non-uniform inflow, Proceedings of the 25th Symposium on Naval Hydrodynamics (2004).
[34] A.E. Raestad, Tip vortex index - an engineering approach to propeller noise prediction, The Naval Architect (1996) 11–16.
[35] J. Bosschers, A semi-empirical method to predict broadband hull pressure, in: Proceedings of the Fifth International Symposium on Marine Propulsors 1 (June). (2017).
[36] A. Asnaghi, R.E. Bensow, and U. Svennberg, Comparative analysis of tip vortex flow using rans and les marine 2017, in: 7th International Conference on Computational Methods in Marine Engineering, MARINE 2017 2017-May (May): (2017) 836–47.
[37] A. Vogel, W. Lauterborn, R. Timm, Optical and acoustic investigations of the dynamics of laser-produced cavitation bubbles near a solid boundary, Journal of Fluid Mechanics (206) (1989) 299–338.

[38] S. Joshi, J.P. Franc, G. Ghigliotti, M. Fivel, SPH modelling of a cavitation bubble collapse near an elasto-visco-plastic material, Journal of the Mechanics and Physics of Solids 125 (2019) 420–439. Available from: https://doi.org/10.1016/j.jmps.2018.12.016.

[39] F.G. Hammitt, Observations on cavitation damage in a flowing system, Journal of Basic Engineering 85 (3) (1963) 347–356. Available from: https://doi.org/10.1115/1.3656601.

[40] C. Flageul, R.F. Patella, and A. Archer, Cavitation erosion prediction by numerical simulations, in: ISROMAC 2012 - 14th International Symposium on Transport Phenomena and Dynamics of Rotating Machinery, (2012) 1–9.

[41] C. Leclercq, A. Archer, R. Fortes-Patella, F. Cerru, Numerical cavitation intensity on a hydrofoil for 3d homogeneous unsteady viscous flows, International Journal of Fluid Machinery and Systems 10 (3) (2017) 254–263. Available from: https://doi.org/10.5293/IJFMS.2017.10.3.254.

[42] M. Tinguely, D. Obreschkow, P. Kobel, N. Dorsaz, A. De Bosset, M. Farhat, Energy partition at the collapse of spherical cavitation bubbles, Physical Review E - Statistical, Nonlinear, and Soft Matter Physics 86 (4) (2012) 1–6. Available from: https://doi.org/10.1103/PhysRevE.86.046315.

[43] S. Schenke, T. Melissaris, T.J.C. van Terwisga, On the relevance of kinematics for cavitation implosion loads, Physics of Fluids 31 (5) (2019). Available from: https://doi.org/10.1063/1.5092711.

[44] S. Schenke, Cavitation Implosion Loads from Energy Balance Considerations in Numerical Flow Simulations, Delft University of Technology, 2020.

[45] S.J. Schmidt, M. Mihatsch, M. Thalhammer, N.A. Adams, Assessment of erosion sensitive areas via compressible simulation of unsteady cavitating flows, in: G.L. Chahine, K.H. Kim, J.P. Franc, A. Karimi (Eds.), Advanced Experimental and Numerical Techniques for Cavitation Erosion Prediction, 106, 2014, pp. 329–344.

[46] A. Vogel, W. Lauterborn, Acoustic transient generation by laser-produced cavitation bubbles near solid boundaries, The Journal of the Acoustical Society of America 84 (2) (1988) 719–731. Available from: https://doi.org/10.1121/1.396852.

[47] Y.-C. Wang, C.E. Brennen, Numerical computation of shock waves in a spherical cloud of cavitation bubbles, Journal of Fluids Engineering 121 (December 1999) (1999) 872. Available from: https://doi.org/10.1115/1.2823549.

[48] T. Melissaris, S. Schenke, N. Bulten, T.J.C. van Terwisga, On the accuracy of predicting cavitation impact loads on marine propellers, Wear 456–457 (June) (2020) 203393. Available from: https://doi.org/10.1016/j.wear.2020.203393.

[49] A. Mantzaris, B. Aktas, P. Fitzsimmons, and M. Atlar. 2015. Establishment and verification of reproducible method for coating propeller blades for erosive cavitation detection, in: Fourth International Conference on Advanced Model Measurement Technologies for the Maritime Industry. Istanbul, Turkey.

[50] O. Usta, B. Aktas, M. Maasch, O. Turan, M. Atlar, and E. Korkut. 2017. A study on the numerical prediction of cavitation erosion for propellers, in: SMP 2017.

[51] M.L. Billet, S.R. Turnock, L. Briancon-Marjollet, B. Jun Chang, J. Firiesch, and L. Wilczynski, The specialist committee on cavitation erosion on propellers and appendages on high powered / high speed ships final report and recommendations to the 24th ITTC, in: Proceeedings of the 24th ITTC, II:509–42. (2005) Edinburgh, Scotland: ITTC.

Further reading

Foeth, E.J. 2008. The Structure of Three Dimensional Sheet Cavitation. TU-Delft, PHD Thesis.

Chapter 4

Numerical Prediction of Cavitation and Related Nuisances in Marine Propulsion Systems

Rickard E. Bensow
Mechanics and Maritime Sciences, Chalmers University of Technology, Gothenburg, Sweden

4.1 Introduction

Cavitation is a major design constraint for marine propulsion systems, whether the concern is vibrations, noise, performance loss, or erosion. It therefore needs to be considered as early as possible in the design process. For an open propeller arrangement, typically a baseline design is chosen without consideration of cavitation by optimization based on empirical relations derived from systematic series. A more detailed blade design is then developed, using potential flow simulation approaches that consider cavitation, with varying degrees of reliability and refinement in the cavitation prediction. When such simulations are used by an experienced blade designer, the behavior of on-blade sheet cavitation is usually acceptably determined. On this basis, a first estimation of hull pressure pulses can be computed, and by visual inspection of the sheet cavity dynamics, the designer can make a

Cavitation and Bubble Dynamics. DOI: https://doi.org/10.1016/B978-0-12-823388-7.00006-0

judgment about the risk of cavitation erosion. For a standard contract, this is often sufficient to develop the final design, which is then verified by experiments. A design turnaround time of a week or two is not uncommon.

For vessels with higher requirements, for example, for cruise ships, research vessels, or navy contracts, more effort is needed. Up to now, this has involved model experiments in a cavitation tunnel. With a time frame of a few months for model production and performing the tests combined with considerable cost, this is performed only in a late stage of the design process, and if problems are discovered, they may delay the delivery. Further, the data that can be extracted from an experiment are quite limited and may suffer strongly from scale effects. Pressure pulse levels for the model-scale system can be well recorded, while radiated noise is more troublesome in the confined space of a cavitation tunnel; cavitation erosion may be assessed by a paint test. Standard video recordings combined with stroboscopic light are used to determine cavity extent, and for more detailed studies of cavity dynamics, in particular in relation to determining risk of erosion, high-speed video is employed. The presence of cavitation prohibits the use of more detailed optical measurement techniques. Scale effects to consider are the shape and dynamics of the wake inflow to the propeller and the presence of laminar flow on the blade; both have a huge impact on cavity development.

Recently, more advanced simulation approaches based on incompressible viscous Reynolds-averaged Navier-Stokes (RANS) methods have become mature enough to be incorporated in commercial design procedures. For a marine propulsion system the large-scale separation between the geometrical scale of the ship and propeller and the small scales governing cavitation and tip vortex development limits the use of scale-resolving modeling and prohibits the use of compressible approaches. Still, reliable quantitative analysis is difficult to achieve in an industrial work flow, in which simulation time and, to some extent, computational resources are limited. For a standard propeller around design operating conditions, performance predictions using RANS are not more reliable than potential flow methods, even when cavitation is considered. However, during off-design conditions, separation or massive cavitation reduces the accuracy of potential flow methods. Also, for propellers with smaller blade areas (to increase efficiency) or thicker blades (for ice-classed vessels), potential flow methods loose reliability, and RANS becomes necessary. For pressure pulse predictions, RANS may allow the prediction of the cavitating tip vortex, at least close to the blade, which is not possible in standard potential flow approaches. Although it is a challenge to propagate the tip vortex farther downstream, the interaction between the cavitating tip vortex and the on-blade sheet cavity has a significant impact on the lower-order pressure pulses and is a strong incentive to promote the use of RANS. When it comes to cavitation erosion, a few approaches to indicate the risk of erosion have just emerged in research, and their reliability and requirements on accuracy in flow predictions have not yet been proven. The published results are promising, and it will be interesting to follow how industrial uptake develops.

This chapter aims to give an overview of numerical prediction of these cavitation issues in both research and industry, with a discussion of the limitations and possibilities of different methods. A certain emphasis will be given to the state of the art as presented in recent scientific literature as an indication of where the topic is heading. This includes, to some extent, performance predictions, that is, thrust and torque over a range of operating conditions; a more detailed discussion of the determination of the cavitation inception point; current level of accuracy in hull pressure pulse predictions, extending into the related area of radiated noise levels; and finally an examination of proposed methods for assessing the risk of cavitation erosion.

4.2 Numerical Methods Used in Predicting Cavitation

For more details on general approaches to simulate cavitating flows, the reader is referred to the corresponding chapter in the book. The discussion here will be more general, focusing on considerations of the suitability of the different methods for applications related to marine propulsion systems.

4.2.1 Potential Flow Methods

As was noted in the introduction, potential flow methods are still very much the workhorse in the design of propellers. They are fast, requiring from less than a minute for a simpler analysis up to half an hour for a more complex cavitation problem when running on a workstation. This makes them well suited for the short design time requirement. Their reliability is reasonably good, at least around the design point, both with respect to performance predictions and for leading edge sheet cavitation.

The method originated from the 19th century work of Rankine and Froude and is based on the idea that the inviscid irrotational flow around a body can be described by using only sources, sinks, vortices, and dipoles. The first application to a marine propeller was published by Lerbs [1], who adopted the classical lifting line theory of Prandtl for a moderately loaded propeller.

To consider cavitation, one needs to increase the complexity and turn to lifting surface models. Here, the blade is modeled from a continuous distribution of sources that have to be determined from boundary conditions. The lattice distribution of these vortex sources that is sometimes used has given the methods the name vortex-lattice methods (VLM). The first application for a propeller was presented by Kerwin [2]. The method was further developed by Cummings [3] to solve steady performance of the propeller and by Lee [4] to predict cavitation and unsteady flow conditions. By introducing a distinct set of source sheets for a blade thickness distribution and leading edge correction [5], the blade thickness effect on the cavitation is captured. Further development includes nonlinear thickness loading, hub effects, midchord cavitation detection, and wake alignment (see, e.g., Kinnas and Fine [6]).

A representation of the actual three-dimensional (3D) blade geometry requires applying a boundary element method (BEM). In this approach, the blade is represented by a number of panels whose boundary and control points are located on the blade surface. A continuous distribution of source sheets and dipoles is computed for the control points to satisfy the flow problem. The computational cost is considerably higher for BEM than for VLM, but BEM has major advantages in accurately representing effects of the important leading edge geometry and the propeller hub.

In computing cavitating flows over the blade, sources are introduced to represent the cavity thickness, and an iterative procedure is needed to adjust the source distribution and vorticity distribution until convergence is reached by satisfying the kinematic and dynamic boundary conditions, the Kutta condition, and a cavity closure condition. Further, to achieve a 3D cavity shape, this is performed in radial strips along the blade span, looping from the root to the tip and back. Because this is a rather costly process, it is often performed for only one blade, the so-called key blade, and all other blades are assumed to develop the same cavity shape. In a transient problem, when the propeller rotates in a ship wake, these iterations needs to be performed for each time step.

As has already been noted, potential flow methods give a usable prediction of the attached sheet cavity on the blade but are basically limited in including tip vortex cavitation or the occurrence of reentrant flow beneath the cavity, inducing shedding. However, it is to some extent possible to capture traveling cavities induced by leading edge desinence. As these methods are so important in the industry, overcoming these deficiencies is still an active research area. Several papers have presented a coupled modeling of the cavitating tip vortex or represented its effect, either by a semiempirical model (e.g., Bosschers [7]) or by a potential flow vortex model (e.g., Berger [8], Kim and Kinnas [9]).

In all cases of practical importance the ship wake needs to be considered in assessing the cavitation behavior due to the varying inflow it creates to the propeller blade during one revolution. In many cases a precomputed wake is given in the setup of the problem, and the influence from the potential of the propeller is computed as part of the problem. It is also possible to couple the potential flow propeller code with a viscous RANS simulation of the flow around the ship, where the RANS feeds the propeller code with an inflow and the acceleration of the flow from the propeller is represented by body forces in the RANS.

4.2.2 Viscous Approaches

Despite the extensive use of potential flow methods, Navier-Stokes-based methods are gaining in usage. For industrial use, this almost exclusively means incompressible RANS methods with a two-equation turbulence model, but in research, incompressible scale-resolving large eddy simulation (LES) and hybrid RANS/LES are also frequently used. In rare cases, compressibility is considered, but owing to the large geometrical scales in marine propulsion systems, this is not feasible in practice and will not be discussed further here.

The industrial interest in RANS methods follows from increasing demands on efficiency and low environmental impact, which leads to a trend toward lighter loaded propellers and reduced blade area. For such designs, potential flow methods are not reliable, as the chord-to-thickness ratio is reduced and viscous effects become important. Further, the real gain in performance comes in considering the system as a whole, including influence of hull, rudder, and other components, and then the viscous flow needs to be considered. This is thus not driven by a need to improve cavitation predictions per se, but it can be expected that once the RANS simulation is set up, also cavitation simulations will be utilized with improvements possible in comparison to pontential flow simulations.

For cavitation simulations within this area a transport equation–based homogeneous mixture approach is almost exclusively used. The actual mass transfer model used varies between codes, but overall they are tuned to yield very similar results and behavior. In some research studies however, a Lagrangian bubble model is coupled to the Eulerian mixture approach (see, e.g., Abdel-Maksoud et al. [10], Ghahramani et al. [11], Hsiao et al. [12]); this can be expected to have a significant benefit when the small-scale bubble behavior becomes important, for example, when inception and water quality is important to consider, or for advanced noise and erosion studies.

4.2.3 Particular Limitations and Issues

As has already been mentioned, the large dimensions of a propeller naturally put practical limitations on the modeling assumptions and resolution in a simulation, even more so considering the often short time allowed in the design cycle. This is obvious for full-scale applications but is relevant to a certain degree also at model scale. Furthermore, the ship wake creating the inflow to the propeller needs to be considered, at least in average.

A problematic aspect in developing the computational methodologies is related to scale effects. The possibilities to do detailed measurements in controlled conditions on the real vessel are limited, and model-scale experiments suffer from a few important effects that are difficult to account for in the simulations. The most important of these is the presence of partly laminar flow on the propeller blades. This leads to a different cavitation behavior compared with fully turbulent flow conditions, and the cavitation models are not able to consider this transitional flow.

4.3 Performance Predictions

Computation of the thrust and torque of a standard propeller in design conditions can be considered fairly reliable in both potential flow methods and RANS approaches; the exception is propellers with relatively thick blades. Limited cavitation does not normally change this situation. The challenge comes when off-design is considered, in particular during high loading when there is a risk of thrust loss due to massive cavitation; this is often referred to as thrust breakdown. In practice, the important prediction is to determine at which operating condition the thrust loss is larger than a certain threshold, such as 1% of design thrust. This is difficult to do computationally for two reasons: first, this massive cavitation is difficult to predict correctly; second, it is expensive to do the large number of simulations required to cover the operating conditions in enough detail.

As an example of this, the data set analyzed at the Second Workshop on Cavitation and Propeller Performance held in connection with the Fourth International Symposium on Marine Propulsors [13] will be used. The propeller used is the open PPTC, and three cases were studied in both cavitating and noncavitating conditions: the design condition (Case 2.2) at $J = 1.269$, a slightly lightly loaded condition (Case 2.3) at $J = 1.408$, and a slightly heavily loaded condition (Case 2.1) at $J = 1.019$. The predictions submitted represent a pool of 17 or 18 simulations (depending on the case) with both potential flow panel codes and viscous approaches from industry, institutes, and academia.

The results are summarized in Fig. 4.1 with the experimental data for a range of cavitation numbers and box plots of the submitted numerical predictions at two cavitation numbers. From the experimental data it is clear that performance is not significantly affected until rather low cavitation numbers are reached, and then there is a relatively fast loss of thrust. Except for a few outliers, the numerical predictions at the noncavitating condition are good with low spread. The predictions are considerably worse at the cavitating condition, with the thrust loss underpredicted in most cases and collectively a large spread in the results.

To visualize the cavitation behavior, the experimental observations and the submission by Chalmers using OpenFOAM are displayed in Fig. 4.2; this CFD submission is quite representative for the others as well. It is worth noticing the bubble cavitation in the experiments, which is predicted to be a sheet cavity in

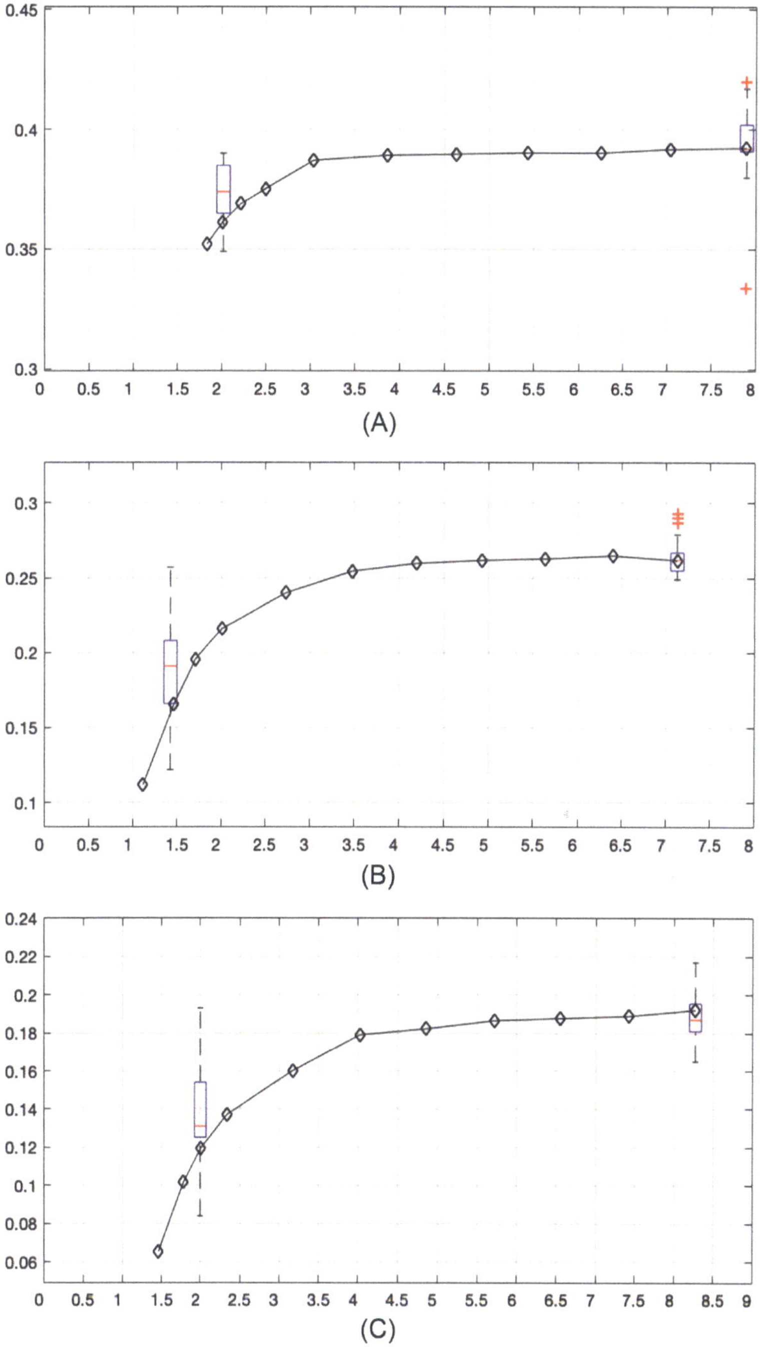

Figure 4.1 Comparison between experimental measurements by SVA Potsdam and computational submissions to the second workshop on cavitation and propeller performance predictions.

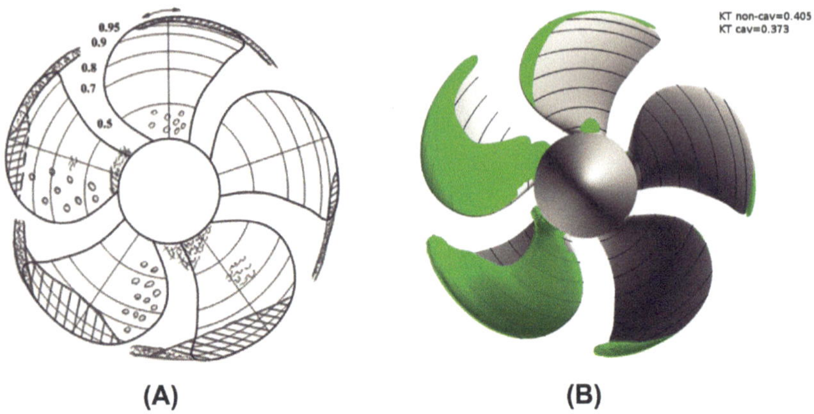

Figure 4.2 Comparison between experimental observations of cavitation extent by SVA potsdam and computational submission by Chalmers to the second workshop on cavitation and propeller performance predictions.

the CFD; this is believed to be an effect of laminar flow on the blade and thus no fixed inception point for the sheet. Whether this is the reason for the difference in thrust prediction has not been investigated.

4.4 Cavitation Inception

Determination of the operating point at which cavitation first appears in a propulsion system is important in design for vessels in which noise and comfort are of importance. This includes a large range of vessels, from naval and research vessels to ferries and yachts to fishing vessels. For a surface vessel this operation point is denoted the cavitation inception speed (CIS). The exact value is very difficult to state, as it strongly depends on the water quality (the number of nucleation sites where cavitation can start) and scale effects, and it is even difficult to define and determine cavitation inception. The design problem is most prominent for open propeller arrangements, in which cavitation most often starts in the tip vortex system, so the discussion to follow will focus on this type of flow. However, many aspects are also relevant to the inception of sheet cavitation.

From a computational perspective, several challenges are involved. First, to correctly predict the tip vortex generation and propagation is demanding. Fortunately, cavitation inception usually starts shortly behind the blade, so the length where we need a correct representation of the vortex is limited. For many of the applications in which CIS is important, the propeller tip is often unloaded, and a fair amount of skew is applied to the blade, so the formation of the vortex on the blade is nontrivial.

Second, a method and criterion for when inception occurs need to be defined. Typically, three different approaches are considered. The first is to consider only the wetted flow and define and monitor a threshold (typically minimum pressure) in the vortex core for when inception occurs. Second, one can transport a nucleus in the vortex core of the (precomputed) wetted flow and track its development; the threshold is then set on the size of the bubble. Last, a fully cavitating flow is simulated in which the cavitation number is successively lowered until it is considered that the vortex cavitates; this can be done for a Eulerian or a Lagrangian modeling of vapor content. If the goal is to rank the behavior of different design changes, it is of less importance which approach is chosen, but if an absolute value is wanted or a comparison with experiments, the response of nucleus to the pressure field will become crucial.

4.4.1 Propeller Tip Vortex Prediction

Even for a rather simple foil geometry, the tip vortex flow becomes quite complicated and poses a challenge for all numerical methods. This holds even without considering cavitation; in fact, the flow often becomes simpler once cavitation is fully developed. A concentrated tip vortex implies very strong velocity gradients across the vortex, putting high requirements on mesh resolution for a viscous modeling approach, and a weaker more diffuse vortex is often associated with the presence of anisotropic turbulence and unsteady flow behavior. A relatively thorough review of the literature on tip vortex flows is given by Asnaghi et al. [14]. Pereira et al. [15] investigated using linear two-equation and Reynolds stress models in RANS and a hybrid RANS/LES approach and showed that only the Reynold stress model could represent the anisotropic flow of a tip vortex. An alternative that has been suggested is to use nonlinear extensions to the standard two-equation models, so-called curvature corrections, and it has been shown that some do give improvements in the predictions [16]. Also a fine computational mesh is required, either by using a manual refinement zone, as done by Asnaghi et al. [14], or more preferably by some adaptive mesh refinement [17,18].

As was noted above, potential flow models, commonly used in propeller design procedures, do not immediately allow for a prediction of the tip vortex. There have been attempts to use semiempirical relations and the computed wake alignment to estimate the strength and dynamics of the tip vortex in order to estimate inception [19,20].

4.4.2 Methods to Determine Inception Point

The numerical methods to determine the cavitation inception can generally be classified into three categories: wetted flow analysis, Eulerian cavitation simulations, and Lagrangian bubble dynamics approaches. In the simplest and computationally cheapest approach, the wetted flow results are used to determine the inception points through some relations or simplified correlations between inception points and the flow properties. The most obvious approach is to monitor the minimum pressure in the vortex, and if that is below vapor pressure (or another threshold), it is considered that cavitation inception occurs [20,21]. This can be expected to give consistent comparisons, but an absolute comparison with experiments will be conservative, as nuclei inertia effects and residence time in the low-pressure zone are not considered.

In the second approach, cavitation simulations have to be conducted to obtain the vapor distribution in each operating condition. Consequently, the method is more costly than the wetted flow approach and yields little benefit compared with wetted flow simulations if standard models for mass transfer are considered.

To include nuclei effects, one can model bubble dynamics behaviors in order to determine inception points through a Lagrangian approach. This approach can provide more details on the interactions of the nuclei and tip vortex, with potentially an extra cost of modeling the nuclei dynamics. One would then define a criterion for inception based on bubble size or growth. In the simplest form, the pressure along a predetermined path in the tip vortex, most naturally the center of the vortex, is extracted, and the Rayleigh-Plesset equation is solved for a bubble exposed to this pressure [21]; the additional cost is then negligible. A more elaborate approach would be to seed a number of bubbles upstream of the tip and trace the trajectories of the bubbles in the precomputed flow [20]. It is still only a one-way coupling, but one could potentially mimic water quality, although determination of bubble size and location is not obvious. This approach also yields a non-neglible amount of additional computational work. The most realistic way to model cavitation inception would be to use a two-way Eulerian-Lagranian coupling between the flow field and the bubbles, as described, for example, by Abdel-Maksoud et al. [10] and Ghahramani et al. [11], although these approaches have not yet been applied to inception studies.

To summarize the behavior of the different approaches, the results of Asnaghi et al. [21] are reviewed. Here, both a foil (Fig. 4.3) and a propeller (Fig. 4.4) were studied with inception determined by the minimum pressure criteria, Eulerian cavitation simulations, and an uncoupled bubble dynamics approach with a range of initial bubble radii for the inserted bubble. For the foil, the different approaches were compared with experimental results in the literature for both weak water and strong water, as summarized in Fig. 4.3. In these simulations the minimum pressure criterion and the Eulerian cavitation simulations give results similar to those of the experiments in weak water of varying quality. Using larger initial bubble radii in the bubble model also gives results as in weak water, but using a very small initial radius would approach the experiments in strong water.

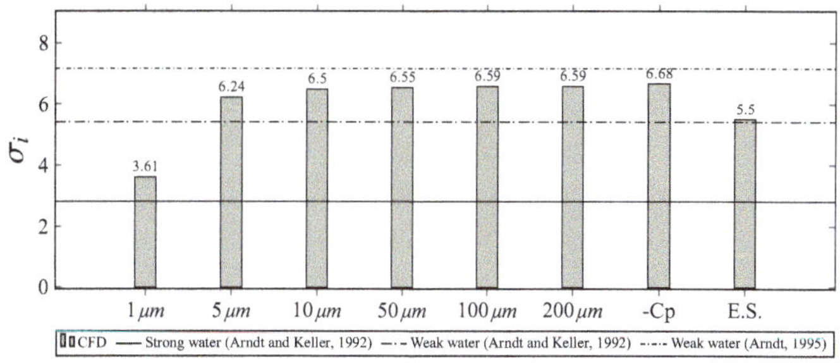

Figure 4.3 Predicted cavitation inception numbers a foil for different initial nuclei radii. $-C_p$, the minimum pressure criterion, *E.S.*, Eulerian simulations. ▯▯CFD — Strong water (Arndt and Keller [22]) -·- Weak water (Arndt and Keller [22]) -··- Weak water (Arndt [23]). *Adapted from A. Asnaghi, U. Svennberg, R. Bensow, Analysis of tip vortex inception prediction methods. Ocean Engineering 167 (2018) 187–203. ISSN 00298018. https://doi.org/10.1016/j.oceaneng.2018.08.053 [20].*

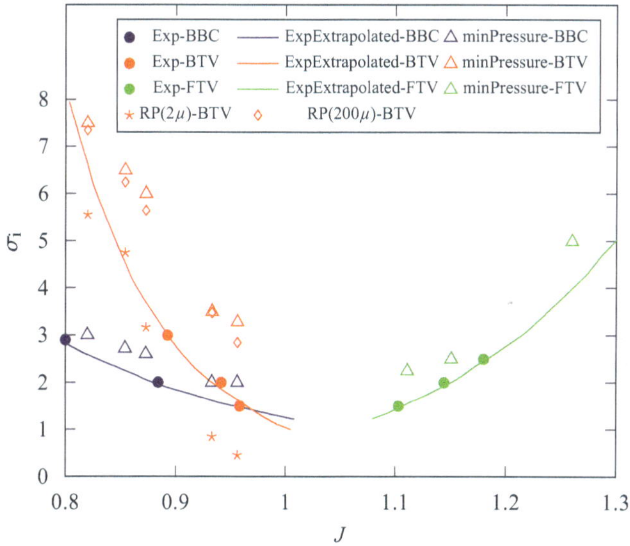

Figure 4.4 Propeller open water cavitation inception diagram, steady simulations. *Adapted from A. Asnaghi, U. Svennberg, R. Bensow, Analysis of tip vortex inception prediction methods. Ocean Engineering 167 (2018) 187–203. ISSN 00298018. https://doi.org/10.1016/j.oceaneng.2018.08.053 [20].*

4.5 Pressure Pulses and Noise

Propeller-induced pressure pulses are normally described and analyzed as harmonics of the blade passing frequency (BPF), overlayed with a broadband contribution. It is generally considered that the first- and second-order BPF fluctuations are caused by the blade thickness and load together with the volume change of propeller sheet cavitation. For third- and higher-order BPF pressure pulses, there are no well-established mechanisms, and knowledge is still developing, but it is believed to be induced by tip vortex cavitation (TVC) related phenomenon and the interaction between the sheet cavity and TVC. This understanding forms the basis for our choice of an appropriate numerical method for assessing the nuisance from cavitation-induced hull pressure pulses.

In a very early design phase, it is still common to use empirical or semiempirical methods, as done by Holden et al. [24]. However, as was noted above, potential flow-based methods give a reasonable prediction of sheet cavity dynamics and are thus commonly used for prediction of pressure pulses of low-order harmonics at a very low computational cost and then combined with a BEM approach to compute the resulting pressure field on the hull. The assessment of higher-order BPF pressure pulses and broadband content requires a tip vortex model [7,8] or a viscous flow model. The methodology and capability of these approaches are active areas of research.

An example of results using a modern potential flow prediction of cavitation-induced pressure pulses is shown in Fig. 4.5 [8]. Here, a BEM propeller model is embedded in a RANS approach for the ship and wake computation; the impact of a TVC model is also evaluated. It is clear that the first- and second-order pressure pulses are in good agreement with the experimental data and only a minor difference can be noted between the cases in which only sheet cavitation is considered and when the TVC model is activated. For this case the TVC model also gives excellent agreement for third- and fourth-order harmonics. This is surely case dependent and will in many cases not be as successful, owing to the limitations of the BEM approach to represent sheet cavity dynamics and the complete interaction between the sheet and the tip vortex cavitation. It clearly indicates the importance of the TVC on pressure pulses.

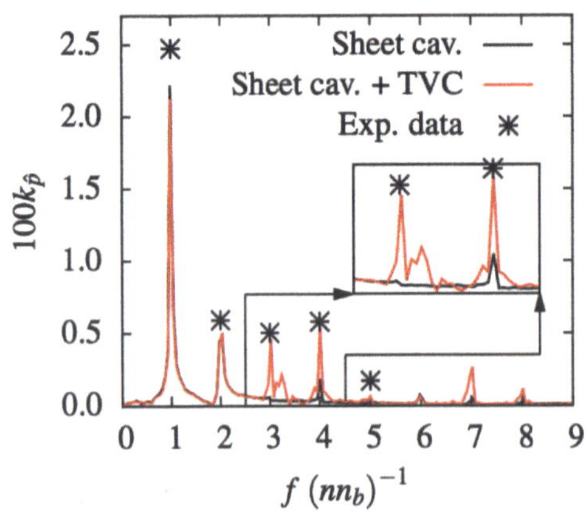

Figure 4.5 Exemplary frequency spectrum observed at the monitoring point directly above the propeller. *From S. Berger, Numerical Analysis of Propeller-Induced Higher-Order Pressure Fluctuations on the Ship Hull (PhD Thesis), Technical Univeristy of Hamburg-Harburg, 2018.*

For some flows, it has been shown that well-resolved RANS approaches may give fairly accurate predictions of high-order BPF, possibly up to fifth-order pulses [25], and indications of broadband content (Figs. 4.6 and 4.7). It is shown that the higher-order contribution comes from the development of the reentrant jet around the convex sheet cavity as the blade moves through the wake peak and how this reentrant jet affects the tip vortex formation. As shown in Fig. 4.8, this leads to a complex vortex breakdown in the wake peak, captured in this RANS

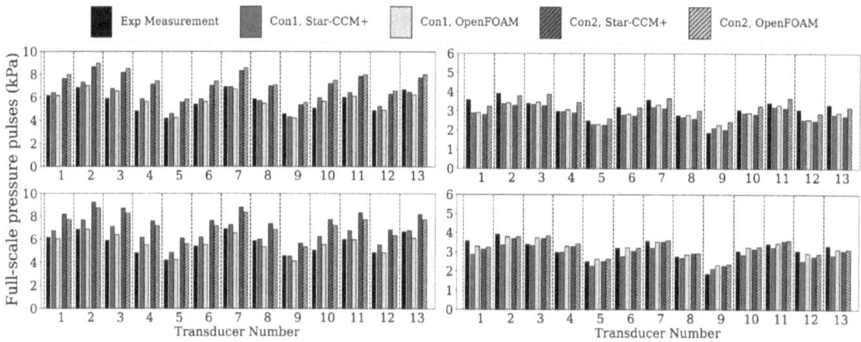

Figure 4.6 Predicted first-order (left column) and second-order (right column) pressure pulse levels using tip refined mesh (upper row) and base mesh (lower row). *Adapted from M. Ge, U. Svennberg, R.E. Bensow, Investigation on RANS prediction of propeller induced pressure pulses and sheet-tip cavitation interactions in behind hull condition, Ocean Engineering 209 (December 2019) (2020) 107503. Available from: https://doi.org/10.1016/j.oceaneng.2020.107503. ISSN 00298018.*

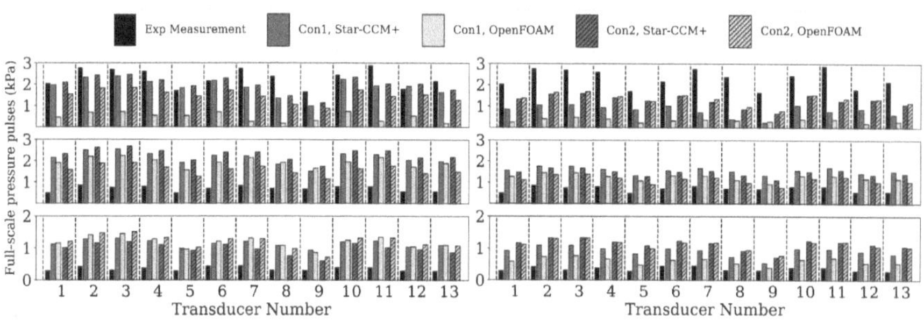

Figure 4.7 Predicted third-order to fifth-order pressure pulse levels using tip refined mesh (left column) and base mesh (right column). *Adapted from M. Ge, U. Svennberg, R.E. Bensow, Investigation on RANS prediction of propeller induced pressure pulses and sheet-tip cavitation interactions in behind hull condition, Ocean Engineering 209 (December 2019) (2020) 107503. Available from: https://doi.org/10.1016/j.oceaneng.2020.107503. ISSN 00298018.*

Figure 4.8 Cavitation patterns with isosurfaces of $Q_n = 1292$, Star-CCM + predictions, condition 1, tip refined mesh. *Adapted from M. Ge, U. Svennberg, R.E. Bensow, Investigation on RANS prediction of propeller induced pressure pulses and sheet-tip cavitation interactions in behind hull condition, Ocean Engineering 209 (December 2019) (2020) 107503. Available from: https://doi.org/10.1016/j.oceaneng.2020.107503. ISSN 00298018.*

simulation. This is related to the closure vortex formation on the sheet cavity, mentioned by Bark and Bensow [26], that is brought up by Bosschers [27] as influencing higher-order pressure pulses and broadband content. These are the kinds of interaction mechanisms that cannot be captured in a BEM approach, as was mentioned above, but are possible in a well-resolved RANS simulation. Despite this being a RANS simulation, the TVC is unsteady, and it can be expected that if the simulation is run long enough, the statistics will reveal a broadband content. Naturally, the reliability will be greatly improved if a hybrid RANS/LES approach can be used.

It should be noted that to achieve an accurate pressure pulse prediction (as well as for erosion), the ship wake needs to be correctly represented as it forms the input to the propeller computation. It is still an open question whether wake unsteadiness contributes significantly to cavitation dynamics and pressure pulse generation, but for most practical applications it is reasonable to expect that an accurate steady wake flow is sufficient, but necessary, to assess cavitation induced pressure pulse nuisance.

There is an increasing interest to assess also the underwater radiated noise (URN) from vessels, in addition to the pressure pulse levels. This comes from an improved awareness of the impact URN has on marine environment. Propeller cavitation is clearly the largest culprit for this type of emission, also indicated above by the importance of prediction of the cavitation inception for many vessels. Prediction of the actual hydro acoustic signal for a far-field receiver is a difficult topic, both numerically and experimentally, related to the propagation properties in water. From a cavitation simulation point of view, it is however reasonable to expect that the discussion on pressure pulses holds also for noise; the acoustically active component of the pressure field is a part of the sources causing hull pressure pulses.

4.6 Cavitation Erosion

The process of cavitation erosion is complex, involving both a large range of scales, both spatial and temporal, and a large number of physical processes, some of which are not completely known or understood. To get a general idea of the requirements for simulating this phenomena, we can very briefly summarize it in the following steps. As a cavity grows in size, it gains in potential energy. Traveling into a higher-pressure region, the cavity starts to undergo collapse, and a part of this energy is transferred into kinetic energy in the liquid that then, during the final stage, is emitted to the material surface by either a shock wave or a microjet, depending on the distance to the surface. If the energy is high enough, it may then affect the material and cause erosion. This description obviously omits several aspects but is sufficient for a discussion on engineering and design aspects relating to marine propulsion systems. Assuming that the material properties are given, we need to simulate or model, to some detail, the generation of the cavity, the hydrodynamics controlling its development during growth and collapse, and the energy that is transferred to the material.

The general methodologies that were outlined previously in the chapter can be assumed to capture the general development of the cavities sufficiently well, provided that the resolution in the simulation is high enough. What is meant by "high enough" is not fully clear, but it is reasonable to believe that a scale-resolving simulation approach is desirable. For more insight into the hydrodynamic mechanisms that govern cavitation erosion, see Bark and Bensow [26] and Bark et al. [28]. Naturally, the final stages of a collapse are not affordable in engineering simulations but need some kind of modeling. Also a compressible flow model, in which the energy transfer and shock waves can be treated in a sound way, is not computationally feasible when it comes to marine propulsion systems.

Some early attempts to estimate the flow aggressiveness were reviewed by Eskilsson and Bensow [29], including the approaches by Fukaya et al. [30] and Ochiai et al. [31] using discrete bubble models in a postprocessing step and by Dular et al. [32] and Li [33], in which the pressure evolution was used in different ways. It was concluded that none of these approaches gave reliable results. Using the pressure could in some cases give an indication of erosion, but it was very sensitive to calibration of the indicator, making it useless in predictive work. The bubble models seemed promising, but it was not possible to get sufficient bubbles to collapse close to the wall. This issue can be resolved by using hybrid mixture-bubble models, in which small cavities are transfered from an Eulerian mixture modeling to Lagrangian discrete bubbles, as was noted by Hsiao et al. [12] and Ghahramani [34]. An erosion assessment based on such a hybrid modeling approach has been presented by Peters and El Moctar [35], in which the energy transfer to the material is modeled by using expressions based on experimental studies of single bubble collapses.

An alternative approach is to estimate the kinetic energy introduced into the liquid during the cavity collapse and use this estimate for the transfer of energy to the material; the principle is presented schematically in Fig. 4.9. This approach has been used by both Schenke and van Terwisga [36] and Arabnejad et al. [38]

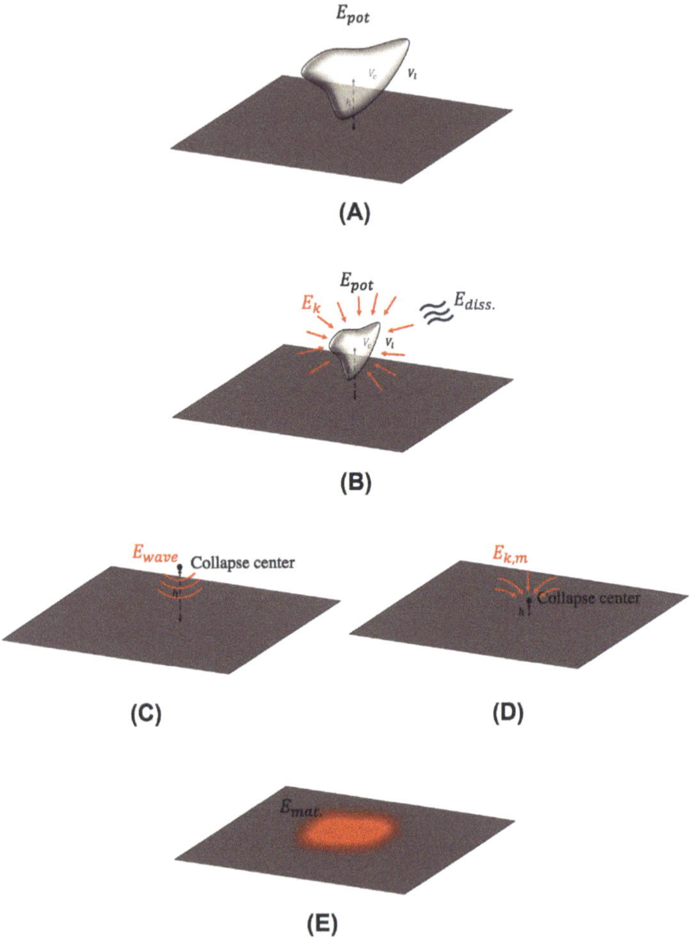

Figure 4.9 Energy-based description of cavitation erosion. (A) Maximum potential energy (E_{Pot}) in the surrounding liquid of a collapsing cavity at the start of the collapse. (B) Conversion of the potential energy to the dissipated energy ($E_{diss.}$) and the kinetic energy (E_k) in the surrounding liquid during collapse. (C) Conversion of the kinetic energy to the acoustic energy carried by shock waves (E_{wave}). (D) Conversion of the kinetic energy to the kinetic energy carried by the microjets ($E_{k,m}$). (E) Energy absorbed by material ($E_{mat.}$) when shock waves or microjets hits the surface.

to independently develop two different ways to assess the risk of erosion in an incompressible flow modeling. Schenke [37] introduces a scalar field representing the cavity-induced kinetic energy in the liquid, based on estimates of the change in potential energy of the cavity, tracing its volume change in a driving pressure field. The kinetic energy is accumulated during the collapse of the cavity, and a focused release to the material is computed. The approach was used for a marine propeller by Melissaris et al. [38]. Arabnejad et al. [38] instead directly computed the induced velocity in the liquid by tracing changes in the vapor-liquid interface; this was applied to a water jet pump by Arabnejad [39]. Although it involves a more complex algorithm and potentially requires higher resolution in the simulation, the latter approach does not, like the former, depend on the driving pressure, which is not straightforward to determine.

To exemplify current capabilities, a brief review is given of the assessment of the risk of cavitation erosion in a water jet pump [39]. What is particularly interesting in this study is that the assessment was performed for two different operating conditions, in which experimental paint tests indicated different level of erosion. This test of the capability to rank conditions has previously not been investigated for any of the proposed erosion indicators for incompressible flows. Simulations were set up in a configuration to mimic the experimental test performed at the Kongsberg Hydrodynamic Research Centre in Kristinehamn, Sweden (see Fig. 4.10). The two flow conditions are given in Table 4.1, where it can be noted that the rotor shaft speed was kept constant and the flow rate was varied. The experiments were documented by video and through a paint test. For the simulations a wall-modeled LES was performed with one blade refined over

Figure 4.10 The geometrical setup for the water jet simulations.

TABLE 4.1 Description of the Studied Flow Conditions.

Conditions	U_{inlet} (m/s)	σ	n_M (rps)
C1	4.58	1.55	20
C2	6.64	0.67	20

the tip to capture tip leakage flow that might influence the cavitation behavior, following the recommendations developed in Arabnejad et al. [40].

The results of the experimental paint tests and the numerical assessment using the approach by Arabnejad et al. [37] are shown in Fig. 4.11. The comparison is favorable in many aspects. For example, the difference between the two conditions is convincingly predicted in the simulations, as are the general shape and location of the region where risk of erosion is indicated. The main differences are the slight shift of location of the midblade erosion for condition C1 and the exaggerated blade tip erosion in both conditions. The reasons for the differences are argued to depend on both issues with the experiments (e.g., uneven paint application and laminar flow effects suppressing cavitation) and the simulations (e.g., lack of statistical convergence and varying resolution on different parts of the blade).

To further support the erosion assessment and to demonstrate the usefulness of simulations to understand the flow behavior, the reason for the higher aggressiveness of the flow in condition C2 was analyzed. It was clear already from the experimental visualizations that the root cavity causing the strong paint removal for C2 was larger than the corresponding root cavity in C1. The simulations could further show that the collapse rate

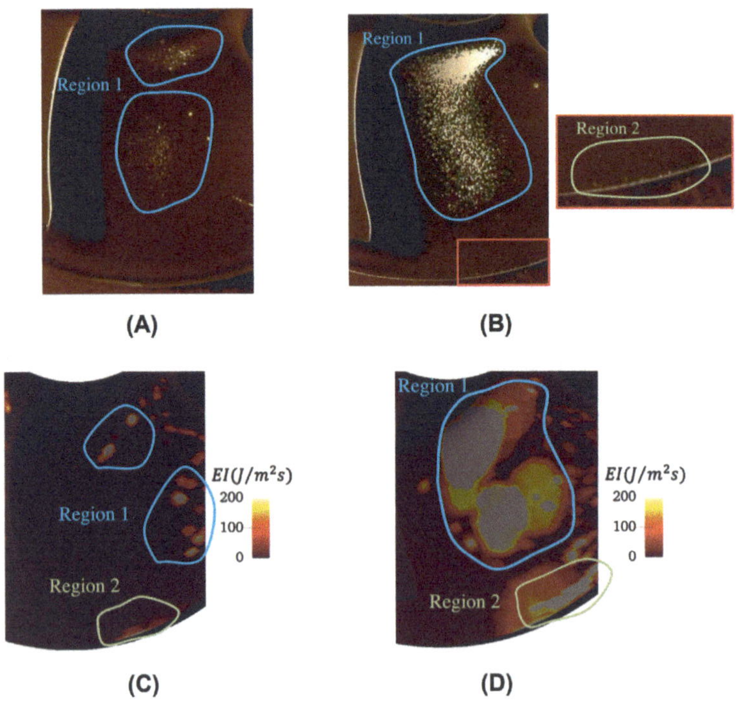

Figure 4.11 Risk of cavitation erosion obtained by the experimental paint test and the presented numerical method in different flow conditions.

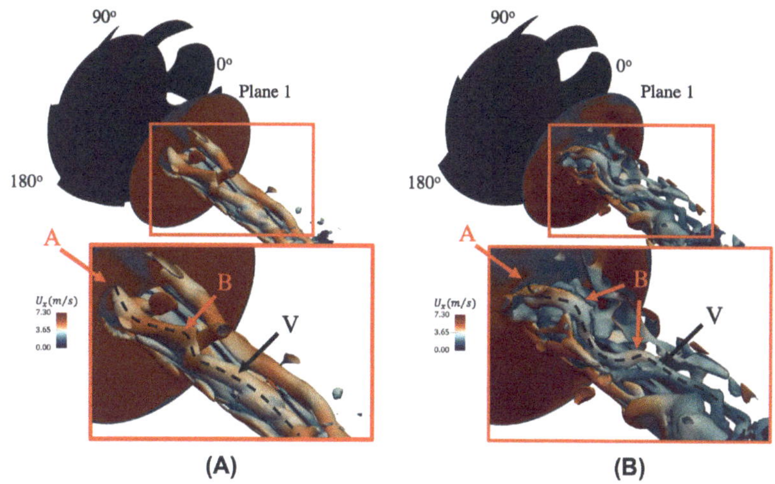

Figure 4.12 Isosurface of Q criteria near the shaft ($Q = 2 \times 10^5$) colored by axial velocity for two studied conditions.

of this vapor structure was significantly higher for C2, indicating a more aggressive collapse. The origin of this behavior was traced to the formation of a vortex structure around the shaft, as shown in Fig. 4.12. At the higher flow rate in C2 a stronger vortex developed on the top of the axis that suffered a stronger vortex breakdown just upstream the rotor. In this case a redesign of the blade shape would have had limited effect on erosion; it would have been more fruitful to try to suppress this vortex breakdown.

4.7 Concluding Remarks

Cavitation nuisance is and will remain a limiting design constraint in marine propulsion systems. As focus turns more and more toward performance at ship scale, possibly over a wide range of operating conditions, assessment of cavitation at ship scale and in off-design is needed, and the use of computational methods will continue to increase. This in turn calls for knowledge about how to do faster and more reliable simulations, which involves the development of modeling approaches and numerical approaches as well as simulation setups that are most suitable for the problem at hand.

Further, some of the physical mechanisms causing nuisance are still not fully understood, so neither are the design requirements to avoid the issues. Experimental methods are developing, but their application is limited to the configurations that it is possible to study. Computational methods complement the development of knowledge on cavitation nuisance mechanisms, in which detailed flow development can be probed to understand the cavitation dynamics and how the efficiency of the propulsion system can be improved without an increased risk of unwanted side effects from cavitation. It is an interesting development.

References

[1] H.W. Lerbs, Moderately loaded propellers with a finite number of blades and an arbitrary distribution of circulation, Transactions of the SNAME, 60 (1952).

[2] J.E. Kerwin, The solution of propeller lifting surface problems by vortex lattice methods (PhD thesis), Massachusetts Institute of Technology, 1961.

[3] D.E. Cummings, Numerical prediction of propeller characteristics, Journal of Ship Research 17 (1973).

[4] C.S. Lee, Prediction of Steady and Unsteady Performance of Marine Propellers With or Without Cavitation by Numerical Lifting-Surface Theory (PhD thesis), Massachusetts Institute of Technology, 1979.

[5] J.E. Kerwin, S.A. Kinnas, M.B. Wilson, Experimental and analytical techniques for the study of unsteady propeller sheet cavitation, in: Symposium on Naval Hydrodynamics, 1986.

[6] S.A. Kinnas, N.E. Fine, A numerical nonlinear analysis of the flow around two- and three-dimensional partially cavitating hydrofoils, Journal of Fluid Mechanics 254 (1993) 151–181. Available from: https://doi.org/10.1017/S0022112093002071.

[7] J. Bosschers, A semi-empirical prediction method for broadband hull-pressure fluctuations and underwater radiated noise by propeller tip-vortex cavitation, Journal of Marine Science and Engineering 60 (49) (2018).

[8] S. Berger, Numerical Analysis of Propeller-Induced Higher-Order Pressure Fluctuations on the Ship Hull (PhD Thesis), Technical Univeristy of Hamburg-Harburg, 2018.

[9] S. Kim, S.A. Kinnas, Prediction of unsteady developed tip vortex cavitation and its effect on the induced hull pressures, Journal of Marine Science and Engineering 80 (2) (2020). Available from: https://doi.org/10.3390/jmse8020114. ISSN 20771312.

[10] M. Abdel-Maksoud, D. Hänel, U. Lantermann, Modeling and computation of cavitation in vortical flow, International Journal of Heat and Fluid Flow 310 (6) (2010) 1065–1074. Available from: https://doi.org/10.1016/j.ijheatfluidflow.2010.05.010. ISSN 0142727X.

[11] E. Ghahramani, M.H. Arabnejad, R.E. Bensow, Realizability improvements to a hybrid mixture-bubble model for simulation of cavitating flows, Computers & Fluids 174 (2018) 135–143.

[12] C.-T. Hsiao, J. Ma, G.L. Chahine, Multiscale tow-phase flow modeling of sheet and cloud cavitation, International Journal of Multiphase Flow 90 (2017) 102–117.

[13] S. Kinnas, M. Abdel-Maksoud, U. Barkmann, L. Lübke, T. Tian, Second workshop on cavitation and propeller performance, in: 4th International Symposium on Marine Propulsors, smp'15, 2015.

[14] A. Asnaghi, U. Svennberg, R.E. Bensow, Large eddy simulations of cavitating tip vortex flows, Ocean Engineering 1950 (November 2019) (2020) 106703. Available from: https://doi.org/10.1016/j.oceaneng.2019.106703. ISSN 00298018.

[15] F.S. Pereira, L. Eça, G. Vaz, Simulation of wingtip vortex flows with Reynolds-averaged Navier-Stokes and scale-resolving simulation methods, AIAA Journal 570 (3) (2019) 932–948. Available from: https://doi.org/10.2514/1.J057512.

[16] A. Asnaghi, U. Svennberg, R.E. Bensow, Evaluation of curvature correction methods for tip vortex prediction in SST k-ω turbulence model framework, International Journal of Heat and Fluid Flow 750 (November 2018) (2019) 135–152.

[17] J. Windt, J. Bosschers, Influence of local and adaptive mesh refinement on the tip vortex characteristics of a wing and propeller, in: ECCOMAS MARINE 2015, Computational Methods in Marine Engineering VI, 2015.

[18] N. Yilmaz, M. Atlar, M. Khorasanchi, An improved mesh adaption and refinement approach to cavitation simulation (MARCS) of propellers, Ocean Engineering 1710 (October 2018) (2019) 139–150. Available from: https://doi.org/10.1016/j.oceaneng.2018.11.001. ISSN 00298018.

[19] J. Hundemer, M. Abdel-Maksoud, Prediction of tip vortex cavitation inception on marine propellers at an early design stage, in: 7th International Symposium on Cavitation, CAV2009, 2009.

[20] Y. Mirsadraee, Development of a Model for Propeller Tip Vortex Cavitation and Analysis of the Radiated Pressure Fluctuations (PhD thesis), Technical University of Denmark, 2018.

[21] A. Asnaghi, U. Svennberg, R. Bensow, Analysis of tip vortex inception prediction methods, Ocean Engineering 167 (2018) 187–203. Available from: https://doi.org/10.1016/j.oceaneng.2018.08.053. ISSN 00298018.

[22] R. Arndt, Fluid Vortices, Springer, Netherlands, 1995.

[23] R. Arndt, A. Keller, Water quality effects on cavitation inception in a trailing vortex, Journal of Fluids Engineering 114 (1992).

[24] K.D. Holden, O. Fagerjord, R. Frostad, Early design stage approach to reducing hull surface forces due to propeller cavitation, SNAME Transactions 88 (1980).

[25] M. Ge, U. Svennberg, R.E. Bensow, Investigation on RANS prediction of propeller induced pressure pulses and sheet-tip cavitation interactions in behind hull condition, Ocean Engineering 209 (December 2019) (2020) 107503. Available from: https://doi.org/10.1016/j.oceaneng.2020.107503. ISSN 00298018.

[26] G. Bark, R. Bensow, Hydrodynamic mechanisms controlling cavitation erosion, International Shipbuilding Progress 600 (1–4) (2013). Available from: https://doi.org/10.3233/ISP-130097. ISSN 0020868X 15662829.

[27] J. Bosschers, Propeller Tip-Vortex Cavitation and Its Broadband Noise (PhD thesis), Twente Technical University, 2018.

[28] G. Bark, N. Berchiche, M. Grekula, Application of Principles for Observation and Analysis of Eroding Cavitation - The EROCAV Observation Handbook. Technical Report, Chalmers University of Technology, 2004.

[29] C. Eskilsson, R.E. Bensow, Estimation of Cavitation Erosion Intensity Using CFD: Numerical Comparison of Three Different Methods, in: 4th International Symposium on Marine Propulsors, smp'15, 2015.

[30] M. Fukaya, Y. Tamura, Y. Matsumoto, Prediction of cavitation intensity and erosion area in centrifugal pump by using cavitating flow simulation with bubble flow model, Journal of Fluid Science and Technology 50 (10) (2010) 305–316.

[31] N. Ochiai, M. Nohmi, Y. Iga, T. Ikohagi, Numerical prediction of cavitation erosion in cavitating flow, in: Proc. of 7th International Symposium on Cavitation. Ann Arbor, MI, 2009.

[32] M. Dular, B. Sirok, B. Stoffel, Experimental and numerical modeling of cavitation erosion, in: Proceedings of 6th International Symposium on Cavitation, Wageningen, The Netherlands, 2006.

[33] Z. Li, Assessment of Cavitation Erosion With a Multiphase Reynolds-Averaged Navier-Stokes Solver (PhD thesis), Delft University of Technology, 2012.

[34] E. Ghahramani, Numerical Simulation and Analysis of Multi-Scale Cavitating Flows Using a Hybrid Mixture-Bubble Model (PhD thesis), Chalmers University of Technology, 2020.

[35] A. Peters, O. El Moctar, Numerical assessment of cavitation-induced erosion using a multi-scale Euler-Lagrange method, Journal of Fluid Mechanics (2020). Available from: https://doi.org/10.1017/jfm.2020.273. ISSN 14697645.

[36] S. Schenke, T.J.C. van Terwisga, An energy conservative method to predict the erosive aggressiveness of collapsing cavitating structures and cavitating flows from numerical simulations, International Journal of Multiphase Flow 111 (2019) 200–218. Available from: https://doi.org/10.1016/j.ijmultiphaseflow.2018.11.016. ISSN 03019322.

[37] S. Schenke, T. Melissaris, T.J.C. Van Terwisga, On the relevance of kinematics for cavitation implosion loads, Physics of Fluids 31 (5) (2019) 10897666. Available from: https://doi.org/10.1063/1.5092711.

[38] T. Melissaris, S. Schenke, N. Bulten, T.J. van Terwisga, On the accuracy of predicting cavitation impact loads on marine propellers, Wear 456-4570 (2020). Available from: https://doi.org/10.1016/j.wear.2020.203393.

[39] M.H. Arabnejad, Assessment of Cavitation Erosion Risk Based on Single-Fluid Simulation of Cavitating Flows (PhD thesis), Chalmers University of Technology, 2020.

[40] M.H. Arabnejad, A. Eslamdoost, U. Svennberg, R.E. Bensow, Scale resolving simulations of the non-cavitating and cavitating flows in an axial water jet pump, in: 33rd Symposium on Naval Hydrodynamics, Osaka, Japan (Virtual), 2020.

Chapter 5

Study and Detection of Cavitation in Centrifugal Pumps

Georgios Mousmoulis, Ioannis Kassanos and Ioannis Anagnostopoulos
Laboratory of Hydraulic Turbomachines, School of Mechanical Engineering, National Technical University of Athens, Athens, Greece

5.1 General Description of Cavitation in Centrifugal Pumps

In this chapter the effect of cavitation development on the operation of centrifugal pumps is presented, along with the experimental and numerical approaches that have been developed and are proposed in the literature for the study and the detection of this phenomenon. Chapter 5 is divided in three parts. In Section 5.1, the various types of vapor formations that can appear in centrifugal pumps are presented, along with the installation considerations for the proper operation of the centrifugal pump. In Sections 5.2 and 5.3, the most seminal experimental and numerical approaches to the study and detection of cavitation development in centrifugal pumps are reviewed. The analysis of the cavitation phenomenon in this chapter is relatively brief. For more details, readers are invited to study the books of Li [1] and Brennen [2,3].

5.1.1 Large-Scale Cavitation Formations in Centrifugal Pumps

Cavitation is the hydrodynamic phenomenon of liquid vaporization that occurs when the local static pressure value, p, becomes lower than the vapor pressure of the fluid,

Cavitation and Bubble Dynamics. DOI: https://doi.org/10.1016/B978-0-12-823388-7.00005-9

p_{sat}. In centrifugal pumps, vapor cavities usually appear close to the suction of the impeller at the leading edge of the blades, where the flow is accelerated as it enters the rotating impeller and the static pressure takes its lowest value. When the vapor bubbles implode close to the solid boundaries of the machine, strong impulsive waves are generated and significantly stress the material structure of the internal walls [1]. The collapse of the cavities is followed by noise that is often masked by the mechanical and hydraulic noises of the machine and the establishment. After a certain amount of operation time under cavitating conditions, the stressing can cause significant wear and material removal at the internal rotating surfaces [2,3] and consequently reduces the efficiency and increases the imbalance of the system, creating additional oscillations that stress the bearings.

Studies of cavitation inside centrifugal pumps approach the phenomenon from a large-scale perspective rather than that of a single bubble, since the two-phase flow area usually consists of a large number of bubbles that form a visually observable two-phase flow area. Four main types of large-scale cavitation have been observed inside centrifugal pumps: (1) attached or sheet cavitation, (2) cloud cavitation, (3) bubble traveling cavitation, and (4) tip vortex cavitation. All types are shown in Fig. 5.1 at the suction side (SS) of the blade of a radial (Fig. 5.1A–C) and an axial (Fig. 5.1D) centrifugal pump.

The attached cavitation is depicted in Fig. 5.1A by a uniform single vapor-filled separation zone that develops at the SS of the blade, usually at large angles of attack. The formation of steady attached cavities changes the hydrodynamic shape of the blade, even for blades with relatively low incidence angle, and creates flow recirculation and separation regions that decrease the overall efficiency of the pump [9]. Under very low suction conditions the extent of the cavity could reach the trailing edge of the blade of the pump.

Cloud cavitation formations are also formed at the SS of the impeller blade (Fig. 5.1B) and usually exhibit a characteristic periodic formation and collapse of vapor bubbles that form a cloud unsteady structure [3]. The complex mechanism that drives the periodicity of the formation and collapse procedure is affected by both the shedding of cavitating vortices and the periodicities imposed by the individual flow characteristics of each application (rotor-stator interaction, etc.). The unsteady nature of the phenomenon may cause more noise, vibration, and damage to the neighboring solid surfaces than in the case of large-scale cavitation types that do not exhibit strong fluctuations in their shape, such as attached two-phase flow formations. The bubble traveling cavitation (Fig. 5.1C) consists of individual bubbles that are formed at the SS and collapse in the higher-pressure areas [10]. Such formations create intensive noise but are considered harmless in terms of material wear, since they usually implode far from the solid surfaces [9]. The tip vortex cavitation type (Fig. 5.1D) appears in areas with extensive vorticity, where the pressure usually takes its minimum value at the vortex core. This type of large-scale cavitation appears at the tip clearance of the blades of open or

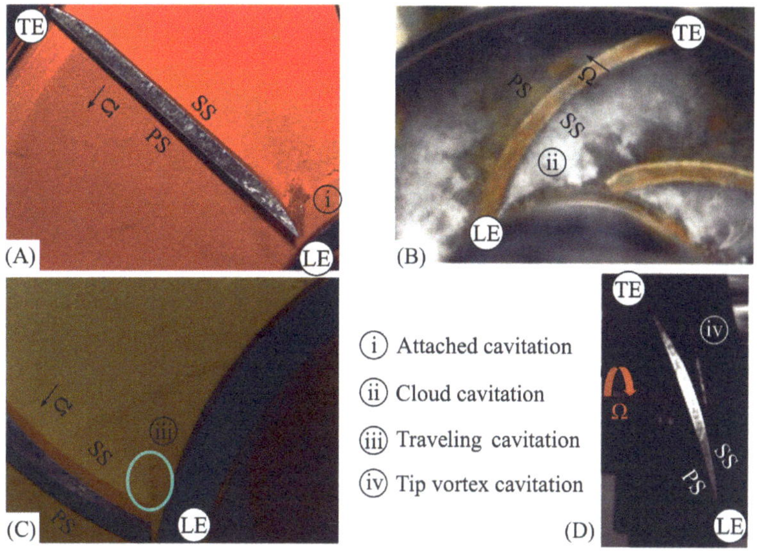

Figure 5.1 Large-scale cavitation types developed in, (A, C) radial centrifugal pump with semi-open impellers [4–6], (B) radial centrifugal pump with closed impellers [7], (D) axial centrifugal pump [8]. *(A, C) Present authors, adapted from G. Mousmoulis, N. Karlsen-Davies, G. Aggidis, J. Anagnostopoulos, D. Papantonis, Experimental analysis of the onset and development of cavitation in a centrifugal pump, Journal of Physics: Conference Series 813 (1) (2017); G. Mousmoulis, N. Karlsen-Davies, G. Aggidis, I. Anagnostopoulos, D. Papantonis, Experimental analysis of cavitation in a centrifugal pump using acoustic emission, vibration measurements and flow visualization, European Journal of Mechanics-B/Fluids 75 (2019) 300–311; G. Mousmoulis, Development of Cavitation Detection Methodology for Hydraulic Turbomachines Using Experimental and Numerical Tools (PhD thesis), National Technical University of Athens, 2020 [4–6]; (B) Reproduced with permission from W. Zhao, G. Zhao, An active method to control cavitation in a centrifugal pump by obstacles, Advances in Mechanical Engineering 9(11) (2017) [7]; (D) Reproduced with permission from D. Zhang, W. Shi, D. Pan, M. Dubuisson, Numerical and experimental investigation of tip leakage vortex cavitation patterns and mechanisms in an axial flow pump, Journal of Fluids Engineering 137 (12) (2015) 121103 [8].*

semi-open rotors and its vortices appear due to the interaction of the end-wall boundary with the main flow and the rotor's rotation. Tip vortex cavitation can cause significant wear in the rotor; nevertheless, it does not reduce the hydraulic efficiency of the machine [8,9,11,12].

5.1.2 Cavitation Effects on Pump Performance

A direct way to depict the effect of cavitation on the operation of a centrifugal pump is by plotting on the same graph (Fig. 5.2) the total head characteristic

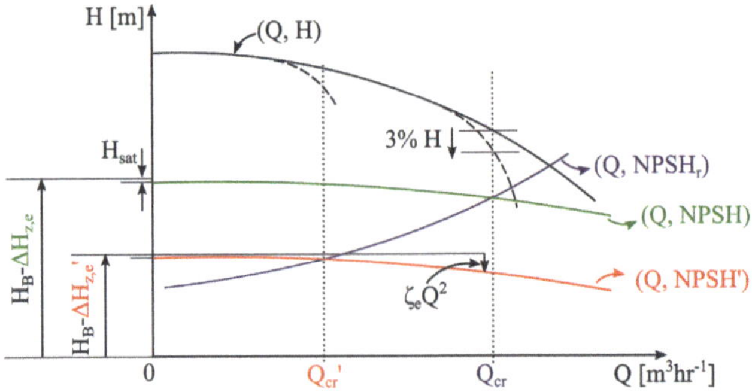

Figure 5.2 Effect of cavitation on the total head characteristic curve (Q, H_{tot}) of a centrifugal pump and effect of pump installation level on NPSH [9]. *Reproduced with permission from D. Papantonis, Hydrodynamic Machines, Hydrodynamic Transmissions, Symeon, 2016. (in Greek) [9].*

curve (Q, H) of the pump when the last works under single-phase conditions in the total loading range (solid black curve) and when vapor formations appear in the pump after a specific flow rate, Q (dashed black curve). Fig. 5.2 depicts the drop of H due to cavitation development and highlights the decrease in the Q range, where the pump operates under the designed and desired H.

In addition, cavitation in the pump establishment is related to the static pressure level (SPL) of the flow. The difference of SPL between the suction of the pump and the p_{sat} defines the net positive suction head (NPSH) available (Eq. 5.1) and associates it with the design characteristics of the installation:

$$\text{NPSH} = H_{\text{oe}} - H_{\text{sat}} = H_B - \Delta H_{z,e} - \zeta_e Q^2 - H_{\text{sat}} \tag{5.1}$$

where H_{oe} is the total energy per unit weight at the SS of the pump, H_{sat} is the saturation pressure, H_B is the pressure at the surface of the suction tank, $\Delta H_{z,e}$ the height difference between the pump suction and the height level of the medium in the suction tank, and the term $\zeta_e Q^2$ represents the hydraulic losses from the suction tank until the pump. For a given installation the terms H_B, $\Delta H_{z,e}$, ζ_e, and H_{sat} are constant, and the NPSH available as a function of flow rate (green solid curve) is also given in Fig. 5.2. Here, it is clear that the operation of the pump at higher flow rates decreases the NPSH available value of the installation and makes the flow in the latter more prone to cavitate.

Moreover, pump manufacturers have set a 3% drop in H (Figs. 5.2 and 5.3A) as a limit for the safe operation of the centrifugal pump [13]. The NPSH value that corresponds to a 3% drop in H defines the critical NPSH$_r$, while the characteristic curve (Q, NPSH$_r$), which is shown by the blue solid curve in Fig. 5.2, characterizes the machine and is provided by the manufacturer of each pump. The intersection point between the NPSH and NPSH$_r$ curves in Fig. 5.2 provides the

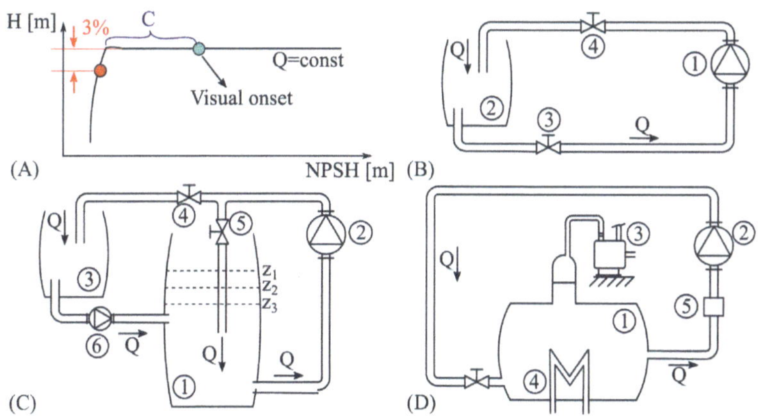

Figure 5.3 (A) Typical cavitation characteristic curve (NPSH, H_{tot}), (B,C,D) Experimental configurations for the cavitation testing of centrifugal pumps: (B) open sump with throttle valve [6], (C) open sump with the level control [9], and (D) closed loop arrangement [9]. *(B) Present authors, adapted from G. Mousmoulis, Development of Cavitation Detection Methodology for Hydraulic Turbomachines Using Experimental and Numerical Tools (PhD thesis), National Technical University of Athens, 2020 [6]; (C,D) Reproduced with permission from D. Papantonis, Hydrodynamic Machines, Hydrodynamic Transmissions, Symeon, 2016. (in Greek) [9].*

critical flow rate, Q_{cr}, which is the maximum flow rate at which the pump can operate without suffering a larger than 3% drop in H [9]. If the characteristics of the installation change, for instance if the pump is installed at a higher level height ($\Delta Hz,e' > \Delta Hz,e$), the new NPSH′ available characteristic curve (the red curve in Fig. 5.2) exhibits lower values. The equivalent effect on the NPSH curve reduces H_B or increases ζ_e and H_{sat}. In practice, the configuration with the new NPSH′ value is characterized by a lower critical flow rate, Q_{cr}' (Fig. 5.2), which reduces the loading range at which the pump can operate safely under the designed H.

5.2 Experimental Investigations

This section reviews the experimental works that aim to study and detect cavitation in centrifugal pumps with the use of experimental tools and techniques. The vast majority of these studies are based on the conduction of cavitation performance tests, according to the standard methods given in ISO 9906:2012 [14]. For that reason the different experimental approaches are initially discussed in Section 5.2.1 before the presentation of the literature results.

5.2.1 Test Configurations

Three different methods for the conduction of cavitation tests in centrifugal pumps are proposed in ISO 9906:2012 [14]: (1) the open sump with throttle valve, (2) the open sump with level control, and (3) the closed loop arrangement. All of the tests aim to derive the cavitation characteristic curves of the pump (NPSH, H_{tot}) that depict the drop of the total head when the available NPSH of the installation decreases (Fig. 5.3A). In practice, the (NPSH, H_{tot}) curve illustrates the effect of cavitation on the ability of the centrifugal pump to provide hydraulic power to the working fluid. The NPSH and H_{tot} parameters are nondimensionalized with the use of Thoma's cavitation number, σ (Eq. 5.2), and the nondimensional total head, Ψ (Eq. 5.3), respectively.

$$\sigma = \frac{\text{NPSH}}{H} \tag{5.2}$$

$$\Psi = \frac{2gH}{(\Omega R_2)^2} \tag{5.3}$$

The cavitation characteristic curves (σ, Ψ) or their dimensional form (NPSH, H) is the basis of the majority of cavitation-related studies in centrifugal pumps. The open sump with throttle valve configuration is shown in Fig. 5.3B, and the installation consists only of the pump (1) and one open tank (2). In this case, the reduction in the NPSH available value is achieved by throttling the suction valve (3) and, consequently, by increasing the losses term ($\zeta_e Q^2$) in Eq. (5.1). Nevertheless, the throttling of the suction valve changes the system's characteristic curve and hence, the flow rate. In order to restore the flow rate for the conduction of the test under constant Q conditions, the discharge valve (4) has to be further regulated. This configuration is the simplest proposed by ISO 9906:2012 [14]. However, the throttling of the suction valve is likely to disturb the axisymmetric flow at the inlet of the impeller.

The open sump with the level control is shown in Fig. 5.3C. The static pressure at the pump inlet is regulated by unloading ($z_1 \rightarrow z_2 \rightarrow z_3$) the open tank (1) that is located at the suction of the tested pump (2) with the aid of a second tank (3) and pump (6) after properly regulating valves (4) and (5). In practice. this method aims to increase the height difference, $\Delta H_{z,e}$ term of Eq. (5.1) by reducing the height level of the fluid in the suction tank. The disadvantage of the aforementioned procedure is that it increases significantly the time duration of the experiment and the number of the hydraulic components (two tanks, feeding pump, valves, etc.) that are used.

Finally, the closed loop arrangement, which is shown in Fig. 5.3D, consists of one closed tank (1) that supplies the centrifugal pump under examination (2). In this case, the SPL of the installation is decreased with the use of a vacuum pump (3) that is mounted on the upper part of the closed tank and regulates the static pressure in the tank. In practice, this method proposes a decrease in H_B in Eq. (5.1) for the reduction of the NPSH available value. The advantage of this testing configuration is that the NPSH available value is controlled accurately and without affecting the flow rate. However, the forced circulation of the working medium in a closed loop configuration increases its temperature; for this reason a heat exchanger (4) is installed to keep the medium's temperature constant. The use of additional mechanical equipment, such as the vacuum pump and the heat exchanger, makes this type of configuration more expensive than the other two already described.

Also, there are cavitation studies on centrifugal pumps that do not follow the aforementioned techniques [15–19]. In these studies, the machine is operated at higher flow rates, where the flow field is accelerated and vapor formations are more likely to appear at the inlet of the pump (Fig. 5.2). In this way, it is possible to obtain measurements and study the phenomenon's characteristics. However, it is impossible to examine the development of two-phase flow area as a function only of the SPL of the installation, since the flow rate also changes.

When cavitation tests are being conducted in centrifugal pumps, it is important to control the amount of air and gas particles in the working medium that circulates in the configuration. The presence of noncondensable gas favors the formation of vapor cavities, since they reduce significantly the tension that the liquid is able to sustain [9,20,21]. Control of the air and dissolved gases is achieved by the use of the heat exchanger (4), shown in Fig. 5.3D. This procedure dissolves the particles, and the heating continues until water with 0.3%–0.5% gas content is obtained [9]. Measurement of the gas content in the fluid is done with the use of special arrays ((5) in Fig. 5.3D), which are included in the configuration. The heating of the fluid can be achieved only with the closed loop arrangements.

5.2.2 Dynamic Pressure Measurements and Flow Visualization

The cavitation phenomenon exhibits unsteady characteristics, and its experimental study demands not only excellent control of the test rig (Section 5.2.1) but also the use of sophisticated measuring techniques and transducers. The studies that examine the vapor characteristics in the flow of centrifugal pumps make use of dynamic static pressure measurements and flow visualization techniques. By definition, cavitation is related to changes in the pressure and the velocity of the flow field. Thus the conduction of static pressure measurements in different positions inside the pump has been found to be effective in identifying the location of bubble formation [22] and their intensity [17,23,24] and in computing the excited frequencies from the phenomenon [23,25,26]. In addition, the incorporation of static pressure sensors on the rotating and stationary surfaces of the centrifugal pump allows for the study of rotating effects on cavitation development. More specifically, the processing of the measured data in the frequency spectrum has been found to highlight the subsynchronous periodicity of rotating cavitation relative to the rotational frequency (RF) of the shaft [27,28].

The vast majority of the cavitation studies are accompanied by flow visualization measurements to illustrate two-phase flow development under various suction conditions (Figs. 5.1 and 5.4B). The results of visual observations are very often used for the validation of the numerical tools that aim to simulate the onset and development of cavitation in centrifugal pumps.

To obtain flow visualization data, parts of the turbomachine are manufactured from transparent material, usually Plexiglas (Fig. 5.4A). Especially for cavitation tests in centrifugal pumps, these parts include the pump casing (Fig. 5.1) [4–6,8,11,12,26,30,31], the suction pipe [4–6,25,29,31], and in some cases the shroud of the closed impeller (Fig. 5.1B) [6,7,27,32–35]. Furthermore, stroboscope light is used to freeze the rotating motion of the impeller, along with high-speed cameras that capture the characteristics of vapor formations.

The visual results illustrate the various types of large-scale vapor formations, which have been already discussed in Section 5.1.1 (Figs. 5.1 and 5.4B), the

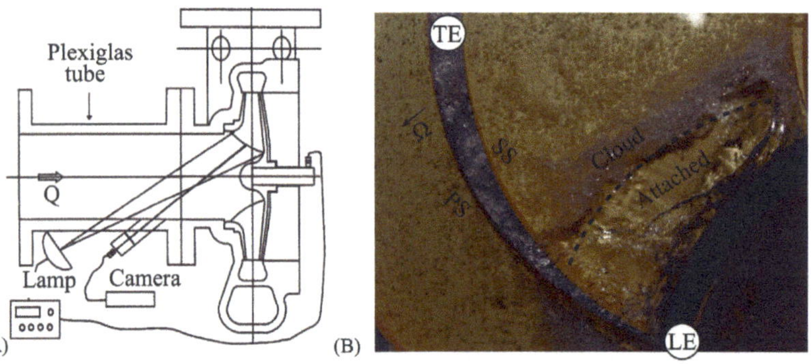

(A) (B)

Figure 5.4 (A) Installation of a Plexiglas configuration for flow visualization at the inlet of a radial pump [29]. (B) Flow visualization of attached and cloud cavitation after total head collapse [5,6]. *(A) Reproduced with permission from A. Baldassarre, M. De Lucia, P. Nesi, Real-time detection of cavitation for hydraulic turbomachines, Real-Time Imaging 4 (6) (1998) 403–416 [29]; (B) Present authors, adapted from G. Mousmoulis, N. Karlsen-Davies, G. Aggidis, I. Anagnostopoulos, D. Papantonis, Experimental analysis of cavitation in a centrifugal pump using acoustic emission, vibration measurements and flow visualization, European Journal of Mechanics-B/Fluids 75 (2019) 300–311; G. Mousmoulis, Development of Cavitation Detection Methodology for Hydraulic Turbomachines Using Experimental and Numerical Tools (PhD thesis), National Technical University of Athens, 2020 [5,6].*

location where they are developed, and their extent. With such pictures it is possible to observe the increase in size and extent of the cavities and the changes in their structure when the static pressure at the impeller inlet is reduced [4–6,8,11,12,25,29,34,35]. Furthermore, the effects of the geometrical characteristics of the blades are assessed when different impellers are tested in the same configuration under similar operating conditions [4–6,34]. More specifically, the visual onset of the phenomenon highlights the formation of small vapor bubbles at the SS of the leading edge of the blades (Fig. 5.1C) [2,4–6,9,13]. These formations are unquestionably very dynamic: The bubbles are created, travel for a short distance in the impeller (Fig. 5.1C), and finally reliquify, owing to the presence of pressure gradient in the impeller. This behavior characterizes a typical traveling bubble cavitation case. Reduction the available NPSH results in an increased number of bubbles that, in the case of impellers that operate at large incidence angles, follows from the formation of attached cavities, as in the case shown in Fig. 5.1A.

When the pump operates in minimum pressure conditions, cavitation development is violent and extended, and it almost blocks the flow passage between the impeller blades (e.g., Fig. 5.4B). In this case, two large-scale cavitation formations are found to coexist in the path between the blades: (1) attached vapor cavities that create a uniform single bubble area and (2) cloud cavitation formations at the outer boundaries of the attached cavitation area. Finally, when cavitation exhibits rotating characteristics, visual tools allow observation of the vapor cells that rotate in the opposite direction of the impeller at lower frequency (subharmonics) of the RF [27,32]. In test configurations, in which the environmental light conditions complicate the identification of vapor bubbles in the impeller of the pump, additional effort has to be made to interpret the obtained photographs. More specifically, the images are processed with the use of statistical algorithms that quantify the brightness of the pixels and aim to identify the two-phase flow area [29,36].

5.2.3 Detection of Cavitation in Centrifugal Pumps

The development and implementation of experimental tools for the detection of cavitation in centrifugal pumps have been taking place for almost a century. The first approaches, which were used until recently, propose the monitoring of the pump total head (H) and detect the presence of vapor formations when H exhibits a drop of 3% (Figs. 5.2 and 5.3A). Although the aforementioned approach provides a detection criterion of general applicability, flow visualization studies have proved that vapor bubbles initiate long before the total head starts to drop (Fig. 5.3A). The operation of the pump in the NPSH area between the head drop and the onset of the phenomenon (area C in Fig. 5.3A) can still wear out the internal solid surfaces and eventually results in additional vibration that stresses the mechanical components of the pump and in decreased performance.

To overcome the inability of the *H* monitoring approach to detect early cavitation stages, the use of noise and vibration measurements has been proposed from several authors and engineers, so as to record the high-intensity pressure waves generated by the bubble implosion. The use of such an approach requires adequate acquisition and processing of the signals to extract the cavitation information from the noisy environment of the pump and the pumping station. In general, the time series obtained from a centrifugal pump include a large number of mechanical and hydraulic excitations that are not related to cavitation, such as the rotor-stator interaction, noise and vibration from the bearings, and turbulence. The traditional signal-processing softwares allow the study of the obtained dynamic measurements in the time domain and frequency spectrum, as well as the analysis of its power content at different discrete frequencies or range of frequencies, with the use of Fast Fourier Transformation, Power Spectral Density, and power band tools. Regarding time domain analysis, only a few studies have managed to identify cavitation appearance, either by presenting the impulses in the raw measured data [33] or by plotting the effects of the impulses in the distribution of the measurements [15]. In the vast majority of the published works, there have not been reported clear strikes in the raw data, even in experiments in which high-frequency measurements are conducted and the low-frequency noise is filtered out [31,37–39].

On the other hand, the impact of two-phase flow development becomes much clearer when the spectrum of the noise and vibration data is examined, since the onset and development of cavitation clearly excites a wide band ranges, as can be observed in Fig. 5.5A [5,6,19,40–44]. The power of those frequency bands is integrated with the use of power bands, and research has found that the noise and vibration values start to increase long before the *H* drop (Fig. 5.5B), and consequently, they are able to detect cavitation at early stages [31,37–40,43,45–49].

The difficulty introduced at this point is the frequencies excited by cavitation change between different testing configurations, machines, sensor-mounting positions, and operating conditions [4,6]. Those irregularities are highlighted after comparing the trends of the various power bands published in the literature [6], and they are mainly related to the stochastic nature of the phenomenon and the random changes that are created in each configuration.

The fact that the vibration power bands quantify the excitation of different ranges in the spectrum introduces additional questions about their use in the identification of vapor bubbles. The literature provides results [50–53] in which impulsive bearing-related faults at some stage of their evolution seem to excite a wide range of high frequencies, up to 20 kHz, in a manner similar to cavitation (blue line in Fig. 5.5A).

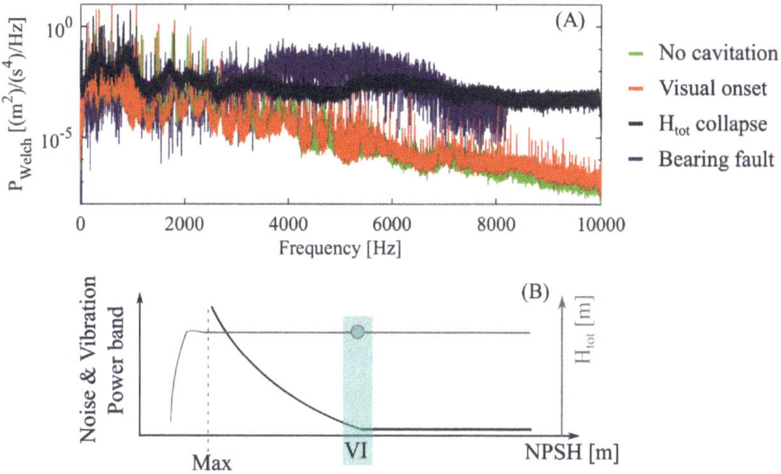

Figure 5.5 Detection of cavitation with the use of (A) the frequency spectrum and (B) power band tools [4–6]. *Present authors, adapted from G. Mousmoulis, N. Karlsen-Davies, G. Aggidis, J. Anagnostopoulos, D. Papantonis, Experimental analysis of the onset and development of cavitation in a centrifugal pump, Journal of Physics: Conference Series 813 (1) (2017); G. Mousmoulis, N. Karlsen-Davies, G. Aggidis, I. Anagnostopoulos, D. Papantonis, Experimental analysis of cavitation in a centrifugal pump using acoustic emission, vibration measurements and flow visualization, European Journal of Mechanics-B/Fluids 75 (2019) 300–311; G. Mousmoulis, Development of Cavitation Detection Methodology for Hydraulic Turbomachines Using Experimental and Numerical Tools (PhD thesis), National Technical University of Athens, 2020 [4–6].*

From the latter it is concluded that when measurements integrate ranges over the audible range (> 20 kHz), as in the case of acoustic emissions sensors, the power band does not include the effect of the bearing-related faults. One way to distinguish cavitation among other faults is to operate the pump under lower (or higher) NPSH conditions so as to study the development of this excitation either in the raw spectrum or with the use of power bands. However, this procedure is not practical in dealing with industrial centrifugal pumps; hence the use of such tools cannot always provide a clear answer regarding the two-phase flow development in the machine.

The above discussion reveals the two main weaknesses of the proposed in the literature noise and vibration measurement approaches. The first weakness is the inability to create a detection cavitation criterion of general application to any testing configuration; the second deals with the confusion that may be caused by the parallel existence of other faults that also exhibit impulsive behavior. As a consequence, additional effort has recently been made to determine a well-rounded characterization of cavitation phenomeon, along with the development of a detection criterion with general applicability.

Toward this direction, recent works [6,54,55] have proposed the use of tools related to spectral kurtosis [56–61] that calculate the band-pass filter characteristics, which are then applied in the raw time series and manage to unmask the impulses due to vapor bubble implosion from the general noise and vibration of the pumping station (Fig. 5.6A). Moreover, the filtered results (Fig. 5.6A) reveal the modulation of the cavitation impulses from the flow field of the turbomachine and, more specifically, from the rotor-stator interaction and the blade-passing frequency. As a result, to further highlight these periodicities, the filtered signals are demodulated with the use of Hilbert transformation, and the frequency spectrum of the envelope is studied, as presented in Fig. 5.6B. Here, the color legends coincide with those presented in Fig. 5.5A.

The impulse appearance in the filtered data and their modulation from the rotor-stator interaction can constitute a general and robust cavitation detection criterion for centrifugal pumps. Those features of the filtered signal are unique, since in the motor- pump system, only the rolling element bearings could excite strikes; however, they are usually modulated by their kinematic frequencies, which are in a different range [61]. Finally, the use of those characteristics for cavitation identification associates the detection criterion with the physical mechanism of

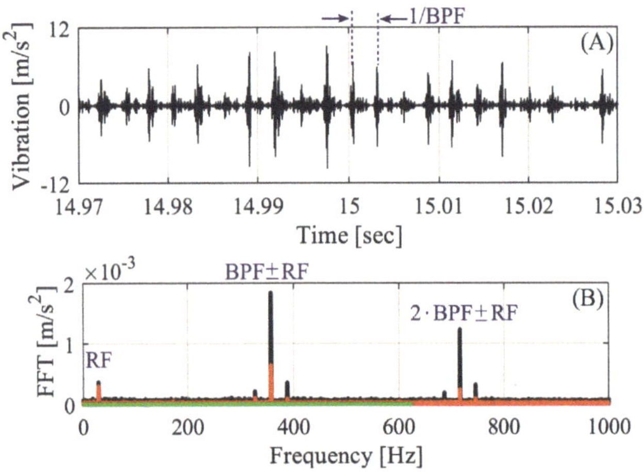

Figure 5.6 (A) Filtered signal after implementation of the fast kurtogram tool. (B) Envelope spectrum of the filtered signals [4–6]. *Present authors, adapted from G. Mousmoulis, N. Karlsen-Davies, G. Aggidis, J. Anagnostopoulos, D. Papantonis, Experimental analysis of the onset and development of cavitation in a centrifugal pump, Journal of Physics: Conference Series 813 (1) (2017); G. Mousmoulis, N. Karlsen-Davies, G. Aggidis, I. Anagnostopoulos, D. Papantonis, Experimental analysis of cavitation in a centrifugal pump using acoustic emission, vibration measurements and flow visualization, European Journal of Mechanics-B/Fluids 75 (2019) 300–311; G. Mousmoulis, Development of Cavitation Detection Methodology for Hydraulic Turbomachines Using Experimental and Numerical Tools (PhD thesis), National Technical University of Athens, 2020 [4–6].*

cavitation, since there is a clear analogy between the impulsive behavior of the filtered signal and the shock pressure mechanism due to vapor bubble implosion as well as between the modulation of the impulses from the rotor-stator interaction with the fact that the shock pressure waves occur in the rotating impeller. More specifically, cavitation bubbles are formed in the impeller, and they implode as the flowpath drives them to the exit of the rotor. The implosion of the vapor bubbles is accompanied by the generation of high-intensity pressure waves that hit the blades of the impeller. The blades rotate and, owing to the pump's nonuniformities, create rotor-stator interaction forces that dominate the flow field and modulate the shock pressure waves.

5.3 Numerical Investigations

Numerical modeling tools have been employed to study the cavitation phenomenon in rotating machinery and understand the onset and development under different operating conditions. The advances and systematic application and validation of these tools have provided new insights into the complex flow physics associated with cavitation in turbomachinery, which would otherwise be difficult to achieve through only laboratory testing. In this section, the various mathematical formulations that have been developed for cavitation modeling and their validation and application in centrifugal pumps are reviewed and discussed.

5.3.1 Modeling Approaches and Mathematical Formulations

The cavitation phenomenon involves several different complex physical processes (phase change, turbulence, multiphase interface, variable fluid properties, relative motion) that need to be modeled simultaneously. In centrifugal pumps, cavitation inception, cavity shape, and dynamics depend also on local flow conditions, streamline curvature, geometrical features, and fluid properties [62,63]. Therefore the development of a generalized model that is applicable to different problems is a task of increased complexity. For this reason, several models have been proposed with different assumptions and simplifications and various levels of complexity. The applicability of these models depends on the nature of the flow and cavitation type being modeled [10] and on the availability of computational resources.

The available cavitation modeling techniques can generally be classified as interface tracking (IT) and interface capturing methods (IC) [64]. In the case of IT methods the interface between the two phases is explicitly modeled [65]. The incompressible flow equations are solved, while the interface between the two phases is treated as an impermeable boundary of constant pressure equal to the vapor pressure of the fluid. An iterative procedure is applied to determine the interface, owing to the nonlinear relationship between the shape of the cavity and the liquid flow field, while an empirical algebraic model is used to model the wake of the cavity where the constant pressure assumption is not valid. IT methods have been applied successfully in the prediction of attached cavities [64,66,67] in two-dimensional hydrofoils and the prediction of pump performance under cavitation inception and heavy cavitating conditions [68,69]. However, limitations imposed by the constant pressure assumption and difficulties in modeling three-dimensional flows using the algebraic wake model have been reported [70,71]; numerical stability issues during the evolution of the cavitation phenomenon have also been observed [72]. Furthermore, IT methods fail to model unsteady cavitation. owing to the inherent inability to model the splitting of large cavities into smaller bubbles associated with the interface definition [73]. For these reasons, as well as owing to the increased computational cost associated with the interface treatment, these methods are limited to simple steady-state problems [64,65,74].

In IC methods the fluid-vapor interface is obtained by solving the complete domain, usually considering a homogenous single-fluid treatment [73], in which the fluid properties are obtained as a function of the vapor volume fraction with the assumptions of no slip between phases and thermodynamic equilibrium. IC methods have been more successful than IT methods in modeling different cavitation types in centrifugal pumps, including bubble, vortex and tip cavitation. In these methods the fluid properties may be determined, on one hand, by using an appropriate equation of state (EOS), such as in the barotropic approach, in which the mixture density is determined by using a barotropic state law [75] and the vapor volume fraction and, on the other hand, by using the transport equation (TEM) based models, in which an additional transport equation is solved for the mass or volume fraction, which is then used to determine the fluid properties. Unlike the TEM approach, in the barotropic approach the mass transfer between the two phases is ignored, and the phase change process is assumed to be instantaneous. The EOS approach has been successfully applied by several authors to study the cavitation around hydrofoils [76–83], ship propellers, centrifugal pumps [36], and pump inducers [84–86] and in hydraulic turbines [87,88]. A significant drawback of the barotropic approach is the inherent inability to resolve vorticity generation due to the misalignment of density and pressure gradients (baroclinicity). This has been shown experimentally to be an important source of vorticity [71], which may result in increased instabilities and cavitation due to the strong interaction between turbulence and cavitation inception [73]. The assumption of a barotropic fluid in which the fluid density is a function of local pressure implies that the baroclinic contributions in the vorticity production and therefore the potential interaction with cavitation generation are ignored. In shear flows, large vortical structures may promote the inception of cavitation within the boundary layer by the entrainment of cavitation nuclei or by microvortices in turbulent jets. The cavitation cloud, which is typically observed downstream of attached cavities, is a result of shed vorticity into the flow field [89] and may be significantly influenced by vorticity generated as a result of baroclinicity. It is therefore apparent that the effects of baroclinic torque are important at the interface of the cavities and the wake region, and they have been shown to significantly affect the vorticity generation in high-pressure gradient areas of the flow while also influencing the cavity shedding, in which a strong interaction between vorticity and cavitation is present [90]. However, for flow problems in which the contribution of the baroclinic torque may be small, the barotropic model may lead to reasonable and comparable predictions with TEM models [83]. Finally, another reported drawback of the barotropic approach is the increased computational requirements compared to TEM models [83].

By contrast, TEM models have been more frequently used in hydraulic machinery applications, using several different models to treat the exchange of mass between the two phases, based on simplified forms of the Rayleigh-Plesset equation (RPE) (Eq. 5.9) for bubble dynamics [91–97]. The RPE relates the vapor bubble radius R to the liquid properties and the difference between the vapor pressure p_v and the liquid pressure p_∞.

Under the assumption of mechanical and thermodynamic equilibrium and of homogenous multiphase fluid flow, the unsteady Reynolds-averaged Navier-Stokes transport equations (continuity Eq. 5.4 and momentum Eq. 5.5) are solved for the mixture (m), where the density ρ and viscosity μ of the fluid are determined as a linear combination of the liquid (l) and vapor (v) phase properties (Eqs. 5.6 and 5.7), while cavitation is modeled by using the mass transfer equation (Eq. 5.8), where α_v and a_l represent the vapor and liquid volume fractions, respectively. In Eqs. (5.4) and (5.5), p represents the pressure field, F represents the external body forces, u_j represents the fluid velocity, and τ_{ij} represents the viscous stress tensor.

$$\frac{\partial \rho_m}{\partial t} + \frac{\partial}{\partial x_j}\left(\rho_m u_j\right) = 0 \tag{5.4}$$

$$\frac{\partial}{\partial t}\left(\rho_m u_i\right) + \frac{\partial}{\partial x_j}\left(\rho_m u_i u_j - \tau_{ij}\right) = -\frac{\partial p}{\partial x_i} + \rho_m g + F \tag{5.5}$$

To model the evaporation and condensation processes, the mass transfer equation includes two different source terms (m^+ and m^-). Various different models have been proposed that offer a different treatment for these terms, using empirical coefficients and simplifications of the RPE for bubble dynamics (Eq. 5.9), in which the effects of noncondensable gasses, surface tension σ, viscosity μ, and second-order effects may be neglected.

$$\rho_m = \rho_v a_v + \rho_l(1 - a_v) \tag{5.6}$$

$$\mu_m = \mu_v a_v + \mu_l(1 - a_v) \tag{5.7}$$

$$\frac{\partial}{\partial t}(a_l\rho_l)+\frac{\partial}{\partial x_j}(a_l\rho_1 u_j)=\dot{m}^-+\dot{m}^+ \tag{5.8}$$

In the following sections the most widely used TEM-based models in centrifugal pumps are discussed.

$$\frac{p_v-p_\infty}{\rho_l}=R\frac{d^2R}{dt^2}+\frac{3}{2}\left(\frac{dR}{dt}\right)^2+\frac{2\sigma}{\rho_l R}+4\frac{\mu_l}{\rho_l R}\frac{dR}{dt} \tag{5.9}$$

5.3.1.1 The Zwart et al. Model

In the Zwart et al. model [95], the mass transfer rates of the cavitation and condensation processes are determined from the RPE by assuming that second-order effects, viscosity, and surface tension are negligible. Furthermore, with the assumptions that there is no bubble interaction and that all bubbles have the same initial spherical size, the number of bubbles contained in the fluid per unit volume are described by using the bubble density n_0. By expressing the vapor volume fraction α_ν as a function of the bubble density and bubble volume (Eq. 5.10), the mass transfer rate during phase change can be derived by combining the truncated Rayleigh-Plesset equation, the rate of change of mass of a bubble, and the bubble density. For the successful implementation of the model, additional empirical terms are included for vaporization and condensation, while the interaction of cavitation bubbles with increasing volume fraction is taken into consideration by assuming that the nucleation site density decreases with increasing volume fraction.

The resulting mass transfer rates relate the evaporation and condensation processes to the densities of the liquid, ρ_l, and vapor phases, ρ_v, the vapor volume fraction α_ν, the nucleation site volume fraction α_{nuc}, the saturation pressure p_s, and the local pressure p.

The above lead to Eqs. (5.11) and (5.12) for the evaporation and condensation mass transfer rates, $\dot{m}^-$ and $\dot{m}^+$, respectively.

$$a_v=n_0\cdot\frac{4}{3}\pi {R_B}^3 \tag{5.10}$$

$$\dot{m}^-=-C_{\text{vap}}\frac{3\alpha_{\text{nuc}}(1-a_v)\rho_v}{R_B}\sqrt{\frac{2(p_{\text{sat}}-p)}{3\rho_l}} \tag{5.11}$$

$$\dot{m}^+=C_{\text{cond}}\frac{3\alpha_v\rho_v}{R_B}\sqrt{\frac{2(p-p_{\text{sat}})}{3\rho_1}} \tag{5.12}$$

The coefficients in these equations, which have been found to be applicable in numerous applications, are the radius of the nucleation site $R_B = 10^{-6}$ m, the nucleation site volume fraction $\alpha_{\text{nuc}} = 5 \times 10^{-4}$, and the evaporation and condenstation empirial coefficients $C_{\text{vap}} = 50$ and $C_{\text{cond}} = 0.01$, respectively.

5.3.1.2 The Kunz et al. Model

In the Kunz et al. [91] cavitation model, surface tension and viscosity effects in the RPE are neglected, leading to appropriate expressions for the source terms in the mass transfer equation (Eqs. 5.13 and 5.14). Specifically, the evaporation mass transfer rate is related to the liquid volume fraction α_l and the difference between local static pressure p and saturation pressure p_{sat}. On the other hand, the condensation mass transfer rate is related to the liquid volume fraction α_l, as well as the noncondensable gases volume fraction α_{ncg}. Both equations are nondimensionalized using a mean flow time scale, while two empirical constants C_{vap} and C_{cond} are employed. Typical values for the empirical coefficients reported in the literature are C_{vap}, $C_{\text{cond}} = 100$.

$$\dot{m}^- = \frac{C_{\text{vap}} \rho_v \alpha_l}{0.5 \rho_l u_\infty^2 t_\infty} (p - p_{\text{sat}}) \tag{5.13}$$

$$\dot{m}^+ = \frac{C_{\text{cond}} \rho_v (\alpha_l - \alpha_{\text{ncg}})^2}{t_\infty} \left(1 - \alpha_l - \alpha_{\text{ncg}}\right) \tag{5.14}$$

Another variation of the Kunz model is the Merkle model [97], in which similarly the source and sink terms for the mass transfer equation are directly related to the pressure difference, in contrast to other TEM models, in which the mass transfer rates are a function of the square root of the pressure difference. The main difference between the Merkle and Kunz models is found in the definition of the condensation mass transfer rate (Eq. 5.14), which in the case of the Kunz model is independent of the difference of local pressure to the saturation pressure.

5.3.1.3 The Schnerr and Sauer Model

Another popular TEM = based model that is used in centrifugal pump applications is the Schnerr and Sauer model [92,98], in which, similar to other TEM models, the Rayleigh-Plesset equation is used to model bubble dynamics without taking into account viscosity, surface tension, and second-order effects. A different definition for the volume fraction is used (Eq. 5.15) that considers the number of bubbles per unit volume of fluid and the equilibrium bubble radius, which influence the development and characteristics of cavities [99]. This model does not incorporate empirical coefficients other than the initial bubble or nuclei concentration and an initial nuclei radius, from which the bubbles grow during the evaporation process. The mass transfer rates and the vapor volume fraction in this model are given by the following equations (Eqs. 5.16 and 5.17):

$$a_v = \frac{n_0 \cdot \frac{4}{3}\pi {R_B}^3}{1 + n_0 \cdot \frac{4}{3}\pi {R_B}^3} \tag{5.15}$$

$$\dot{m}^- = -\frac{\rho_v \rho_l}{\rho_m}\alpha_v(1-\alpha_v)\frac{3}{R_B}\sqrt{\frac{2(p_{\text{sat}}-p)}{3\rho_l}} \tag{5.16}$$

$$\dot{m}^+ = \frac{\rho_v \rho_l}{\rho_m}\alpha_v(1-\alpha_v)\frac{3}{R_B}\sqrt{\frac{2(p-p_{\text{sat}})}{3\rho_l}} \tag{5.17}$$

where n_o is defined as the number of bubbles per unit volume of liquid and R_B as the nucleation site radius, with typical values of $n_o = 10^8$ and $R_B = 30\ \mu\text{m}$.

5.3.1.4 The Singhal et al. Model

In contrast to the other cavitation models based on the RPE to describe the growth of bubbles in multiphase flows, the Singhal cavitation model [93] incorporates the noncondensable gas mass fraction f_{ncg} as well as the effects of surface tension σ and turbulent pressure fluctuations by incorporating the turbulence kinetic energy k in the expressions for mass transfer rates (Eqs. 5.18 and 5.19). The empirical constants in this model should be equal to $C_{\text{vap}} = 0.02$ and $C_{\text{cond}} = 0.01$, as proposed in the original derivation of the equations:

$$\dot{m}^- = -C_{\text{vap}}\frac{\sqrt{k}}{\sigma}\rho_v\rho_l\sqrt{\frac{2(p_{\text{sat}}-p)}{3\rho_l}}(1-f_v-f_{\text{ncg}}) \tag{5.18}$$

$$\dot{m}^+ = C_{\text{cond}}\frac{\sqrt{k}}{\sigma}\rho_l\rho_l\sqrt{\frac{2(p-p_{\text{sat}})}{3\rho_l}}f_v \tag{5.19}$$

5.3.1.5 Effects of Turbulence Model, Compressibility, and Cavitation Model Parameters

There is a strong interrelation between turbulence and cavitation. Turbulence contributes to the inception process, but equally, cavitation can be considered a mechanism for turbulence generation [73,89,100]. The process of cavity collapse involves large vorticity production and a substantial increase in turbulence and boundary layer thickness. As vapor cavities collapse, large eddies are generated, while vortices are formed in the shear layer between the cavity closure region and the recirculating flow [71]. On the other hand, cavities that are generated in vortex cores generate additional vorticity, which demonstrates the strong interaction between turbulence and cavitation [90]. Therefore turbulence modeling and the accurate prediction of the dynamics of vortical structures are essential for the proper modeling of cavitation and unsteady effects [101].

Linear eddy viscosity models used to solve the Reynolds-averaged Navier-Stokes (RANS) equations have been widely used to model turbulent fluid flow problems with good success. However, for cavitating flows, the application is limited to steady-state problems, owing to the inherent inability to accurately model turbulent viscosity at the cavity interface and the downstream recirculation zone [102,103]. The resulting increased turbulent stresses cause the reentrant jet to lose momentum and reduce the penetration towards the leading edge of the cavity, eventually causing its premature termination, leading to the formation of a stable cavity. In RANS turbulence modeling, the small-scale turbulent fluctuations are averaged out. Therefore fluid flow characterized by high vorticity generation cannot be accurately captured [102]. Moreover, the overestimation of turbulent viscosity leads to an underestimation of the cavity extent and may affect the prediction accuracy for stable cavitation [63]. The inability to accurately compute the eddy viscosity is attributed to the fluid mixture compressibility, which influences the turbulent characteristics of the flow.

To improve the performance of linear eddy viscosity models to simulate the dynamic characteristics of cavities and to account for compressibility effects, an arbitrary correction (density correction model, or DCM) has been introduced that limits the mixture turbulent viscosity by using a function based on the vapor and liquid phase density (Eq. 5.20):

$$f(\rho) = \rho_v + \left(\frac{\rho_v - \rho}{\rho_v - \rho_l}\right)^n (\rho_l - \rho_v) n >> 1 \tag{5.20}$$

where ρ is the density and the subscripts v and l denote the vapor phase and liquid phase, respectively, while n is an empirical coefficient, typically set as $n = 10$. This modification has been applied in numerous applications and has been shown to improve the prediction capabilities in both stable and unstable cavitation flow problems [63,104–109]. In stable cavitation the DCM model leads to larger predicted cavities, in line with experimental observations, by decelerating the cavity collapse process, while in unstable cavitation the model allows the reentrant jet to maintain momentum and reach the leading edge of the cavity [108].

In centrifugal pumps, the DCM correction has been used in the past and shown to marginally improve the prediction characteristic of the pump breakdown curve. As the onset of pump head breakdown depends predominantly on the development of a stable sheet cavity that blocks the flow passage and not on the development of unsteady cavitation phenomena, the turbulent viscosity corrections are unlikely to greatly influence the predicted cavitation breakdown coefficient [110]. In the prediction of the pump cavitation curve, large-scale cavitation structures are of interest, while unsteady smaller-scale structures, such as bubbles and cavitation clouds, may be ignored [87].

In flows with recirculation zones and increased streamline curvature there is an imbalance between the production and dissipation of turbulent kinetic energy. Therefore the choice of the turbulence model can affect the accuracy of the predicted cavities [111].

To improve the performance of RANS models in such flows, a modification was introduced by Spalart and Schur [112] and validated by Schur et al. [113] in curved two-dimensional and three-dimensional channels. This modification introduces a correction source term in the turbulent kinetic energy transport equation, allowing improvements in the predicted velocity profiles in flows characterized by strong rotation and streamline curvature. The addition of this source term in the turbulent kinetic energy transport equation allows for the limitation of the predicted turbulent viscosity and of turbulence stresses that have been found to interfere with the development of the reentrant jet and thus suspend the development of unsteady cavitation shedding. The incorporation of this modification in modeling cavitating flows has been found to adequately capture the interaction of large vortical structures downstream of the attached cavity and the reentrant jet, allowing the accurate capture of the periodic characteristics of cavity shedding [114]. In another approach, to include the effects of rotation and improve the prediction capabilities in centrifugal pumps, the homogenous cavitation model may be modified by including additional terms in the definition of the evaporation and condensation mass transfer rates to account for rotation and flow passage geometry [115].

Even when suitable eddy viscosity limiters are included following the RANS approach, the shedding frequency and hydrodynamic load are still not well predicted, highlighting the requirement for a more refined turbulence modeling approach [116]. To better capture the dynamics of transient medium- to small-scale structures and the development of unsteady cavitation, another alternative is the large eddy simulation (LES) approach [117]. In this approach, the larger turbulent structures are computed, while smaller-scale turbulent structures, which are expected to exhibit isotropic behavior, are modeled. LES can provide insights into the mechanisms of cavitation and its interaction with turbulence while at the same time providing added accuracy, though at the expense of increased computational requirements [100]. However, even with LES, analyses in stationary hydrofoils show that the cavity extent and hydrodynamic loads may be underpredicted, while a better accuracy for the cavity dynamics can be achieved [116]. The sensitivity of the LES approach allows the accurate prediction of different flow patterns and hydrodynamic fluctuations as a function of geometry modifications [118,119], while it is notable that generally good accuracy is achieved even using standard cavitation model parameters [116].

Equally important in turbulence modeling is the proper spatial discretization of the computational domain using local mesh refinement, where deemed necessary, to properly resolve vortex cores and properly resolve the development of the reentrant jet facilitating cavity break-off [65,74,117,120]. Otherwise, the associated pressure fluctuations may be underpredicted, and the model may fail to predict cavitation. Furthermore, mesh resolution effects are also important in evaluating the effect of small geometry variations on cavitation performance [118].

Apart from adopting a suitable turbulence modeling approach to account for the associated flow physics in different flow problems, the empirical parameters used in the various TEM models need to be carefully tuned. These parameters control the rate of mass transfer between the phases and influence more the numerical results in the closure region of the cavity, where large density variation are observed [111,121]. They have been determined for each TEM model after rigorous experimentation in different applications and are known to greatly influence the predicted shape of the cavity [80], producing different compressibility effects. As was previously discussed, the mixture compressibility is related to the computed turbulent viscosity and the turbulent kinetic energy, which is produced at the trailing edge of the cavity. These are more important in time-dependent problems, in which accurate prediction of the turbulent viscosity strongly influences the development and shedding of cavities [121]. The applicability of these empirical parameters has been found to be problem specific [122], so in cases in which the physical processes defining the flow regime being studied differ from those used to determine the baseline values, a careful calibration is necessary in conjunction with the selected turbulence model [123]. An analysis of the RPE has indicated that these empirical coefficients can be thought of as representating compressibility effects [124]. This highlights the need to properly establish the relationship between them and the associated flow physics [80].

In numerous studies, the flow over a stationary hydrofoil or a Venturi geometry is typically used as a benchmark case to fine-tune the condensation and evaporation rate coefficients, which are then applied to model the cavitation development in hydraulic machinery. Depending on the specific application and flow regime studied, many different values can be found that allow improved accuracy [80,122,123,125,126,127] (Table 5.1).

Recently, in an effort to produce a single cavitation model based on the RPE, various cavitation models were compared, and a new formulation was proposed that allowed switching from one model to the other by modifying the model coefficients accordingly [124]. It was shown that by appropriately modifying the coefficients of each model, virtually the same results can be achieved as the models

TABLE 5.1 Typical cavitation model parameters used in the literature in general and pump cavitation computations.

Model	Coefficient	Default	Frika et al. [80]	Morgut et al. [123], Biluš et al. [126]	Limbach and Skoda [128]	Liu et al. [129]	Salvadori et al. [149]	Zhou et al. [150]	Senocak and Shyy [74]	Vaidyanathan et al. [111]
Zwart	C_v	50	—	300	150	500	Default	—	—	—
	C_c	0.01	—	0.03	0.03	0.0001	Default	—	—	—
	R_B	1×10^{-6}	—	—	—	—	—	—	—	—
	α_{nuc}	5×10^{-4}	—	—	—	—	4×10^{-3}	—	—	—
Kunz	C_v	100	10	4100	—	—	—	3323	3×10^{4}	4×10^{4}
	C_c	100	0.7	455	—	—	—	4328	9×10^{5}	8×10^{5}
Shnerr	n_o	1×10^{8}	—	—	—	—	—	—	—	—
Singhal	C_v	0.2	4×10^{-4}	0.4	—	—	—	—	—	—
	C_c	0.01	1.4	2.3×10^{-4}	—	—	—	—	—	—
Method	—	—	Hydrofoil	Hydrofoil, Venturi	Pump	Pump	Pump	Hydrofoil	Blunt object	Hydrofoil
Turbulence model	RNG k-e	k-e sc. wf	k-w sst	RNG k-e	k-w	k-e	k-e	neq. k-e		

become homologous. The validity of the models and the corresponding coefficients varies with the flow conditions being modeled, so there is not a unique set of model coefficient that applies to any condition being modeled.

5.3.2 Validation and Implementation of Modeling Approaches in Centrifugal Pumps

5.3.2.1 The Zwart et al. Model

In the derivation of their model, Zwart et al. [95] compared the performance of the model in various test cases, including a steady cavitation in a hydrofoil, inducer cavitation at different flow rates, and the unsteady cavitation in a Venturi geometry. For the steady sheet cavitation the model was able to predict the cavity length and cavitation breakdown with relatively good accuracy. However, this was not the case in unsteady cavitation, in which a careful tuning of the model constants along with the inclusion of compressibility effects is required for proper correlation. At partial discharge conditions, the Zwart model was shown to slightly underpredict the cavitation breakdown point of the pump inducer, while the opposite was at design and larger discharges. However, the overall qualitative behavior of the model is well correlated with experimental data. In the case of the Venturi cavitation problem the shedding frequency and average cavity length was in good agreement with experiments after fine-tuning of the model constants and turbulence model modifications to include compressibility effects. In subsequent analyses by Bakir et al. [130] and Mejri et al. [131,132], similar conclusions were drawn for cavitation simulations in various pump inducers. These validation studies and the implementation of this model in most commercial solvers have contributed to its popularity in numerous investigations related to hydraulic machinery.

The cavitation performance of a low specific speed centrifugal pump is evaluated at different operating conditions by using the Zwart et al. model by Limbach et al. [128,133]. Their results showed that at low flow conditions, the critical cavitation coefficient is well predicted, while higher inaccuracies are achieved with increasing flow conditions. The differences were attributed to the model cavitation coefficients and a requirement for proper calibration was highlighted. Furthermore, Limbach et al. [128] observed that for the accurate prediction of the head losses and head drop curve, the proper resolution of the spiral casing tongue is of utmost importance. Sufficiently refined meshes are required to resolve the boundary layer and capture the effects of flow incidence at various operating conditions. Unstable cloud cavitation will also form in the volute tongue, and the accuracy of the model at overload conditions is strongly related to the resolution, as head losses may be in excess of the conventional 3% limit used for the identification of the critical cavitation number. The inclusion of surface roughness was seen to improve the correlation with experimental observations of the predicted head drop curve by Limbach et al. [128], with the prediction of an unsteady cavity shed from the tongue of the spiral casing. Although in general the model performed well across deferent operating conditions, an underprediction of the head drop point was found at overload conditions. Liu et al. [129] were able to improve the correlation to measurements by modifying the model parameters controlling the mass transfer rates and found that improved performance is achieved by a reduction of the condensation coefficients and an increase of the evaporation coefficient. The importance of properly resolving the flow in the vicinity of the tongue was also highlighted by Tang et al. [134]

Recently, Mousmoulis et al. [135] studied both numerically and experimentally the inception and development of cavitation in a low specific speed pump using the Zwart et al. model. By analyzing numerical flow patterns in conjunction with experimental flow visualizations, the extent of the cavity was identified and compared at various flow conditions (Fig. 5.7). Overall, good qualitative and quantitative agreement between observations and predictions was achieved. Particularly, the predicted cavitation extent and inception location at the tip of the leading edge was in agreement with the experimental observations (Fig. 5.7). On the other hand, the critical cavitation point was slightly underpredicted at the various flow conditions studied; this is consistent with similar studies in centrifugal pumps. Finally, at heavy cavitating conditions in which the head had begun to drop and the flow passage was completely blocked by water vapor, the numerical model was able to capture all the relevant flow features (Fig. 5.8).

5.3.2.2 The Kunz et al. Model

The Kunz et al. model has been previously been validated for super-cavitating and submerged body flow problems, and subsequently applied in turbomachinery. Kunz et al. and Medvitz et al. [30,136] used this model for the prediction of cavitation performance of a seven-blade centrifugal pump and incorporated experimental measurements to determine the performance characteristics, while visual observations allowed the identification of cavitation onset. An overall good agreement between numerical and experimental results was achieved, as the authors were able to sufficiently predict the head drop curve and the critical cavitation coefficient, both qualitatively and quantitatively. The prediction of the cavitation breakdown point was also accurately calculated for the nominal flow conditions, while the analysis of the computed flow features provided insights into the mechanisms of cavitation breakdown in centrifugal pumps. Similarly, Pouffary et al. in [137] and [138] used the Kunz model for the predication of centrifugal pump and inducer cavitation performance. According to their results, the cavitation breakdown point in both occasions agrees qualitatively

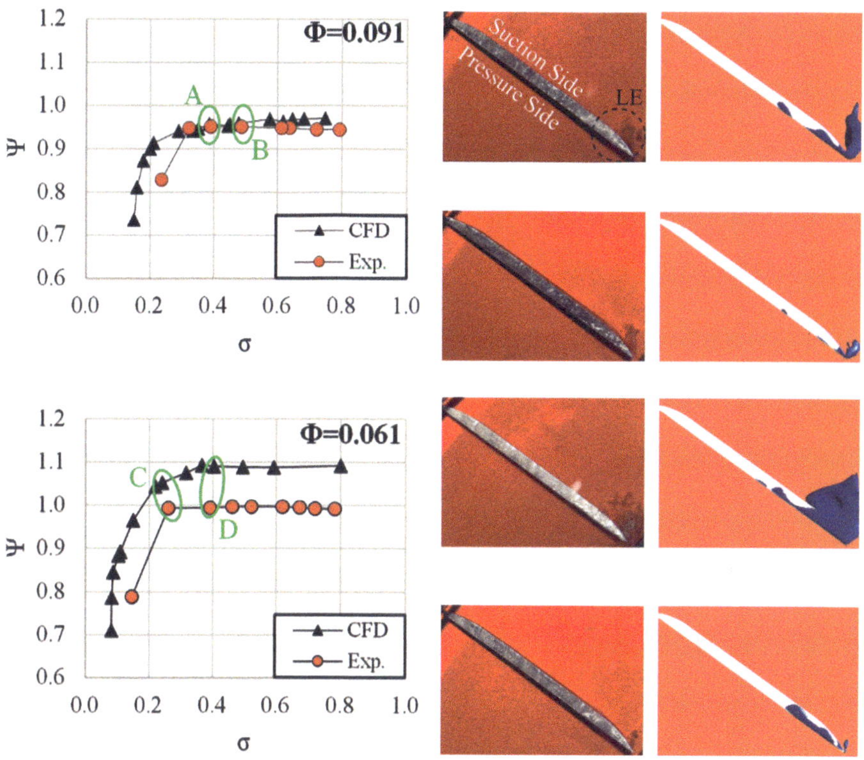

Figure 5.7 Experimental and numerical $\sigma - \Psi$ curves for different Φ values at incipient and developed cavitating conditions. *Present authors, adapted from G. Mousmoulis, Development of Cavitation Detection Methodology for Hydraulic Turbomachines Using Experimental and Numerical Tools (PhD thesis), National Technical University of Athens, 2020 [6]; G. Mousmoulis, I. Kassanos, G. Aggidis, I. Anagnostopoulos, Numerical simulation of the performance of a centrifugal pump with a semi-open impeller under normal and cavitating conditions, Applied Mathematical Modelling 89 (2) (2021) 1814–1834 [135].*

with experimental results, while a comparison with a simpler barotropic cavitation model is also provided. Both models are shown to predict comparable flow fields and volume fractions for steady sheet cavitation. The analysis of the flow field at cavitating conditions enabled the identification of the developed flow patterns in the transition from noncavitating conditions to full breakdown of the pump head.

Wang et al. [139] studied the cavitation performance of a centrifugal pump with diffuser blades using the Zwart and Kunz models, including the effects of turbulent pressure fluctuations on the flow. Both models underpredicted the critical cavitation coefficient, however the Kunz model showed better agreement to experimental measurements, albeit with a slight over-prediction of the extent of the cavity. This shortcoming is likely to influence the prediction of the unsteady

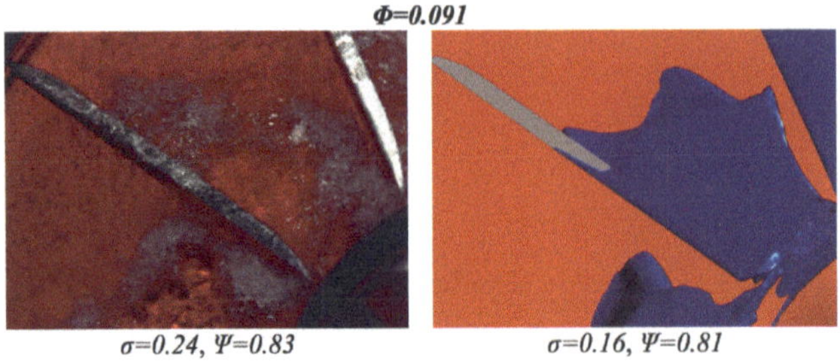

Figure 5.8 Experimental and numerical visualizations of the cavity extent within an impeller passage at the Ψ collapse point for a given Φ value. *Present authors, adapted from G. Mousmoulis, Development of Cavitation Detection Methodology for Hydraulic Turbomachines Using Experimental and Numerical Tools (PhD thesis), National Technical University of Athens, 2020 [6]; G. Mousmoulis, I. Kassanos, G. Aggidis, I. Anagnostopoulos, Numerical simulation of the performance of a centrifugal pump with a semi-open impeller under normal and cavitating conditions, Applied Mathematical Modelling 89 (2) (2021) 1814–1834 [135].*

characteristics of the pump at unsteady cavitation conditions. A similar pump configuration was used by Liang et al. [140], who in conjunction with experimental flow visualizations, were able to characterize the flow through various phase of cavitation development and head breakdown.

5.3.2.3 The Schnerr and Sauer Model

Limited application of the Schnerr and Sauer model in centrifugal pumps can be found in the literature. In the original derivation of the model [98], the authors were able to predict the shedding character of the unsteady cavitation over a hydrofoil. The model seemed to capture the reentrant jet, which travels up to the leading edge of the cavity, and promoted the cavity shedding. Qualitatively, their results compared well with available experimental observations [92]. Lu et al. [141] used the Schnerr-Sauer model following the LES approach for the prediction of sheet cavitation and the unsteady flow characteristics on a NACA0015 hydrofoil to study the cavitation dynamics at different cavitation numbers and angles of attack. The authors highlight the importance of mesh resolution to properly capture the cavitation dynamics and accurately model the flow in the vicinity of the walls.

The first application of the Schnerr and Sauer model in centrifugal pumps was carried out by Frobenious et al. [99] to determine the head drop characteristic curve of a centrifugal pump at nominal operating conditions and at off-design conditions. The predicted vapor volume showed good qualitative agreement with experimental measurements, while the calculated critical and incipient cavitation coefficients were systematically underpredicted. Sato et al. [142] used the Schnerr and Sauer model to determine the onset of vortex cavitation in a double volute centrifugal pump. In double suction pumps that incorporate a baffle plate at the entrance of the impeller, an inlet vortex is induced, which may cavitate if the cavitation number is low enough. The authors carried out both numerical and experimental analyses to elucidate the relationship between vibration and noise as a result of the baffle vortex cavitation. They observed that vibration and noise are modulated by the pump rotational velocity, while the observed pressure pulsation is found to have a period corresponding to the blade-passing frequency. The pressure pulsation amplitude increased under cavitating conditions, while the vortex was also present under noncavitating conditions, albeit with lower amplitudes. The comparison with experimental data showed good qualitative correlation, being able to correctly predict the dominant frequency of the vortex pressure pulsation and reproduce the increase in amplitude at cavitating conditions. However, the cavitation model predicted an additional peak at the first harmonic of the pressure pulsation, which was not present in the experiment. No explanation about the origin of this harmonic was offered by the authors.

5.3.2.4 The Singhal Model

The Singhal model, also known as the full cavitation model, was validated for several different cases, including hydrofoil cavitation, diffuser cavitation, and orifice flow, by in the work of Singhal et al. [93], who, along with the mathematical formulation, determined the optimum cavitation model coefficients by considering a large number of different test cases.

The first application of the Singhal model in centrifugal pumps was carried out by Athavale et al. [143], who studied numerically the flow in a water jet propulsion centrifugal pump and an inducer. Their study did not include experimental measurements; however, the predicted flow patterns have been found to be correct from a physical standpoint. In their results, the predicted cavitation initiates at the SS of the blade in the vicinity of the leading edge and grows with decreasing cavitation number. The authors showed that the inclusion of the noncondensable gases affects the predicted head, with lower head values achieved, while there is also an important effect on the cavitation characteristics with smaller cavitation zones predicted with increasing concentration. The incipient cavitation coefficient was predicted with better accuracy at design and high load conditions, while the accuracy is slightly deteriorated at reduced loads. The model was qualitatively validated by Mostafa et al. [144] on an axial flow pump, including the tip clearance. Visualizations of the predicted flow field showed good agreement with experimental observations regarding the location and extent of the cavity. Amezcuca et al. [145] used the model in two different inducer geometries, and good qualitative agreement was found without taking into consideration the effects of tip clearance. Their steady-state results showed decent agreement with experimental head drop curves for partial flow rates, where the cavitation coefficient was overpredicted. Furthermore, they showed that unsteady inducer instabilities may initiate numerically after a significant number of revolutions has been completed, underpinning the requirement for substantial computational resources for such simulations. Various other authors have used the Singhal model in similar centrifugal pump applications and were able to validate its applicability to predict with good accuracy the head drop curve [146–148]. The inclusion of noncondensable gas concentration was shown to improve the correlation [146]; however, there were larger inaccuracies when looking at the incipient cavitation coefficient. Furthermore, the predicted head drop curve was strongly dependent on the turbulence model used, with the RNG k-e model failing to achieve cavitation breakdown conditions, owing to convergence difficulties.

5.4 Summary

In this chapter, emphasis has been placed first on the weaknesses and limitations of the available cavitation detection techniques in centrifugal pumps. The main difficulties associated with cavitation detection are the lack of general applicability of a particular method to different pump designs and testing configurations and the difficulty of distinguishing cavitation from other mechanical faults of the pump-motor system. A new detection tool is presented that is able to isolate in the raw signal the impulses due to vapor bubble collapse from noise or other sources, facilitating the efficient detection of cavitation at early stages after its development.

This chapter has also reviewed the various numerical modeling techniques that have been developed to predict cavitation and related flow instabilities in centrifugal pumps. These numerical tools have promoted the understanding of the flow mechanisms of cavitation under different operating conditions of centrifugal pumps and can be used with relatively good accuracy for the prediction of cavitation, provided that the various model parameters are properly tuned to capture the underlying physical processes.

Abbreviations

DCM density correction method
EOS equation of state
IC interface capturing methods
IT interface tracking methods
LES large eddy simulation
NPSH net positive suction head

RANS	Reynolds-averaged Navier-Stokes
RF	rotational frequency
RPE	Rayleigh-Plesset equation
SPL	static pressure level
SS	suction side
TEM	transport equation-based models

Nomenclature

σ	Thoma's cavitation number (−)
a_{ncg}	noncondensable gas volume fraction (−)
a_v	vapor mass fraction (−)
a_{nuc}	nucleation site volume fraction
Q	flow rate (m^3/s)
Q_{cr}	critical flow rate (m^3/s)
ζ_e	energy loss coefficient
f_{ncg}	noncondensable gas mass fraction (−)
f_v	Vapor mass fraction (−)
ρ	fluid density (kg/m^3)
σ	surface tension (N/m)
m^-	evaporation mass transfer rate
m^+	condensation mass transfer rate
t_∞	mean flow time scale (s)
u_∞	characteristic velocity (m/s)
H_{sat}	saturation pressure (m)
H_{tot}	total head (m)
H_B	pressure at the surface of the suction tank (Pa)
H_{oe}	total energy per unit weight
p	pressure (Pa)
p_{sat}	vapor pressure of fluid (Pa)
n_o	bubble density ($1/m^3$)
R_B	nucleation site/bubble radius
k	turbulent kinetic energy (m^2/s^2)
C_{vap}	evaporation mass transfer coefficient
C_{con}	condensation mass transfer coefficient

References

[1] S.C. Li, Cavitation of Hydraulic Machinery, vol. 1, World Scientific, 2000.
[2] C.E. Brennen, Fundamentals of Multiphase Flow, Cambridge University Press, 2005.
[3] C.E. Brennen, Cavitation and Bubble Dynamics, Cambridge University Press, 2014.

[4] G. Mousmoulis, N. Karlsen-Davies, G. Aggidis, J. Anagnostopoulos, D. Papantonis, Experimental analysis of the onset and development of cavitation in a centrifugal pump, Journal of Physics: Conference Series 813 (1) (2017). Available from: https://doi.org/10.1088/1742-6596/813/1/012044.
[5] G. Mousmoulis, N. Karlsen-Davies, G. Aggidis, I. Anagnostopoulos, D. Papantonis, Experimental analysis of cavitation in a centrifugal pump using acoustic emission, vibration measurements and flow visualization, European Journal of Mechanics-B/Fluids 75 (2019) 300–311. Available from: https://doi.org/10.1016/j.euromechflu.2018.10.015.
[6] G. Mousmoulis, Development of Cavitation Detection Methodology for Hydraulic Turbomachines Using Experimental and Numerical Tools PhD thesis), National Technical University of Athens, 2020. Available from: https://dspace.lib.ntua.gr/xmlui/handle/123456789/50836?locale-attribute = en
[7] W. Zhao, G. Zhao, An active method to control cavitation in a centrifugal pump by obstacles, Advances in Mechanical Engineering 9 (11) (2017). Available from: https://doi.org/10.1177/1687814017732940.
[8] D. Zhang, W. Shi, D. Pan, M. Dubuisson, Numerical and experimental investigation of tip leakage vortex cavitation patterns and mechanisms in an axial flow pump, Journal of Fluids Engineering 137 (12) (2015) 121103. Available from: https://doi.org/10.1115/1.4030914.
[9] D. Papantonis, Hydrodynamic Machines, Hydrodynamic Transmissions, Symeon, 2016. (in Greek).
[10] C.E. Brennen, Hydrodynamic of Pumps, Cambridge University Press, 2011.
[11] D. Zhang, W. Shi, B.B. van Esch, L. Shi, M. Dubuisson, Numerical and experimental investigation of tip leakage vortex trajectory and dynamics in an axial flow pump, Computers & Fluids 112 (2015) 61–71. Available from: https://doi.org/10.1016/j.compfluid.2015.01.010.
[12] D. Zhang, L. Shi, W. Shi, R. Zhao, H. Wang, B.B. van Esch, Numerical analysis of unsteady tip leakage vortex cavitation cloud and unstable suction-side-perpendicular cavitating vortices in an axial flow pump, International Journal of Multiphase Flow 77 (2015) 244–259. Available from: https://doi.org/10.1016/j.ijmultiphaseflow.2015.09.006.
[13] J.F. Gülich, Centrifugal Pumps, Springer, 2008.
[14] International Organization for Standardization, Rotodynamic Pumps: Hydraulic Performance Acceptance Tests: Grades 1, 2 and 3, ISO 9906:2012, International Organization for Standardization, Geneva, 2012.
[15] S.A. Al Hashmi, Statistical analysis of vibration signals for cavitation detection, IEEE Symposium on Industrial Electronics & Applications 1 (2009) 78–82.
[16] F. Al Thobiani, F. Gu, A. Ball, The monitoring of cavitation in centrifugal pumps based on the analysis of vibro-acoustic measurements, in: The Seventh International Conference on Condition Monitoring and Machinery Failure Prevention Technologies, 2010.
[17] M.M. Stopa, B.J. Cardoso Filho, C.B. Martinez, Incipient detection of cavitation phenomenon in centrifugal pumps, IEEE Transactions on Industry Applications 50 (2013) 120–126.
[18] A.R. Al-Obaidi, Investigation of effect of pump rotational speed on performance and detection of cavitation within a centrifugal pump using vibration analysis, Heliyon 5 (6) (2019) e01910. Available from: https://doi.org/10.1016/j.heliyon.2019.e01910.
[19] J. Lu, X. Liu, Y. Zeng, B. Zhu, B. Hu, S. Yuan, H. Hua, Detection of the flow state for a centrifugal pump based on vibration, Energies 12 (16) (2019) 3066. Available from: https://doi.org/10.3390/en12163066.
[20] R.E. Arndt, Cavitation in fluid machinery and hydraulic structures, Annual Review of Fluid Mechanics 13 (1) (1981) 273–326. Available from: https://doi.org/10.1146/annurev.fl.13.010181.001421.

[21] J.P. Best, The formation of toroidal bubbles upon the collapse of transient cavities, Journal of Fluid Mechanics 251 (1993) 79–107. Available from: https://doi.org/10.1017/S0022112093003349.

[22] K. Kobayashi, Y. Chiba, Computational fluid dynamics of cavitating flow in mixed flow pump with closed type impeller, International Journal of Fluid Machinery and Systems 3 (2) (2010) 113–121. Available from: https://doi.org/10.5293/IJFMS.2010.3.2.113.

[23] T. Cencîc, M. Hoĉevar, B. Ŝirok, Study of erosive cavitation detection in pump mode of pump-storage hydropower plant prototype, Journal of Fluids Engineering 136 (5) (2014) 051301. Available from: https://doi.org/10.1115/1.4026476.

[24] C.A. Lima, M.M. Stopa, B.J. Cardoso Filho, L.S. Miranda, A.S.A. Luiz, C.B. Martinez, Cavitation detection in centrifugal pumps driven by frequency inverter fed motors, in: IEEE Industry Applications Society Annual Meeting, 2017.

[25] T. Watanabe, H. Sato, Y. Henmi, H. Horiguchi, Y. Kawata, Y. Tsujimoto, Rotating choke and choked surge in an axial pump impeller, International Journal of Fluid Machinery and Systems 2 (3) (2009) 232–238. Available from: https://doi.org/10.5293/IJFMS.2009.2.3.232.

[26] S. Hatano, D. Kang, S. Kagawa, M. Nohmi, K. Yokota, Study of cavitation instabilities in double-suction centrifugal pump, International Journal of Fluid Machinery and Systems 7 (3) (2014) 94–100. Available from: https://doi.org/10.5293/IJFMS.2014.7.3.094.

[27] J. Friedrichs, G.N. Kosyna, Rotating cavitation in a centrifugal pump impeller of low specific speed, Journal of Fluids Engineering. 124 (2) (2002) 356–362. Available from: https://doi.org/10.1115/1.1457451.

[28] G. Pace, D. Valentini, A. Pasini, R. Hadavandi, L. d'Agostino, Analysis of flow instabilities on a three-bladed axial inducer in fixed and rotating frames, Journal of Fluids Engineering 141 (4) (2019) 041104. Available from: https://doi.org/10.1115/1.4041731.

[29] A. Baldassarre, M. De Lucia, P. Nesi, Real-time detection of cavitation for hydraulic turbomachines, Real-Time Imaging 4 (6) (1998) 403–416. Available from: https://doi.org/10.1006/rtim.1997.0100.

[30] R.B. Medvitz, R.F. Kunz, D.A. Boger, J.W. Lindau, A.M. Yocum, L.L. Pauley, Performance analysis of cavitating flow in centrifugal pumps using multiphase CFD, Journal of Fluids Engineering 124 (2) (2002) 377–383. Available from: https://doi.org/10.1115/1.1457453.

[31] L. Dong, Y. Zhao, C. Dai, Detection of inception cavitation in centrifugal pump by fluid-borne noise diagnostic, Shock and Vibration 2019 (3) (2019) 1–15. Available from: https://doi.org/10.1155/2019/9641478.

[32] M. Hofmann, B. Stoffel, J. Friedrichs, G. Kosyna, Similarities and geometrical effects on rotating cavitation in two scalted centrifugal pumps, 2001. http://resolver.caltech.edu/cav2001:sessionB8. 001.

[33] S. Duplaa, O. Coutier-Delgosha, A. Dazin, O. Roussette, G. Bois, G. Caignaert, Experimental study of a cavitating centrifugal pump during fast startups, Journal of Fluids Engineering 132 (2) (2010) 021301. Available from: https://doi.org/10.1115/1.4000845.

[34] S.D. Kyparissis, D.P. Margaris, Experimental investigation and passive flow control of a cavitating centrifugal pump, International Journal of Rotating Machinery 2012 (2012) 8 p. Available from: https://doi.org/10.1155/2012/248082.

[35] S.D. Kyparissis, D.P. Margaris, Experimental investigation of cavitation in a centrifugal pump with double-arc synthetic blade design method, International Review of Mechanical Engineering 5 (5) (2011) 884–892.

[36] O. Coutier-Delgosha, R. Fortes-Patella, J.L. Reboud, M. Hofmann, B. Stoffel, Experimental and numerical studies in a centrifugal pump with two-dimensional curved blades in cavitating

condition, Journal of Fluids Engineering 125 (6) (2003) 970–978. Available from: https://doi.org/10.1115/1.1596238.

[37] S. Gopalakrishnan, Modern cavitation criteria for centrifugal pumps, in: Proceedings of the 2nd International Pump Symposium, 1985.

[38] G.D. Neill, R.L. Reuben, P.M. Sandford, E.R. Brown, J.A. Steel, Detection of incipient cavitation in pumps using acoustic emission, Proceedings of the Institution of Mechanical Engineers, Part E: Journal of Process Mechanical Engineering 211 (1997) 267–277. Available from: https://doi.org/10.1243/0954408971529737.

[39] L. Alfayez, D. Mba, Detection of incipient cavitation and the best efficiency point of a 2.2 MW centrifugal pump using acoustic emission, Journal of Acoustic Emission 22 (2004) 77–82.

[40] J. Cernetic, J. Prezelj, M. Chudina, Use of noise and vibration signal for detection and monitoring of cavitation in kinetic pumps, The Journal of the Acoustical Society of America 123 (2008) 3316. Available from: https://doi.org/10.1121/1.2933777.

[41] M. Chudina, J. Prezelj, Detection of cavitation in operation of kinetic pumps. Use of discrete frequency tone in audible spectra, Applied Accoustics 70 (4) (2009) 540–546. Available from: https://doi.org/10.1016/j.apacoust.2008.07.005.

[42] S.N. Ganeriwala, V. Kanakasabai, Using vibration signatures analysis to detect cavitation in centrifugal pumps, Rotating Machinery, Structural Health Monitoring, Shock and Vibration 5 (2011) 307–324. Available from: https://doi.org/10.1007/978-1-4419-9428-8_41.

[43] N. Zhang, M. Yang, B. Gao, Z. Li, Vibration characteristics induced by cavitation in a centrifugal pump with slope volute, Shock and Vibration 2015 (2015) 10 p. Available from: https://doi.org/10.1155/2015/294980.

[44] A.M. Abdulaziz, A. Kotb, Detection of pump cavitation by vibration signature, Australian Journal of Mechanical Engineering 15 (2017) 103–110. Available from: https://doi.org/10.1080/14484846.2015.1093261.

[45] M. Chudina, Noise as an indicator of cavitation in a centrifugal pump, Acoustical Physics (2003) 463–474. Available from: https://doi.org/10.1134/1.1591303.

[46] D.P. Sloteman, Cavitation in high energy pumps-detection and assessment of damage potential, in: Proceedings of the 23 rd International Pump Users Symposium, 2007.

[47] P.J. McNulty, I.S. Pearsall, Cavitation inception in pumps, Journal of Fluids Engineering 104 (1982) 99–104. Available from: https://doi.org/10.1115/1.3240865.

[48] M. Chudina, J. Prezelj, Detection of cavitation in situ operation of kinetic pumps: effect of cavitation on the characteristic discrete frequency component, Applied Accoustics. 70 (9) (2009) 1175–1182. Available from: https://doi.org/10.1016/j.apacoust.2009.04.001.

[49] J. Cernetič, M. Chudina, Cavitation noise phenomena in centrifugal pumps, in: 5th Congress of Alps-Adria Acoustics Association, 2012.

[50] P.D. McFadden, J.D. Smith, Vibration monitoring of rolling element bearings by the high-frequency resonance technique – a review, Tribology International 17 (1984) 3–10. Available from: https://doi.org/10.1016/0301-679X(84)90076-8.

[51] H. Prashad, M. Ghosh, S. Biswas, Diagnostic monitoring of rolling-element bearings by high-frequency resonance technique, ASLE Transactions 28 (4) (1985) 439–448. Available from: https://doi.org/10.1080/05698198508981641.

[52] L. Saidi, J. Ben Ali, M. Benbouzid, E. Bechhoefer, The use of spectral kurtosis as a trend parameter for localized bearing fault diagnosis in induction machines, ISA Transactions 63 (2016) 436–447. Available from: https://doi.org/10.1016/j.isatra.2016.02.019.

[53] C. Yiakopoulos, K. Gryllias, I. Antoniadis, Spectral effects of the slip motion in defective rolling element bearings, Advances in Vibration Engineering 4 (2005) 199–211.

[54] G. Mousmoulis, C. Yiakopoulos, I. Kassanos, I. Antoniadis, J. Anagnostopoulos, Vibration and acoustic emission monitoring of a centrifugal pump under cavitating operating conditions, in: IOP Conference Series, Earth and Environmental Science (2019). Available from: https://doi.org/10.1088/1755-1315/405/1/012003.

[55] N. Chu, L. Wang, L. Yu, C. He, L. Cao, B. Huang, D. Wu, An adaptive autogram approach based on a cfar detector for incipient cavitation detection, Sensors 20 (8) (2020) 27 p., doi:10.3390/s20082303.

[56] R.F. Dwyer, A technique for improving detection and estimation of signals contaminated by under ice noise, The Journal of the Acoustical Society of America 74 (1983) 124–130. Available from: https://doi.org/10.1121/1.389729.

[57] R.F. Dwyer, Use of the kurtosis statistic in the frequency domain as an aid in detecting random signals, IEEE Journal of Oceanic Engineering 9 (2) (1984) 85–92, doi:10.1109/JOE.1984.1145602.

[58] J. Antoni, The spectral kurtosis: a useful tool for characterising non-stationary signals, Mechanical Systems and Signal Processing. 20 (2006) 282–307. Available from: https://doi.org/10.1016/j.ymssp.2004.09.001.

[59] J. Antoni, Fast computation of the kurtogram for the detection of transient faults, Mechanical Systems and Signal Processing. 21 (2007) 108–124. Available from: https://doi.org/10.1016/j.ymssp.2005.12.002.

[60] R.B. Randall, J. Antoni, Rolling element bearing diagnostics – a tutorial, Mechanical Systems and Signal Processing. 25 (2011) 485–520. Available from: https://doi.org/10.1016/j.ymssp.2010.07.017.

[61] R.B. Randall, Vibration-Based Condition Monitoring: Industrial, Aerospace and Automotive Applications, John Wiley & Sons, 2011.

[62] G. Ardizzon, G. Pavesi, Theoretical evaluation of the effects fo the impeller entrance geometry and of the incident angle on cavitation inception in centrifugal pumps, Proceedings of the Institute of Mechanical Engineers, Part C: Journal of Mechanical Engineering Science 209 (1) (1995) 29–38. Available from: https://doi.org/10.1243/PIME_PROC_1995_209_119_02.

[63] O. Coutier-Delgosha, J.L. Reboud, Y. Delannoy, Numerical simulation of the unsteady behaviour of cavitating flows, International Journal for Numerical Methods in Fluids 42 (2003) 527–548. Available from: https://doi.org/10.1002/fld.530.

[64] Y. Chen, S.D. Heister, A numerical treatment for attached cavitation, Journal of Fluids Engineering 116 (1994) 613–618. Available from: https://doi.org/10.1115/1.2910321.

[65] J.H. Ferziger, M. Peric, Computational Methods for Fluid Dynamics, Springer-Verlag, Berlin Heidelberg, 2002.

[66] M. Deshpande, J. Feng, C.L. Merkle, Cavity flow predictions based on the Euler equations, Journal of Fluids Engineering. 116 (1) (1994) 36–44. Available from: https://doi.org/10.1115/1.2910238.

[67] L. Liu, J. Li, Z. Feng, A numerical method for simulation of attached cavitation flows, International Journal Numerical Methods in Fluids 52 (2006) 639–658. Available from: https://doi.org/10.1002/fld.1192.

[68] R. Hirschi, P. Dupont, F. Avellan, J.N. Favre, J.F. Guelich, E. Parkinson, Centrifugal pump performance drop due to leading edge cavitation: numerical predictions compared with model tests, Journal of Fluids Engineering. 120 (4) (1998) 705–711. Available from: https://doi.org/10.1115/1.2820727.

[69] Y.A. Bouziad, M. Farhat, F. Guennoun, J.-L. Kueny, F. Avellan, Physical modelling and simulation of leading edge cavitation, application to an industrial inducer, in: Fifth International Symposium on Cavitation (CAV2003), Osaka, Japan, November 2003.

[70] Y. Chen, S.D. Heister, Modelling hydrodynamic non-equilibrium in cavitating flows, Journal of Fluids Engineering. 118 (1) (1996) 172–178. Available from: https://doi.org/10.1115/1.2817497.

[71] S. Gopalan, J. Katz, Flow structure and modelling issues in the closure region of attached cavitation, Physics of Fluids 12 (4) (2000) 895–911. Available from: https://doi.org/10.1063/1.870344.

[72] T. Michael, J. Yang, F. Stern, A sharp interface approach for cavitation modelling using volume of fluid and ghost fluid methods, Journal of Hydrodynamics 29 (6) (2017) 917–925. Available from: https://doi.org/10.1016/S1001-6058(16)60806-5.

[73] A. Kubota, H. Kato, H. Yamaguchi, A new modelling of cavitating flows: a numerical study of unsteady cavitation on a hydrofoil section, Journal of Fluid Mechanics 240 (1992) 59–96. Available from: https://doi.org/10.1017/S002211209200003X.

[74] I. Senocak, W. Shyy, A pressure-based method for turbulent cavitating flow computations, Journal of Computational Physics 176 (2) (2002) 363–383. Available from: https://doi.org/10.1006/jcph.2002.6992.

[75] Y. Dellanoy, J.L. Kueny, Two phase flow approach in unsteady cavitation modeling, ASME FED 98 (1990) 153–158.

[76] Y. Ventikos, G. Tzabiras, A numerical method for the simulation of steady and unsteady cavitating flows, Computers & Fluids 29 (1) (2000) 63–88. Available from: https://doi.org/10.1016/S0045-7930(98)00061-9.

[77] C. Pascarella, V. Salvatore, Numerical study of unsteady cavitation on a hydrofoil section using a barotropic model, in: Proc. Fourth Int. Symp. Cavitation, 2001, p. 12.

[78] Q. Qin, C.C.S. Song, R.E.A. Arndt, A virtual single-phase natural cavitation model and its application to CAV2003 hydrofoil, in: Proceedings of CAV2003—Fifth International Symposium on Cavitation, Osaka (Japan), November 2003.

[79] E. Rapposelli, L. d'Agostino, A barotropic cavitation model with thermodynamic effects, in: 5th International Symposium on Cavitation CAV2003, Osaka, Japan, 2003.

[80] S. Frikha, O. Coutier-Delgosha, J.A. Astolfi, Influence of the cavitation model on the simulation of cloud cavitation on 2D foil section, International Journal of Rotating Machinery 2008 (2008) 12 p. Available from: https://doi.org/10.1155/2008/146234.

[81] S. Barre, J. Rolland, G. Boitel, E. Goncalves, R.F. Patella, Experiments and modeling of cavitating flows in venturi: attached sheet cavitation, European Journal of Mechanics-B/Fluids 28 (3) (2009) 444–464. Available from: https://doi.org/10.1016/j.euromechflu.2008.09.001.

[82] M. Blume, R. Skoda, Numerical 3D analysis of cloud cavitation shedding frequency on a circular leading edge hydrofoil with a barotropic cavitation model, Journal of Physics: Conference Series 656 (1) (2015) 012146IOP Publishing. Available from: https://doi.org/10.1088/1742-6596/656/1/012146.

[83] K. Hejranfar, E. Ezzatneshan, K. Fattah-Hesari, A comparative study of two cavitation modeling strategies for simulation of inviscid cavitating flows, Ocean Engineering 108 (2015) 257–275. Available from: https://doi.org/10.1016/j.oceaneng.2015.07.016.

[84] O. Coutier-Delghosa, R. Fortes Patella, J.-L. Reboud, B. Pouffary, 3D numerical simulation of pump cavitating behavior, in: ASME FEDSM, Montreal, Canada, July 2002. 10.1115/FEDSM2002-31188.

[85] O. Coutier-Delgosha, R. Fortes-Patella, J.L. Reboud, N. Hakimi, C. Hirsch, Numerical simulation of cavitating flow in 2D and 3D inducer geometries, International Journal for Numerical Methods in Fluids 48 (2005) 135–167. Available from: https://doi.org/10.1002/fld.820.

[86] M.V. Salvetti, E. Sinibaldi, F. Beux, Towards the simulation of cavitating flows in inducers through a homogeneous barotropic flow model, in: L. d'Agostino, M.V. Salvetti (Eds.), Fluid Dynamics of

Cavitation and Cavitating Turbopumps. CISM International Centre for Mechanical Sciences, vol. 496, Springer, Vienna, 2007, pp. 317–351. Available from: https://doi.org/10.1007/978-3-211-76669-9_9

[87] U. Jese, R. Fortes-Patella, S. Antheaume, High head pump-turbine: pumping mode numerical simulations with a cavitation model for off-design conditions, IOP Conference Series: Earth and Environmental Science 22 (3) (2014) 032048IOP Publishing. Available from: https://doi.org/10.1088/1755-1315/22/3/032048.

[88] P. Limbach, T. Müller, R. Skoda, Application of a compressible flow solver and barotropic cavitation model for the evaluation of the suction head in a low specific speed centrifugal pump impeller channel, Journal of Physics: Conference Series 656 (1) (2015) 012065IOP Publishing. Available from: https://doi.org/10.1088/1742-6596/656/1/012065.

[89] R.E.A. Arndt, Cavitation in vortical flows, Annual Review of Fluid Mechanics 34 (1) (2002) 143–175. Available from: https://doi.org/10.1146/annurev.fluid.34.082301.114957.

[90] B. Ji, X.W. Luo, R.E.A. Arndt, X. Peng, Y. Wu, Large Eddy simulation and theoretical investigations of the transient cavitating vortical flow structure around a NACA66 hydrofoil, International Journal of Multiphase Flow 68 (2015) 121–134. Available from: https://doi.org/10.1016/j.ijmultiphaseflow.2014.10.008.

[91] R.F. Kunz, D.A. Boger, D.R. Stinebring, T.S. Chyczewski, J.W. Lindau, H.J. Gibeling, S. Venkateswaran, T.R. Govindan, A preconditioned Navier–Stokes method for two-phase flows with application to cavitation prediction, Computers & Fluids 29 (8) (2000) 849–875. Available from: https://doi.org/10.1016/S0045-7930(99)00039-0.

[92] G.H. Schnerr, J. Sauer, Physical and numerical modeling of unsteady cavitation dynamics, in: Fourth international conference on multiphase flow, vol. 1, ICMF New Orleans, 2001.

[93] A.K. Singhal, M.M. Athavale, H. Li, Y. Jiang, Mathematical basis and validation of the full cavitation model, Journal of Fluids Engineering 124 (3) (2002) 617–624. Available from: https://doi.org/10.1115/1.1486223.

[94] Y. Saito, I. Nakamori, T. Ikohagi, Numerical analysis of unsteady vaporous cavitating flow around a hydrofoil, in: Proc. 5th Int. Symp. Cavitation (CAV2003), Osaka, 2003.

[95] P.J. Zwart, A.G. Gerber, T. Belamri, A two-phase flow model for predicting cavitation dynamics, in: Fifth International Conference on Multiphase Flow, vol. 152, Yokohama, Japan, 2004.

[96] D. Dauby, P. Queutey, A. Leroyer, M. Visonneau, Computation of 2D cavitating flows ant tip vortex flows with an unstructured RANS solver, 2006.

[97] C.L. Merkle, J.Z. Feng, P.E.O. Buelow, Computational modeling of the dynamics of sheet cavitation, in: Proceedings of the 3rd International Symposium on Cavitation, Grenoble, France, 1998.

[98] J. Sauer, G.H. Schnerr, Unsteady cavitating flow—a new cavitation model based on a modified front capturing method and bubble dynamics, in: Proceedings of 2000 ASME Fluid Engineering Summer Conference, Boston, MA, June 11–15, 2000.

[99] M. Frobenius, R. Schilling, J. Friedrichs, G. Kosyna, Numerical and experimental investigations of the cavitating flow in a centrifugal pump impeller, in: Proceedings of the ASME 2002 Joint U.S.-European Fluids Engineering Division Conference. Volume 1: Fora, Parts A and B. Montreal, Quebec, Canada. July 14–18, ASME, 2002, pp. 361–368. Available from: https://doi.org/10.1115/FEDSM2002-31006.

[100] B. Huang, Y. Zhao, G. Wang, Large eddy simulation of turbulent vortex-cavitation interactions in transient sheet/cloud cavitating flows, Computers & Fluids 92 (2014) 113–124. Available from: https://doi.org/10.1016/j.compfluid.2013.12.024.

[101] P. Koukouvinis, H. Naseri, M. Gavaises, Performance of turbulence and cavitation models in prediction of incipient and developed cavitation, International Journal of Engine Research 18 (4) (2017) 333–350. Available from: https://doi.org/10.1177/1468087416658604.

[102] O. Coutier-Delgosha, R. Fortes-Patella, J.L. Reboud, Simulation of unsteady cavitation with a two-equation turbulence model including compressibility effects, Journal of Turbulence 3 (2002) 20 p. Available from: https://doi.org/10.1088/1468-5248/3/1/058.

[103] O. Coutier-Delgosha, R. Fortes-Patella, J.L. Reboud, Evaluation of the turbulence model influence on the numerical simulations of unsteady cavitation, Journal of Fluids Engineering 125 (1) (2003) 38–45.

[104] B. Pouffary, R.F. Patella, J.-L. Reboud, Numerical simulation of cavitating flow around a 2D hydrofoil: barotropic approach, in: Fifth International Symposium on Cavitation (CAV2003), Osaka, Japan, November 2003.

[105] D.-q Li, M. Grekula, P. Lindell, Towards numerical prediction of unsteady sheet cavitation on hydrofoils, Journal of Hydrodynamics, Series B 22 (5 Suppl. 1) (2010) 741–746. Available from: https://doi.org/10.1016/S1001-6058(10)60024-8.

[106] E. Goncalvès, Numerical study of unsteady turbulent cavitating flows, European Journal of Mechanics – B/Fluids 30 (1) (2011) 26–40ISSN 0997–7546. Available from: https://doi.org/10.1016/j.euromechflu.2010.08.002.

[107] J. Decaix, E.G. da Silva, Compressible effects modelling in turbulent cavitating flows, European Journal of Mechanics – B/Fluids 39 (2013) 11–31. ff10.1016/j.euromechflu.2012.12.001ff.

[108] T.D. Tran, B. Nennemann, T.C. Vu, F. Guibault, Numerical simulation of unsteady sheet/cloud cavitation, IOP Conference Series: Earth and Environmental Science 22 (5) (2014) 052012. Available from: https://doi.org/10.1088/1755-1315/22/5/052012.

[109] G. Chen, G. Wang, C. Hu, B. Huang, Y. Gao, M. Zhang, Combined experimental and computational investigation of cavitation evolution and excited pressure fluctuation in a convergent–divergent channel, International Journal of Multiphase Flow 72 (2015) 133–140. Available from: https://doi.org/10.1016/j.ijmultiphaseflow.2015.02.007.

[110] H. Liu, D. Liu, Y. Wang, X. Wu, J. Wang, Application of modified k-ω model to predicting cavitating flow in centrifugal pump, Water Science and Engineering 6 (2013) 331–339. Available from: https://doi.org/10.3882/j.issn.1674-2370.2013.03.009.

[111] R. Vaidyanathan, I. Senocak, J. Wu, W. Shyy, Sensitivity evaluation of a transport-based turbulent cavitation model, Journal of Fluids Engineering 125 (3) (2003) 447.

[112] P.R. Spalart, M. Shur, On the sensitization of turbulence models to rotation and curvature, Aerospace Science and Technology 1 (5) (1997) 297–302. Available from: https://doi.org/10.1016/S1270-9638(97)90051-1.

[113] M.L. Shur, M.K. Strelets, A.K. Travin, P.R. Spalart, Turbulence modeling in rotating and curved channels: assessing the Spalart-Shur correction, AIAA Journal 38 (5) (2000) 784–792. Available from: https://doi.org/10.2514/2.1058.

[114] Y. Zhao, G. Wang, B. Huang, A curvature correction turbulent model for computations of cloud cavitating flows, Engineering Computations 33 (1) (2016) 202–216. Available from: https://doi.org/10.1108/ec-01-2015-0026.

[115] W. Jian, W. Yong, L. Houlin, S. Quaorui, D. Matevz, Rotating corrected-based cavitation model for a centrifugal pump, Journal of Fluids Engineering 140 (2018) 111301. Available from: https://doi.org/10.1115/1.4040068.

[116] C. Tian, T. Chen, T. Zou, Numerical study of unsteady cavitating flows with RANS and DES models, Modern Physics Letters B 33 (20) (2019) 1950228. Available from: https://doi.org/10.1142/S0217984919502282.

[117] R. Bensow, Simulation of the unsteady cavitation on the the Delft Twist11 foil using RANS, DES and LES, in: 2nd International Symposium on Marine Propulsors, Workshop, 2011.

[118] N.X. Lu, R.E. Bensow, G. Bark, Large eddy simulation of cavitation development on highly skewed propellers, Journal of Marine Science and Technology 19 (2014) 197–214. Available from: https://doi.org/10.1007/s00773-013-0240-3.

[119] h Changli, G.-H. Chen, L. Yang, G. Wang, Large eddy simulation of turbulent attached cavitating flows around different twisted hydrofoils, Energies 11 (2018) 2768. Available from: https://doi.org/10.3390/en11102768.

[120] A. Asnaghi, A. Feymark, R.E. Bensow, Numerical investigation of the impact of computational resolution on shedding cavity structures, International Journal of Multiphase Flow 107 (2018) 33–50. Available from: https://doi.org/10.1016/j.ijmultiphaseflow.2018.05.021.

[121] I. Senocak, W. Shyy, Interfacial dynamics-based modelling of turbulent cavitating flows, Part-1: model development and steady-state computations, International Journal for Numerical Methods in Fluids 44 (2004) 975–995. Available from: https://doi.org/10.1002/fld.692.

[122] M.P. Kinzel, J.W. Lindau, R.F. Kunz, A unified homogenous multiphase CFD model for cavitation, in: Proceedings of the ASME 2017 Fluids Engineering Division Summer Meeting. Volume 1B, Waikoloa, Hawaii, USA. V01BT11A020. ASME, July 30–August 3, 2017. Available from: https://doi.org/10.1115/FEDSM2017-69363.

[123] M. Morgut, E. Nobile, I. Bilus, Comparison of mass transfer models for the numerical prediction of sheet cavitation around a hydrofoil, International Journal of Multiphase Flow 37 (2011) 620–626. Available from: https://doi.org/10.1016/j.ijmultiphaseflow.2011.03.005.

[124] M. Kinzel, J. Lindau, R. Kunz, An assessment of CFD cavitation models using bubble growth theory and bubble transport modeling, Journal of Fluids Engineering 141 (2019) 041301. Available from: https://doi.org/10.1115/1.4042421.

[125] M. Morgut, E. Nobile, Numerical predictions of cavitating flow around model scale propellers by CFD and advanced model calibration, International Journal of Rotating Machinery 2012 (2012) 61818011 pages. Available from: https://doi.org/10.1155/2012/618180.

[126] I. Biluš, M. Morgut, E. Nobile, Simulation of sheet and cloud cavitation with homogenous transport models, International Journal of Simulation Modelling 12 (2013) 94–106, doi:10.2507/IJSIMM12(2)3.229.

[127] Y. Zhang, X. Ren, Y. Wang, X. Li, Y. Ito, C. Gu, Investigation of the cavitation model in an inducer for water and liquid nitrogen, Proceedings of the Institution of Mechanical Engineers, Part C: Journal of Mechanical Engineering Science 233 (19–20) (2019) 6939–6952. Available from: https://doi.org/10.1177/0954406219869980.

[128] P. Limbach, R. Skoda, Numerical and experimental analysis of cavitating flow in a low specific speed centrifugal pump with different surface roughness, ASME Journal of Fluids Engineering 139 (10) (2017) 101201. Available from: https://doi.org/10.1115/1.4036673.

[129] H.-l Liu, J. Wang, Y. Wang, H. Zhang, H. Huang, Influence of the empirical coefficients of cavitation model on predicting cavitating flow in the centrifugal pump, International Journal of Naval Architecture and Ocean Engineering 6 (1) (2014) 119–131. Available from: https://doi.org/10.2478/IJNAOE-2013-0167.

[130] F. Bakir, R. Rey, A. Gerber, T. Belamri, B. Hutchinson, Numerical and experimental investigations of the cavitating behavior of an inducer, International Journal of Rotating Machinery 10 (2004) 11 p. Available from: https://doi.org/10.1155/S1023621X04000028.

[131] I. Mejri, F. Bakir, R. Rey, T. Belamri, Comparison of computational results obtained from a homogeneous cavitation model with experimental investigations of three inducers, Journal of Fluids Engineering-Transactions of The ASME 128 (2006) 1308–1323. Available from: https://doi.org/10.1115/1.2353265.

[132] F. Bakir, I. Mejri, S. Kouidri, R. Rey, Hub shape effects on the inducers performance under cavitation, Proceedings of the Institution of Mechanical Engineers, Part A: Journal of Power and Energy 220 (2003).

[133] P. Limbach, M. Kimoto, C. Deimel, R. Skoda, Numerical 3D simulation of the cavitating flow in a centrifugal pump with low specific speed and evaluation of the suction head, Proceedings of the ASME Turbo Expo 2 (2014). Available from: https://doi.org/10.1115/GT2014-26089.

[134] X. Tang, M. Zou, F. Wang, X. Li, X. Shi, Comprehensive numerical investigations of unsteady internal flows and cavitation characteristics in double-suction centrifugal pump, Mathematical Problems in Engineering 2017 (2017) 501382613 pages. Available from: https://doi.org/10.1155/2017/5013826.

[135] G. Mousmoulis, I. Kassanos, G. Aggidis, I. Anagnostopoulos, Numerical simulation of the performance of a centrifugal pump with a semi-open impeller under normal and cavitating conditions, Applied Mathematical Modelling 89 (2) (2021) 1814–1834. Available from: https://doi.org/10.1016/j.apm.2020.08.074.

[136] R.F. Kunz, D.R. Stinebring, T.S. Chyczewski, D.A. Boger, H.J. Gibeling, T.R. Govindan, Multi-phase CFD analysis of natural and ventilated cavitation about submerged bodies, in: Proceedings of the 1999 3rd ASME/JSME Joint Fluids Engineering Conference, FEDSM'99, San Francisco, CA, July 18–23, 1999.

[137] B. Pouffary, R.F. Patella, J.-L. Reboud, Unsteady cavitating flows in turbomachinery: comparison of two numerical models and applications, in: ECCOMAS2004 Symposium, Jyväskylä, Finland, July 2004.

[138] B. Pouffary, R.F. Patella, J.-L. Reboud, P.-A. Lambert, Numerical simulation of 3D cavitating flows: analysis of cavitation head drop in turbomachinery, Journal of Fluids Engineering 130 (6) (2008) 061301. Available from: https://doi.org/10.1115/1.2917420.

[139] Y. Wang, H. Liu, D. Liu, S. Yuan, J. Wang, L. Jiang, Application of the two-phase three-component computational model to predict cavitating flow in a centrifugal pump and its validation, Computers & Fluids 131 (2016). Available from: https://doi.org/10.1016/j.compfluid.2016.03.022.

[140] D. Liang, Z. Yuqi, C. Dai, Research on cavitation acoustic characteristics of centrifugal pump based on fluid-acoustic field coupling method, Advances in Mechanical Engineering 10 (2018). Available from: https://doi.org/10.1177/1687814018773665. 168781401877366.

[141] N.-X. Lu, R.E. Bensow, G. Bark, Mesh resolution effects in simulating unsteady sheet cavitation, Ship Technology Research 60 (2) (2013) 86–97. Available from: https://doi.org/10.1179/str.2013.60.2.004.

[142] T. Sato, T. Nagahara, K. Tanaka, M. Fuchiwaki, F. Shimizu, Vortex cavitation and oscillation in a double-suction volute pump, in: IOP Conference Series: Earth and Environmental Science, vol. 12, 25th IAHR Symposium on Hydraulic Machinery and Systems 20–24 September "Politehnica" University of Timişoara, Timişoara, Romania, 2010.

[143] M.M. Athavale, H.Y. Li, Y. Jiang, A.K. Singhal, Application of the full cavitation model to pumps and inducers, International Journal of Rotating Machinery 8 (2002) 65687012 pages. Available from: https://doi.org/10.1155/S1023621X02000052.

[144] N.H. Mostafa, M.A., Numerical and experimental investigation of cavitation in axial pumps, in: 2006 8th International Congress of Fluid Dynamics and Propulsion, 2006, pp. 14–17.

[145] R. Campos-Amezcua, S. Khelladi, Z. Mazur-Czerwiec, F. Bakir, A. Campos-Amezcua, R. Rey, Numerical and Experimental Study of Mass Transfer Through Cavitation in Turbomachinery, Mass Transfer – Advanced Aspects, IntechOpen, Hironori Nakajima, 2011. Available from: http://doi.org/10.5772/20006.

[146] H. Ding, F. Visser, Y. Jiang, A practical approach to speed up npshr prediction of centrifugal pumps using CFD cavitation model, in: American Society of Mechanical Engineers, Fluids Engineering Division (Publication), FEDSM. 1, 2012. 10.1115/FEDSM2012-72282.

[147] W.G. Li, Validating full cavitation model with an experimental centrifugal pump, TASK Quarterly: scientific bulletin of Academic Computer Centre in Gdansk 18 (1) (2014) 81–100.

[148] W. Li, Modeling viscous oil cavitating flow in a centrifugal pump, ASME Journal of Fluids Engineering 138 (1) (2015) 011303. Available from: https://doi.org/10.1115/1.4031061.

[149] A. Marini, S. Salvadori, C. Bernardini, M. Insinna, F. Martelli, A. Nicchio, A. Piva, Numerical prediction of cavitation inception in centrifugal impellers, In 9th European Turbomachinery Conference (2011) 21–25.

[150] H. Zhou, M. Xiang, P.N. Okolo, Z. Wu, G.J. Bennett, W. Zhang, An efficient calibration approach for cavitation model constants based on OpenFOAM platform, Journal of Marine Science and Technology 24 (4) (2019) 1043–1056. Available from: https://doi.org/10.1007/s00773-018-0604-9.

Chapter 6

Cavitation in Hydraulic Turbines

Ioannis Kassanos and Ioannis Anagnostopoulos
Laboratory of Hydraulic Turbomachines, School of Mechanical Engineering, National Technical University of Athens, Athens, Greece

6.1 Introduction

Among the various renewable energy sources, hydropower has played a dominant role in meeting energy demands throughout the world. The abundance of this available resource in many parts of the world, the capability to cover base load demand, and the great reliability and longevity of hydropower plants have contributed to its popularity over the years. Despite the significant presence of hydropower in the energy mix, a large untapped potential remains, particularly for small-scale installations. On the other hand, the ever-increasing energy demand and the requirement for versatility and sustainability in power generation have highlighted the need for the mass deployment of renewable energy power plants. However, other renewable energy sources are generally characterized by large variability with a significant imbalance between supply and demand, which poses a great challenge for the stability of electricity grids. On the contrary, hydropower constitutes a versatile and predictable energy source that can also cover peak load demand and contribute to the stability of the electricity network by the quick regulation of supplied power. In this framework, hydropower can play a critical role in the efficient integration of renewable energy sources and in meeting the goals of sustainable and clean energy generation.

Cavitation and Bubble Dynamics. DOI: https://doi.org/10.1016/B978-0-12-823388-7.00001-1

Hydraulic turbines constitute the main components in hydroelectric power plants, where the potential energy of an available water resource is converted to mechanical power in the turbine shaft and subsequently to electrical power in the generator. Depending on the hydraulic characteristics of the resource (available head H in meters and flow rate Q in cubic meters per second), different types of turbines and different configurations may be used (Fig. 6.1). The selected size may also vary, but the ultimate goal is to determine an optimal balance between energy conversion efficiency and manufacturing cost. Depending on the type and configuration used, the turbine usually consists of a rotating runner, which is the main component where energy conversion takes place; stationary flow diversion components; regulation components such as guide vanes and injectors; a draft tube, which connects the turbine outlet to a downstream reservoir or river; and a casing that guides the flowing water to the runner and supports all the turbine components. Depending on the principle of operation, hydraulic turbines can be classified as reaction turbines and impulse turbines (Fig. 6.1). In reaction turbines the runner is fully submerged in water, and the flow results in the development of a pressure field on the runner blades, which creates a net force or reaction and, consequently, torque on the shaft. Apart from the reaction force, a portion of the available energy is converted into a jet of water, which interacts with the runner blades. A draft tube is also employed, connecting the turbine to the tail race water channel, which allows a portion of the available kinetic energy that would otherwise dissipate to be recovered, while also allowing the turbine reference level to be set irrespective of the tail water level without loss in efficiency. In impulse turbines the energy exchange occurs in atmospheric pressure in air by converting the

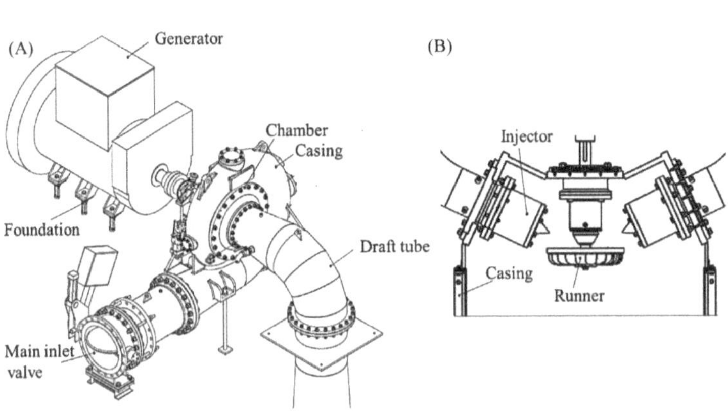

Figure 6.1 Different turbine arrangements. (A) A typical reaction-type Francis turbine, including the spiral casing, the draft tube, the casing chambers, the wicket gate and control actuator, the main inlet valve, and the generator. (B) A typical impulse-type Turgo turbine, showing the runner and injectors.

hydraulic energy to a high-speed jet of water that impinges on specially shaped buckets. The impulse generated by the interaction of the jet with the buckets translates to a tangential force on the runner and allows the development of torque on the shaft.

The selection of turbine type is driven by the relationship between the available head H, flow rate Q, and selected rotational velocity n of the turbine, which may be expressed by a characteristic number, that is, the specific speed n_q:

$$n_q = n\frac{Q^{1/2}}{H^{3/4}} \tag{6.1}$$

As the head increases in proportion to the flow rate, the shape of the runner becomes more radial (Fig. 6.2A,B), until friction losses become substantial, owing to the increase in flow velocity. A point is reached at which it becomes inefficient to maintain a fully submerged arrangement. Therefore a change in configuration is required so that only part of the runner interacts with water at any given time (Fig. 6.2C). By contrast, for a given hydraulic power, if the flow rate is proportionally larger than the head, it is more efficient to keep the runner fully submerged in water and allow all parts to contribute to the energy conversion process.

Therefore the specific speed can be thought of as a characteristic property of hydraulic turbines that dictates their shape. Another intrinsic characteristic of hydraulic turbines is the propensity for cavitation development within the flow path at certain operating conditions. In regions where the flow velocity may rise excessively or a significant drop in pressure is observed, the local static pressure

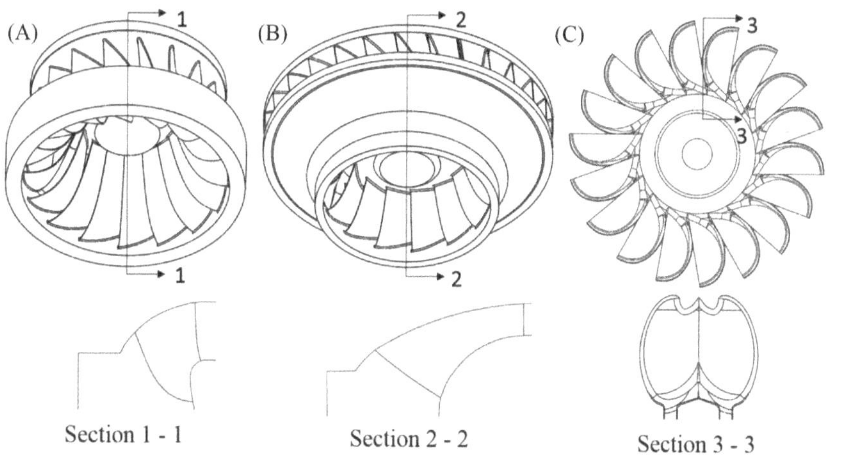

Figure 6.2 Variation in the shape of the turbine runner with specific speed. (A) A high-speed/low-head Francis runner consisting of the hub, shroud, and blade. (B) A low-speed/high-head Francis runner. (C) A Pelton runner with buckets on a central hub.

may reduce to a point below the vapor pressure p_{sat} of the fluid. If this condition is met, the liquid phase cannot be sustained and evaporation is initiated, appearing as small bubbles dispersed in the main flow. A further reduction in pressure and a further increase in the low pressure zones results in growth of the vaporous regions and the appearance of large cavities. As these bubbles and cavities are entrained in regions of the flow field where the local pressure increases again, they will reliquify in a process that is associated with the generation of noise, vibration, and erosion on the metal surfaces of the machine. During normal operation the flow conforms to the hydraulic profile of the turbine components, and the propensity for cavitation is low. However, when there is a departure from normal operating conditions, the flow becomes more turbulent, there is scope for the development of flow separations, and the propensity for cavitation rises.

From this discussion, as turbines are called to operate more frequently in off-design conditions in which the flow may no longer conform to the hydraulic profile, the risk of operation in cavitating and damaging conditions is increased. Therefore to allow the effective and smooth penetration of renewable energy sources in the electrical network and to fully exploit the advantages of hydropower, the in-depth study of cavitation in hydraulic turbines is crucial. In this chapter, cavitation phenomena observed in hydraulic turbines along with experimental and numerical tools applied for their study are presented.

6.2 Cavitation Types and Effects in Hydraulic Turbines

Cavitation in hydraulic turbines appears in several different forms, depending on the flow conditions, the geometry of the turbine components [1,2], and the location of inception. Cavitation can develop in all types of hydraulic turbines and can significantly affect the lifespan of the hydromechanical equipment [3]. The mechanism for the inception and development of cavitation is the same in hydraulic turbines and centrifugal pumps. As was discussed in Chapter 5, when vapor bubbles under an adverse pressure gradient reach a position in the flow path where the local static pressure exceeds the vapor pressure, they will collapse releasing strong impulsive waves that travel through the water and turbine structure [4]. If the bubble collapse occurs near solid surfaces, a liquid microjet may form that impinges on the structure, potentially leading to erosion. The erosive potential of the imploding bubbles may vary significantly [5], leading to minor pitting after a significant amount of operating hours of the machine, or can quickly lead to a significant loss of material from the surface of the structure and ultimately to failure. The damaged region may initially appear as a frosted surface, with visible pits growing over time (Fig. 6.3).

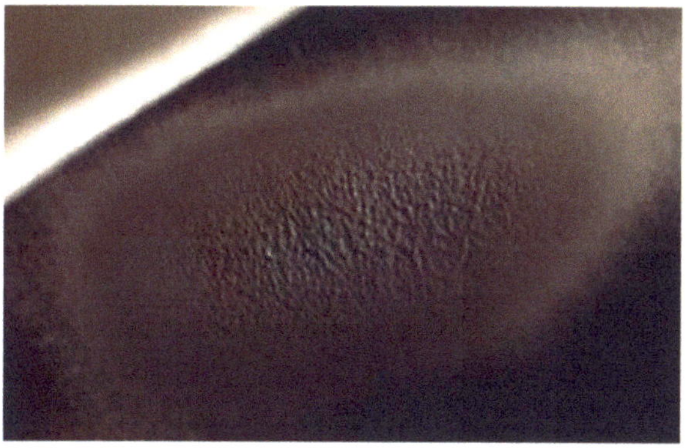

Figure 6.3 Typical pitting and frosting developed on a metal surface affected by cavitation.

The extent of the pitted area may be an indication of the severity of the developed cavitation and may provide useful information about the conditions that are most likely responsible for its inception, since the minimum pressure level in the runner is a strong function of the operating conditions [6]. If the cavities grow significantly, owing to extensive reduction of the pressure level, they can modify the flow structure and influence the dynamic characteristics of the flow [5,7]. The above may ultimately impair the energy conversion process and lead to a reduction in the hydraulic efficiency of the turbine. Apart from the potential erosion on the metal surfaces and efficiency degradation, cavitation may lead to excessive vibration and noise, with detrimental effects on bearings and seals.

Similarly to centrifugal pumps, cavitation in hydraulic machinery may manifest as sheet cavitation, cloud cavitation, traveling bubble cavitation, and vortex cavitation [3]. Depending on the operating conditions and the turbine type, cavitation may be observed at the inlet of the runner as leading edge cavitation (LEC), at the trailing edge cavitation (TEC) as traveling bubbles, in the draft tube as vortex cavitation (DTVC), between consecutive runner blades as interblade vortex cavitation (IBVC), between the blade tips and the discharge ring as tip leakage cavitation (TLC), in the wake of blades as von Kármán vortex cavitation [3,8], and in the splitter and tips of impulse turbine buckets and spear tips as sheet cavitation. Figs. 6.4 and 6.5 show the typical locations and cavitation types than can develop in hydraulic turbines.

More specifically, LEC appears as an attached cavity originating at the leading edge of the runner promoted by flow separation due to the flow approaching with positive or negative incidence. It is an aggressive form of cavitation, causing severe damage on the blades, and is the most frequently encountered type of cavitation [6,8,9]. The turbine is susceptible to this cavitation type when it is called to operate at a higher or lower head than the design head.

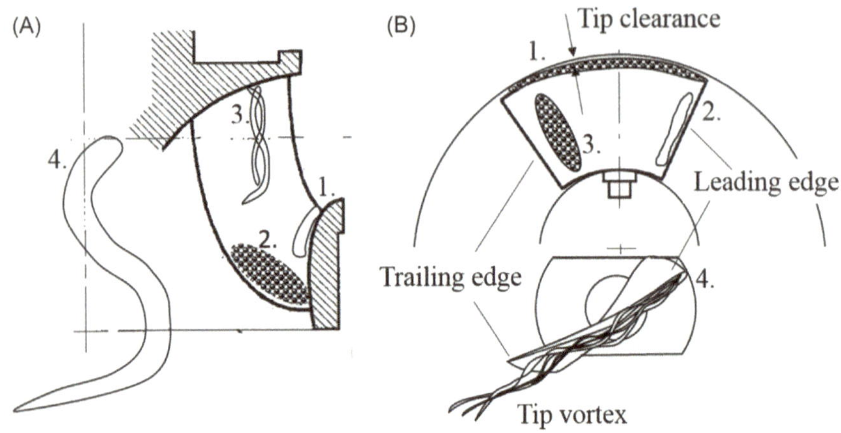

Figure 6.4 Cavitation types in reaction turbines. (A) Typical cavitation types in Francis turbines (1) Leading edge-sheet cavitation. (2) Trailing edge traveling bubble cavitation. (3) Interblade vortex cavitation. (4) Draft tube part load helical vortex. (B) Typical cavitation types in axial turbines (1) Tip clearance cavitation. (2) Leading edge-sheet cavitation. (3) Trailing edge cavitation. (4) Tip vortex cavitation.

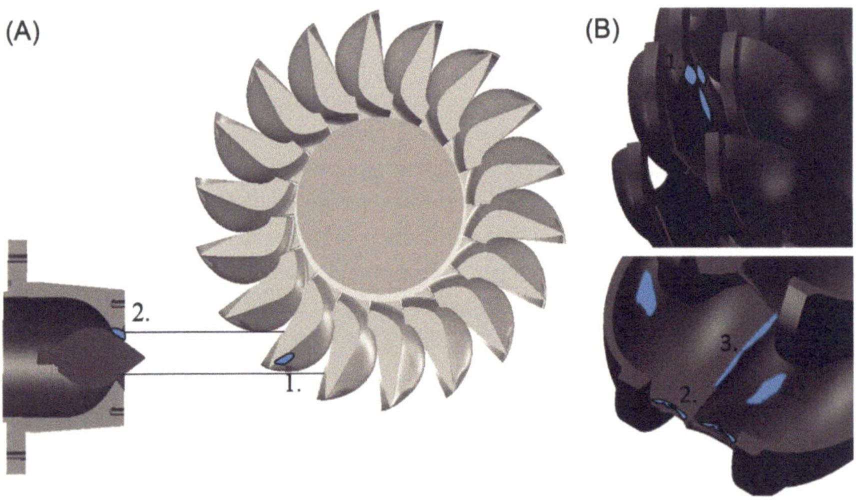

Figure 6.5 Cavitation types in impulse machines. (A) Typical locations for cavitation development in Pelton turbines (sheet cavitation). (1) Cavitation on the buckets. (2) Cavitation on the spear tip of the injector. (B) Cavitation development (1) on the back of the Pelton bucket near the splitter tip and under the splitter, (2) on the bucket pressure side near the splitter tip, (3) along the splitter on the front side of the bucket, (4) downstream of the splitter, and (5) toward the outlet edge of the bucket.

In reaction turbines, TEC appears as traveling bubbles [10] on the blade suction side toward the runner outlet. This cavitation type may develop even under design conditions if the overall pressure level in the turbine is low enough, and it is responsible for the eventual efficiency degradation. Since it is very sensitive to the variation in minimum pressure in the flow path as well as the nuclei content, it is used to determine the setting height of the machine [7] relative to the tail water level. With increasing flow rate, the cavity extent also increases, and the efficiency of the turbine begins to drop when the cavity extends all the way to the outlet. This efficiency reduction is attributed to large-scale modification of the pressure field by the bubble dynamics, interfering with the development of torque on the turbine shaft.

Vortex cavitation develops in highly vortical flows and is initiated at the vortex center where the minimum pressure is observed. In reaction turbines this type of cavitation takes the form of DTVC, IBVC, and TVC.

The vortex cavitation in the draft tube of reaction turbines develops at off-design conditions in which the flow leaving the runner is characterized by excessive or residual swirl (Fig. 6.6). At near optimum flow conditions, the inlet flow velocity is aligned with the runner blades and the flow exits the blade with zero swirl. At part load however, the energy exchange with the blades is not optimum and the flow exits with residual swirl, which promotes the development of a helical vortex also known as vortex rope. The vortex rope initiates at the runner hub and extends downstream in the suction cone of the draft tube. This is a highly unstable phenomenon and leads to substantial pressure fluctuations and potentially excessive vibration in the hydraulic

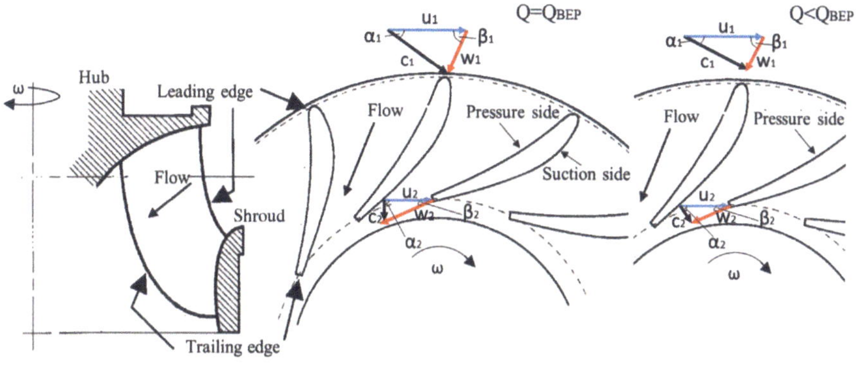

Figure 6.6 Meridional and plan cross-sectional views of a Francis runner showing the variation of the velocity vectors near optimum and part load conditions. At part load (left) there is residual swirl; that is, the tangential component of absolute velocity is not zero. Note that u is the peripheral velocity, c is the absolute velocity, w is the relative velocity, ω is the angular velocity, and subscripts 1 and 2 denote the inlet and outlet, respectively.

system. The vortex rope follows a complex precessing motion (Fig. 6.7A) that depends on the flow rate and the overall pressure level in the draft tube [11]. It can be decomposed into a synchronous component and an asynchronous component [12], the first generating synchronous pressure pulses that are transmitted through the system and the latter generating strong subsynchronous pulsations with a frequency analogous to the frequency of precession. The level of swirl downstream of the runner may be expressed by the swirl number *S*, defined as the ratio of the flux of moment of momentum to the flux of the axial momentum [Eq. (6.2)], where *c* is flow velocity, subscripts *t* and *a* denote the tangential and axial velocity components, respectively, *r* is the radial coordinate, and *R* is the runner outlet radius [12];

$$S = \frac{\int_0^R c_t c_a r^2 dr}{R \int_0^R c_a{}^2 r dr} \tag{6.2}$$

The swirl number is representative of the flow conditions at the outlet of the runner and may be used to infer the pulsation frequency [13] and flow characteristics. The dynamic characteristic of the part load vortex depends on the operating conditions, the shape of the draft tube, the design of the runner, the specific speed, and the overall pressure level in the draft tube [14–17].

On other hand, at overload conditions there is excessive swirl that leads to the formation of a large conical cavity that is rotating in direction opposite that of the runner at synchronous speed and generates high-amplitude periodic pressure pulsations [18]. The size of the cavity is a function of the pressure level in the draft tube and the level of swirl exiting the runner. These highly unstable operating conditions can be considered to be intrinsic characteristics of reaction turbines with fixed blades that cannot be avoided if the turbine is operated at an extended envelope. As the various forms of the DTVC develop away from metal surfaces, there is little

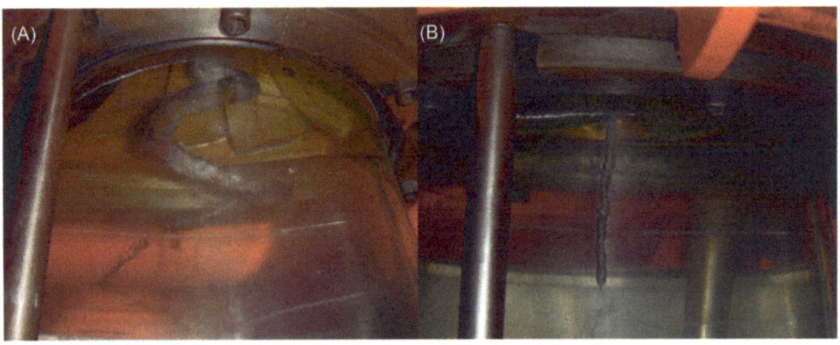

Figure 6.7 Draft tube vortex cavitation in a Francis turbine. (A) Part load vortex with characteristic helical form (B) Vortex core with conical form at overload conditions near the design head.

scope for cavitation erosion. Fig. 6.7 shows typical examples of the part load vortex rope in the draft tube of a high specific speed Francis turbine with a precessing frequency approximately equal to 0.3 times the runner rotation frequency and the overload vortex with the rotating frequency equal to the runner rotation frequency.

The IBVC forms between successive runner blades and is a result of secondary flow patterns developed in the channel between blades due to a spanwise variation of flow velocity angle. It appears as a rotating cavity that is attached on the hub of the runner between two blades and extend downward toward the shroud and the runner exit. Depending on the flow conditions, it may be related to a strong backflow in the suction cone extending in the blade channels [19] or due to large flow separation originating at the runner hub [20] and has been shown to magnify the pressure fluctuations in the runner, generating broadband noise [21]. In this case, cavitation erosion may appear on the runner hub where the vortex is attached (Fig. 6.8B). If

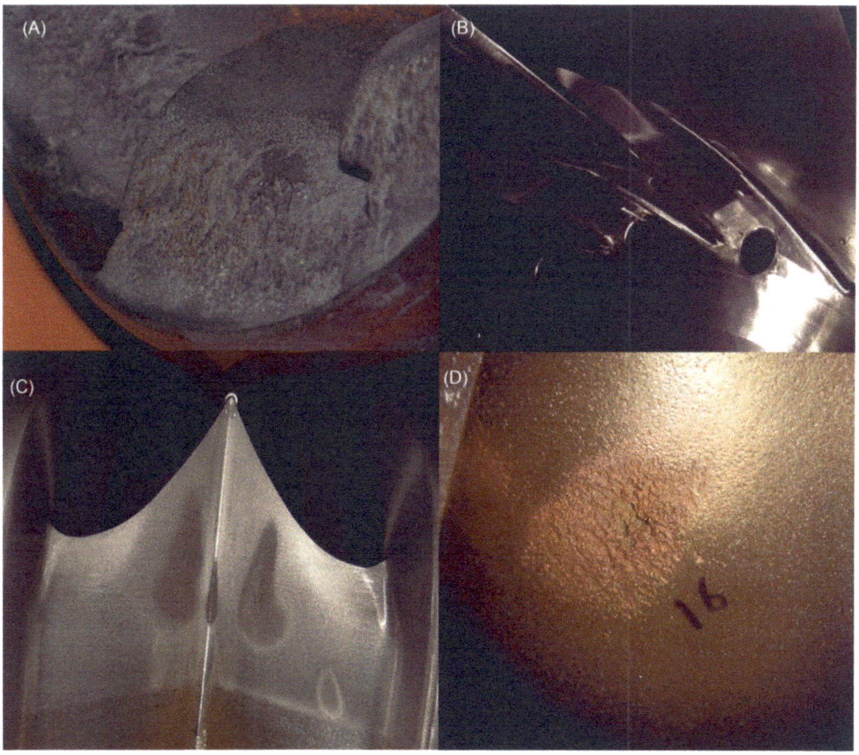

Figure 6.8 Cavitation erosion in hydraulic turbines. (A) Traveling bubble cavitation erosion on the suction side of a Francis runner blade. (B) Interblade vortex cavitation erosion on the hub of a Francis runner. (C) Sheet cavitation erosion on the back side and splitter on the bucket of a Pelton runner. (D) Sheet cavitation erosion on the back side of the bucket of a Turgo turbine.

the cavity is large enough or crosses the blades as it travels to the outlet of the runner, erosion on the suction or pressure side of the blades may also appear.

TVC appears in shroudless reaction turbines and is created in the gap between runner blades and the casing walls by a fast-moving jet that forms as the flow leaks through the small clearance [3,10]. This high-velocity jet forms a shear layer downstream of the runner with high vorticity originating at the blade leading edge. The tip leakage vortex forms as the shed vorticity rolls up, and cavitation may develop at the core of the vortex where the minimum pressure is observed. Cavitation damage may appear on suction side of the blade near the blade tip and the discharge ring [22].

Similarly to TVC, TLC appears in the gap between runner blades and the casing walls as a result of the accelerated flow in the narrow area of the tip. The pressure differential between the suction and pressure sides of the blade in shroudless reaction turbines promotes a recirculation in which a portion of the flow from the pressure side leaks through to the suction side at high velocity. It can cause significant cavitation damage in the runner blades and the discharge ring of reaction turbines but does not affect the hydraulic efficiency. Tip leakage cavitation may appear even under design conditions without a substantial reduction in the overall pressure level of the turbine.

On the trailing edge of stationary or rotating blades, periodic vortex shedding may occur, which may lead to severe periodic pressure pulsations and vibrations that may fatigue the runner blades in the vicinity of the trailing edge [22]. The frequency of vortex shedding is a function of flow velocity, and there is a strong interaction between cavitation and vortex shedding frequency [23]. Furthermore, an increase in the vortex induced vibration amplitude is observed, owing to the increase in vorticity at cavitating conditions.

Finally, in impulse machines, sheet cavitation can develop on the back of the buckets on the splitter (Fig. 6.5A) as well as on the back and front of the cutout, owing to a difference in the relative flow angle and the blade angle in the vicinity of the cutout during the jet cut process [24]. If there is substantial deviation in the flow angle from the bucket inlet and outlet angles, cavitation may appear toward the deepest portion of the bucket close to the splitter or toward the outlet edge (Fig. 6.5B). As the bucket begins to cut into the jet and the jet attaches on the back of the bucket, a cavity develops in that region, which begins to break up and is replaced by air as the jet interacts with the bucket and water begins to flow out [25,26]. Cavitation may also appear on the spear tip and nozzle of the injector as a result of the high velocity and locally low pressure that may develop owing to geometrical characteristics. In Turgo turbines, sheet cavitation may form on the back side of the bucket, where the flow may be accelerated owing to the increased curvature and flow blockage. Depending on the erosive potential in the cavitated regions as well as geometrical features, significant damage in these positions will commence. Especially in Pelton turbines in which sharp edges in the splitter and cutout may be quickly eroded (Fig. 6.8C), leading to substantial reductions in hydraulic efficiency, the jet cut process will be impaired. In Turgo turbines, although cavitation may also lead to significant erosion on the back of the blades (Fig. 6.8D), efficiency degradation will be observed only after a significant portion of the bucket has been eroded.

6.3 Experimental Methods and Investigations

Cavitation tests are typically carried out in the laboratory to determine the conditions of cavitation inception and development and to evaluate the effects on turbine performance. Through these tests, the required setting height and erosion risks are established and correlated to an allowable operating envelope. In this section the experimental methods that aim to study and detect cavitation hydraulic turbines are presented. Subsequently, the application of these methods and tools for the analysis of cavitation phenomena and the characterization of the cavitation performance of hydraulic turbines are discussed.

6.3.1 Laboratory Cavitation Tests

Determination of the cavitation characteristics of a given turbine is carried out in laboratory tests in a highly controllable environment, according to established international guidelines, such as IEC60193 [27]. The prototype turbine is scaled down to proper model dimensions, taking care to maintain hydraulic similarity. Measurements are carried out at different operating conditions by modifying the operating parameters of the test rig to determine the performance characteristics of the turbine under test conditions. By maintaining hydraulic similarity between the model and prototype, it is possible to determine dimensionless performance parameters, which allow the description of the performance of the turbine independent of size. Typical dimensionless parameters that are used are the speed factor n_{ED}, the discharge factor Q_{ED}, and the power factor P_{ED}. These parameters [Eqs. (6.3–6.6)] relate the rotational speed n, head H, and flow rate Q to the power output P and the hydraulic efficiency of the machine η_h [Eq. (6.7)]:

$$n_{\mathrm{ED}} = \frac{nD}{E^{0.5}} \tag{6.3}$$

$$Q_{\mathrm{ED}} = \frac{Q}{D^2 E^{0.5}} \tag{6.4}$$

$$P_{ED} = \frac{P}{\rho D^2 E^{1.5}} \tag{6.5}$$

$$E = \frac{p_{abs1} - p_{abs2}}{\rho} + \frac{v_1^2 - v_2^2}{2} + g(z_1 - z_2) = gH \tag{6.6}$$

$$\eta_h = \frac{T\omega}{\rho E Q} \tag{6.7}$$

The specific hydraulic energy E [Eq. (6.6)], which the turbine is called to convert to mechanical energy, expresses the available energy per unit mass of available water between the inlet 1 and outlet 2 sections of the turbine (Fig. 6.1). In Eqs. (6.3–6.7), H is the turbine net head, ω is the angular velocity, n is the rotational velocity, Q is the flow rate, P is the power output, T is the torque developed on the turbine shaft, ρ is the fluid density, D is the turbine reference diameter, p_{abs} is the absolute pressure at a given location, v is the velocity of water at a given location, and g is the gravitational acceleration.

$$\text{NPSE} = \frac{p_{abs2} - p_v}{\rho} + \frac{v_2^2}{2} - g(z_r - z_2) \tag{6.8}$$

$$\sigma = \frac{\text{NPSE}}{E} \tag{6.9}$$

Hydraulic similarity can be achieved by ensuring that the model and prototype machines are geometrically similar and the ratio of the flow velocity components is equal throughout the flow path of the machine. However, full similarity between model and prototype requires that dynamic similarity is also fulfilled, which may not be achievable, owing to difficulty in maintaining the same Reynolds number between the two scales. For this reason, friction losses in the model scale become more significant; therefore specific corrections need to be applied to obtain the prototype scale hydraulic efficiency [27].

Typically, turbine performance measurements are carried out in a closed loop arrangement (Fig. 6.9) consisting of a high-pressure tank, a low-pressure tank downstream of the model turbine, a vacuum pump, and a centrifugal pump. The centrifugal pump draws water for the low pressure tank and discharges it at the high pressure vessel. By adjusting the speed of the pump, the flow rate and delivery pressure can be controlled, as required by the test specifications. The pressure vessel can act as a buffer, dampening pressure fluctuations originating from the pump operation, ensuring a constant pressure at the inlet of the model turbine. During performance tests, the flow rate, torque, pressure differential, and speed are measured to determine the hydraulic efficiency of the turbine across different noncavitating conditions.

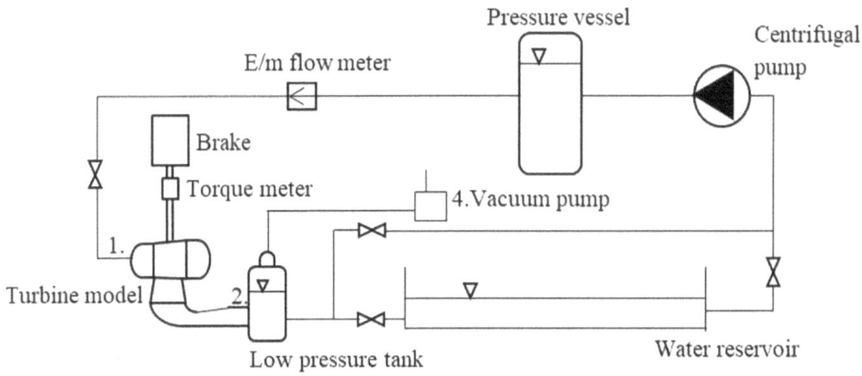

Figure 6.9 Typical closed loop arrangement for turbine performance and cavitation tests.

To determine the cavitation limits of the machine under test conditions, the dimensionless number σ is introduced (Thoma number), which expresses the difference of the outlet hydraulic energy and the pressure energy of water vapor as a fraction of the available specific hydraulic energy [Eqs. (6.8–6.9)]. This number is used to relate the pressure at the outlet of the turbine to the minimum pressure observed within the turbine flow path. The purpose of cavitation tests is to determine the minimum outlet pressure across different operating conditions in which the minimum pressure within the turbine becomes lower than vapor pressure, leading to cavitation and potential erosion on the turbine components.

During cavitation tests, the turbine is operated at a given flow and output setting, while the pressure in the downstream tank is gradually reduced by means of the vacuum pump until an efficiency drop of typically 1% is observed (Fig. 6.10A), allowing the identification of the critical Thoma number σ_1. Cavitation tests are repeated for different guide vane openings, flow rates, and test heads to establish the turbine's cavitation limits across different operating conditions (Fig. 6.10B). By incorporating suitably located transparent windows, the inception point σ_i as well as the development of cavitation can be determined.

The closed loop circuit may also incorporate a heat exchanger to maintain the fluid temperature constant, a deaerator to control the air content, and a resorber tank to reabsorb the released nuclei in the course of the cavitation tests. Alternatively, an active injection system for nuclei may be used [28,29]. This is particularly important as the inception of cavitation may be significantly influenced by the concentration of air or gas bubbles in water, leading to different cavitation characteristics.

Through these tests on the model scale, it is possible to determine the performance variation of the turbine at different operating conditions and to identify the

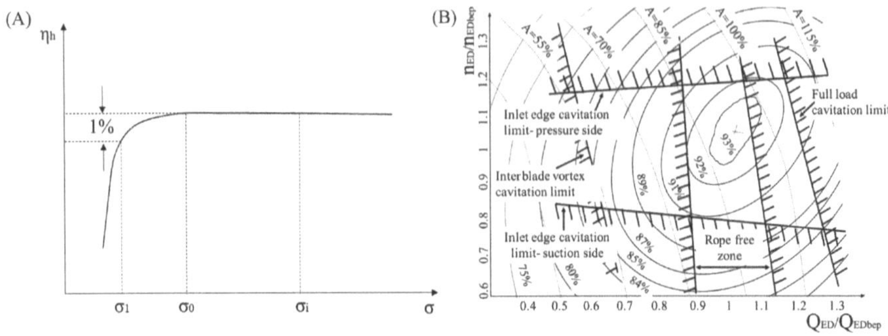

Figure 6.10 (A) Typical variation of turbine hydraulic efficiency η_h during cavitation tests where σ_i is the incipient Thoma number, σ_0 is the Thoma number zero, and σ_1 is the critical Thoma's number where efficiency of the turbine has dropped by 1%. (B) Typical Francis turbine performance chart with highlighted cavitation limits, where A is the guide vane opening given is a percentage of the rated value and the speed and discharge factors are normalized with the best efficiency point (bep) values.

conditions in which different types of cavitation may develop. By determining the nondimensional parameters, it is possible to transpose the model performance to the prototype scale, provided that the various similarity conditions are fulfilled. For cavitation tests, particularly in cases in which the dimensions of the runner are of similar magnitude to the net head or large cavitated regions are to be expected and the effect of gravity forces may become proportionally greater, the extent of the developed cavities may differ between the model and prototype scales. On such occasions it is preferable to also maintain Froude similitude, expressed as the ratio of inertia forces to gravity forces. However, this may not always be possible, as it could result in excessively large model dimensions or excessively low test heads, violating standard specifications [27]. In this case, special considerations should be made to achieve Thoma number equality for the proper transposition of the cavitation performance from model to prototype scale [6,27].

6.3.2 Experimental Techniques and Flow Visualization

Cavitation is a highly dynamic phenomenon in which the process of cavity collapse causes vibration and noise that is transmitted from the fluid to the mechanical components of the turbine. Therefore for the detection of cavitation, accelerometers and acoustic emission sensors installed in suitable locations on the turbine are typically used [8]. Furthermore, to study the cavitation phenomenon in hydraulic turbines, dynamic pressure measurements and advanced flow visualizations techniques are employed.

To determine the dynamic stresses on runner blades at different operating conditions and establish the fatigue characteristics, techniques to install miniature pressure sensors and strain gauges on turbine runner blades have been developed [30–32]. These miniature sensors are flush mounted and glued onto the instrumented blade, while the associated cabling is routed through channels machined on the blade and runner hub. The channels are then typically filled with water-resistant polymer compound, and the blades are dressed so that the original hydraulic profile is maintained. The required instrumentation can be installed on the hub in a perfectly sealed compartment, and the signal can be transmitted by using either slip rings or a telemetry system [32,33]. The induced pressure and strain fluctuations strongly depend on the operating conditions. Therefore the measured fluctuations and root mean square value can be used to infer the appearance of cavities.

One of the key challenges of cavitation detection is the identification of the erosive potential of the developed cavities [32]. If the cavities implode far from the turbine's components or the energy of the imploding bubbles is below a certain threshold, depending on the materials used for the turbine components, then erosion can be avoided. Vibration measurements on the turbine bearings or shaft can be employed to determine the erosive potential of cavitation. The advantage of measuring vibration on the rotating frame lies in the ability to isolate other excitation sources that may introduce noise and interfere with the dynamic characteristics of the bearings [33].

In cavitation tests, transparent windows are typically employed to visually confirm the inception of cavitation, determine the extent and position of cavities, and monitor the development of the phenomenon [34,35]. Furthermore, cavitation visualizations within the turbine flow path may be used to qualitatively validate results from numerical analyses and develop numerical tools for the accurate prediction of cavitation [19,36–38]. Usually, the suction cone of the draft tube, which immediately follows the runner, is made of transparent Plexiglas, allowing the visual observation of the flow field. By obtaining high-speed images at different operating conditions and different Thoma numbers, valuable insights into the physical mechanism involved with different cavitation types can be gained [39,40]. This is usually achieved by using monochrome charge-coupled device (CCD) cameras targeting a single blade at a given rotation angle [34,36,39,41,42]. To obtain good-quality images, the acquisition system needs to be coupled to an illumination system usually employing a stroboscope light synchronized with the runner rotation. Other techniques, particularly involving the study of vortex-induced cavitation in the runner and draft tube, employ dynamic pressure measurements as the triggering signal [41]. In this way it is possible to target specific runner blades or specific instances within the cycle of periodic oscillations. The resulting observations in conjunction with vibroacoustic measurements can connect the variation of vibration amplitude to the cavitation number and the physical extent of the cavities. Moreover, image-processing techniques in which the variation of the mean and standard deviation of gray level from a set of images is monitored can allow the determination of the vapor volume fraction as well as the dynamic characteristics of cavitation [19,34,42]. Other visualization techniques that are used to determine the extent of the cavities employ the use of fluorescent polymer particles released in the test circuit that scatter light at specific wavelengths when illuminated by a laser source [18,41].

6.3.3 Velocity Measurements

The measurement of the velocity filed at specific locations of interest has been employed to help elucidate the physical mechanism of different types of cavitation in hydraulic turbines. Measurements are carried out spanning a full period of the observed pressure oscillation to identify the dynamic characteristics of the velocity field and increase understanding of its relationship with the inception and development of cavitation.

A typical location for such measurement is the draft tube of hydraulic turbines utilizing the transparent suction cone downstream of the runner, which also allows for visual access to the runner. For such measurements, two different techniques can be used: laser Doppler velocimetry (LDV) and particle image velocimetry (PIV).

6.3.3.1 Laser Doppler Velocimetry in Hydraulic Turbines

LDV is a nonintrusive velocity measurement method through which the local instantaneous fluid velocities can be measured at a high temporal resolution [43]. It can therefore be used to quantify velocity fluctuations and turbulence characteristics of a flow field [44]. A typical LDV measurement system (Fig. 6.11) consists of a laser probe, an optical signal transmition system, the detection system, and a synchronization system so that the measurements are carried out at specific runner positions [45]. The measurement is based on the principle of the observed frequency shift of scattered light as a result of flow velocity. A monochromatic laser is split into two beams, both pointed at the measurement location. The region where the two beams overlap forms the measurement volume, where a characteristic interference pattern is developed (fringe). Seeding particles are introduced into the system, and as they pass through the measurement volume, they scatter the laser light at a different frequency, proportionally to the flow velocity [46]. The characteristics of the interference pattern depend on the laser light wavelength and the angle between the beams, while the frequency shift is related to the velocity and detector position. A photodetector is used to convert the optical signal to an analog signal, which is then analyzed in the frequency domain to determine the frequency of the backscattered light. The velocity components

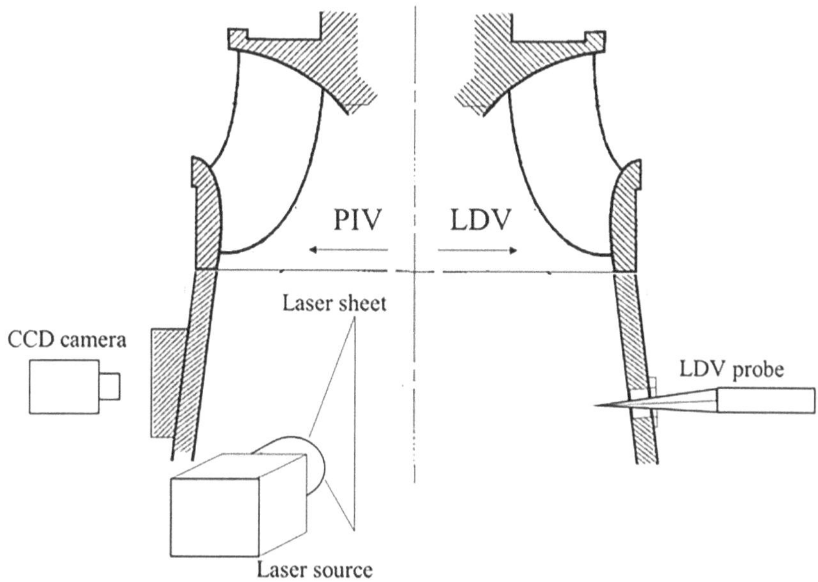

Figure 6.11 Schematic representation of PIV and LDV measurement techniques applied in the suction cone of a Francis turbine. *PIV*, Particle image velocimetry; *LDV*, laser Doppler velocimetry.

normal to the laser beam can then be determined as the product of the frequency and the fringe pattern spacing. Different velocity components can be measured by modifying the orientation of the laser beam [47]

Furthermore, to obtain the velocity field at various locations, a traversing system is incorporated that allows accurate movement in the radial and vertical directions. Typically, measurements can be obtained with a spatial resolution of 0.1 mm.

The LDV measurement technique is typically applied in hydraulic turbines for the study of the different forms of vortex cavitation in the draft tube of reaction turbines. Ciocan and Avellan [45] used LDV to determine the velocity distribution downstream of the runner of model Francis turbines at various operating conditions and relate the distributions to the observed flow field. Recently, Müller et al. [39] carried out LDV measurements in the suction cone of a Francis turbine to study the full-load pressure surge coupled to dynamic pressure measurements and flow visualizations. Measurements were obtained at two different cross sections to evaluate the swirl intensity and relate it to the observed cavitation patterns. By combining flow visualization, dynamic pressure measurements, and velocity measurements, the interaction between the developed runner blade cavitation and the variation of the swirl number was identified as a mechanism influencing the dynamic characteristics of the full-load surge. Favrel et al. [11] investigated the effects of hydroacoustic resonance on the velocity field in the draft tube of a Francis turbine model. The effect of cavitation on the precessing vortex characteristics and of resonance in the velocity field was studied. To capture velocity fluctuations with increased accuracy, the LDV technique is selected, enabling velocity measurements with increased temporal discretization. The variation in pressure fluctuation amplitude and precessing frequency was observed with different Thoma's numbers until resonance conditions were achieved. This was carried out by modification of the pressure level in the draft tube, influencing the vapor volume and conversely modifying the natural frequency of the system. If the precessing frequency matches the natural frequency of the system, resonance is achieved, whereby the amplitude of the synchronous pressure fluctuation is maximized. A reduction of Thoma number leads to a reduction in fluctuation amplitudes and precession frequency, owing to the decreasing vortex trajectory with increasing vapor volume. On the other hand, it was shown that resonance conditions have no effect on the velocity fluctuations, and this does not lead to substantial discharge fluctuations. Similarly, Skirpin et al. [48] carried out LDV measurements on a model simulating the vortical flow in the draft tube of a reaction turbine. By modifying the system parameters, it was possible to modify the swirl intensity and observe different patterns of the vortex core. Interestingly, it was observed that at a relatively low swirl number, an increase in the precession frequency was observed, while at the same time the spatial wavelength was reduced.

6.3.3.2 Particle Image Velocimetry in Hydraulic Turbines

An alternative technique to measure the instantaneous velocity field in hydraulic turbines is the PIV method. Similarly to LDV, it is a nonintrusive technique that incorporates a laser light source and spherical and cylindrical lenses to produce a two-dimensional sheet while also employing a digital camera for image acquisition (Fig. 6.11). In contrast to LDV, in the PIV technique a laser light sheet is used to illuminate a planar region in the flow field where small and naturally buoyant particles have been released and carried by the flow. The displacement of the illuminated particles is monitored by obtaining successive images over a given time period where a second pulse of the laser sheet is repeated. By comparing the images between two successive laser sheet pulses, the flow velocity components can be determined [49]. Of great importance for the elimination of measurement errors is the proper calibration of the system and the appropriate selection of the seeding particles. The main advantage of PIV is that large portions of the velocity field can be mapped simultaneously, since the developed laser sheet illuminates a large number of particles in the measurement plane. By contrast, with LDV, which is a point-by-point technique, separate measurements are required for each position within the measurement plane.

For this reason the PIV technique has been more frequently applied in investigations covering the transient characteristics and cavitation in hydraulic turbines.

In particular, Illiescu et al. [50] used the PIV technique to investigate the flow field at different operating conditions and to elucidate the flow physics behind the observed fluctuations in the presence of the precessing vortex rope. To carry out measurements in the two-phase flow observed in the draft tube of a Francis turbine, two different digital cameras were employed. One camera measured the in-plane velocity field, and a second camera was used to determine the volume of the vapor region through image-precessing techniques. Optical distortions imposed by the suction cone material and fluid were accounted for by proper system calibration. The system triggering was carried out by monitoring the vortex rope precession through wall pressure measurements, and the phase-averaging technique was applied to determine periodic velocity components. Illiescu et al. [51] carried out stereoscopic PIV measurements to determine the three-dimensional velocity field, coupled to image-processing techniques to determine the shape of the vortex rope. The velocity components in all three dimensions were obtained by using two different cameras in stereoscopic configuration, both focusing on the laser sheet at different orientations. The diameter and shape of the rope were determined by background illumination, using a light-emitting diode (LED) panel in conjunction with tracer fluorescent particles, taking advantage of the shadow effect of the vapor region blocking the optical access of the camera. Similar technique, were applied by Tridon et al. and Iliescu et al. [41,52] to study the velocity field and induced instabilities at part load conditions in a Francis turbine and a Kaplan turbine, respectively. In both studies, an asymmetry in the radial velocity components and the overall flow characteristics was identified, which was attributed to the interaction of the flow with the downstream draft tube bend and the developed secondary flow.

Favrel et al. [36] carried out PIV measurements at various horizontal sections in conjunction with dynamic pressure measurements in noncavitating conditions to determine the effect of the flow rate on the precession frequency and the pressure fluctuation amplitude of the synchronous component. The evolution of the velocity field over one period of the periodic cycle was recorded, and the vortex core boundaries, the trajectory and position were identified. The analysis that was carried out distinguished three different flow regimes, depending on the flow conditions, where the pulsation frequency may remain constant or vary linearly with decreasing flow rate, with a simultaneous increase in the precession radius. This increase was related to an increase in pulsation amplitude and was attributed to a flow interaction between the precessing vortex and secondary flow and separation developed downstream of the bend. Finally, as was discussed by Nishi and Liu [12], the synchronous component was attributed to the interaction of the flow with the draft tube bend. Subsequently, the same authors, [53] used PIV measurements in the draft tube of a Francis turbine to extract the vortical characteristics at part-load conditions. By measuring the velocity field in three different planes at different operating conditions, the variation in the vortex dynamics were determined. Specifically, with increasing flow rate, the tangential velocity distribution was found to be quasiaxisymmetric, and a circular cross section in the vortex core was observed. As the flow rate was reduced, the shape became elliptical, while the tangential velocity component lost its axial symmetry. A relationship between the operating condition and the vorticity and the precession radius was also established. Goyal et al. [54] used the PIV technique in the draft tube of a high head Francs turbine to analyze the physical mechanism for the developed vortex rope. By decomposition of the rotating vortex rope into a synchronous and an asynchronous component [12], a phase shift between the two components was observed, with the synchronous component preceding the asynchronous by approximately 0.8 s. A shear layer developed between the recirculation region in the draft tube centerline and the swirling flow exiting the runner, which rolled up, forming the vortex rope, inducing large axial velocity gradients. Finally, the authors showed that the radial velocity component influences the asynchronous pressure fluctuation, while the axial velocity component is responsible for variations in the synchronous pressure fluctuations.

6.3.4 Detection of Cavitation in Hydraulic Turbines

Cavitation detection techniques are common in hydraulic machinery, owing to similarities in the various cavitation types and the associated impact on various machine components. Specifically, for centrifugal pump applications, cavitation detection has been the focus of numerous studies in the past. The main cavitation techniques that have been developed and applied in centrifugal pumps are presented in Chapter 5. Although the implications associated with the effects of cavitation may be larger in hydraulic turbines, application of the available detection techniques and the development of monitoring systems to ensure the reliable operation of hydropower plants is a relatively recent practice [8]. Owing to the physical processes involved in the development of vapor cavities in the hydraulic components of the turbine, cavitation detection mainly involves the use of vibration sensors and acoustic emission sensors. Coupled to a suitable analysis technique, the aims of cavitation detection are to extract the dynamic characteristics of cavitation and to determine the most critical factors influencing its inception and development. A prerequisite for the successful detection of cavitation is isolation of the signal components from other sources of noise that may be present in the system. This also requires identification of the physical mechanisms involved in different cavitation types and the determination of the individual contributions of various components in the development of the phenomenon [55]. In practical applications this is difficult to achieve, as it may be challenging to evaluate the effects of the transmission path in prototype machines, while it is usually not possible to correlate the measurements with a certain cavitation type, since no optical access is available. Furthermore, the excitation characteristics of the observed cavitation is not a priori known, and the frequency bands of focus for each sensor are difficult to determine [56].

Apart from this, of critical importance is verification of the erosive potential of different types of cavitation and the connection to different observations and measured signals. As was discussed in Chapter 5, the available analysis techniques involve amplitude demodulation of the measured signals focused on a given frequency band, which is thought to carry the excitation signal in a more representative manner. It has been found that by applying this method, cavitation may be detected if the observed pressure fluctuations are modulated by characteristic frequencies of the system, such as the blade-passing frequency (BPF) [57]. Typically, cavitation monitoring in hydraulic turbines involves the use of vibration and acoustic emissions measurements by specific sensors installed at different location on the stationary or rotating components, taking into consideration factors influencing signal transmissibility from the excitation source to the location of the sensor.

The earliest attempt to identify the relationship between vibration signals and erosive potential was carried out by Bourdon et al. [58]. Their focus was to be able to distinguish nonerosive cavitation from turbulence which may appear simultaneously in the selected frequency band. They found that the erosive power was connected to the flow velocity and cavity length and observed the largest damage in the closure region of the cavity.

Bajic and Keller [59], applied their technique on vibroacoustic signals and reported occasions where erosive cavitation was presented even at the absence of signal modulation by the BPF. Their technique involved the measurement of noise emitted from each guide vane in a reaction turbine and at the same time removal of the background environmental noise from the power spectrum. The authors were able to successfully detect cavitation and were able to use their technique to distinguish between different types of cavitation.

Escaler et al. [60] were able to relate vibration measurements to erosion intensity, highlighting the importance of sensor location with a consideration to signal transmissivity. In particular, they were able to determine a correlation between the eroded area and the amplitude of impact forces and proposed measurement of vibration as a means to evaluate cavitation erosion. In a later study, Escaler et al. [61] highlighted the importance of sensor placement by comparing the signals from different sensors installed in various locations on the turbine under different operating conditions. For this purpose, specific frequency bands were singled out that seemed to be excited by cavitation and were used to extract and compare the resulting modulating frequencies. This was again stressed by Escaler et al. [62], who, applying the amplitude demodulation technique, identified a frequency band suitable for cavitation detection in their setup and the turbine model that they considered. The generalized applicability of the proposed frequency band to different turbines should be evaluated.

To relate the observed cavitation patterns to the different vibroacoustic characteristics of the measured signals, Rus et al. [34] carried out vibration and acoustic emissions measurements on a Kaplan turbine at different Thoma numbers until full cavitation breakdown conditions were reached. A single blade was targeted and visualized by using a digital camera so that the inception and development of cavitation could be progressively observed. The authors saw that at the critical cavitation number, the vibration amplitude was maximized. Furthermore, they observed that in a Kaplan runner, cavitation was initiated near the hub and the tip clearance between the runner and discharge ring, while cloud cavitation with large erosive potential developed at lower Thoma numbers. However, it was difficult to infer the aggressiveness of cavitation from the measured signals, as a decrease in the amplitude was observed with reducing Thoma number. This effect is attributed to the increasing vapor volume in the region between the sensors and the cavities, which leads to an attenuation of the impulse generated by the collapsing cloud. In a later study, Faria et al. [63] were able to identify a positive relationship between the measured acoustic emissions and the level of erosion on the runner blades of a Francis turbine. Similar to Francis turbines, the importance of sensor placement is highlighted in the study of Schmidt et al. [64], who also noticed the dampening effect of the increasing vapor volume as the flow rate was increased.

Previously, the frequency demodulation technique was applied for the detection of traveling bubble and cloud cavitation in Francis turbines. Escaler et al. [65] applied the same technique to detect the full-load draft tube surge and confirmed that the highest risk for cavitation erosion is found at full-load conditions in which cavitation is modulated by the BPF. Valentin et al. [56] aimed to address the known challenges of cavitation detection in the field by incorporating a simplified test rig, using a rotating disk in a closed container filled with water. By controlling the excitation source and simulating a cavitation event, it was possible to determine the signal transmission characteristics of a given position. It was shown that on-board sensors were the optimum choice for high-frequency excitation measurements on the rotor. Furthermore, it was shown that microphones and acoustic emissions sensors are not the best option in measuring vibration in the stationary frame. On the other hand, Hamstuchi et al. [66] showed that a similar response could be observed from vibration and a strain sensor installed on the stationary and rotating frames.

Presas et al. [67] concluded that it has still not been determined whether the appearance of the BPF in the demodulated signal is an indication of highly erosive cavitation. Considering signal transmissivity and sensor placement, they concluded that for the detection of erosive cavitation an acoustic emission sensor on the rotating shaft should be used.

6.4 Numerical Investigations

In recent years the development of new computational fluid dynamics (CFD) tools and the availability of computational resources have facilitated reliable simulations focusing on the analysis of complex flow patterns in hydraulic machinery. Several studies have used CFD to simulate the three-dimensional turbulent two-phase flow conditions that may be observed in hydraulic turbines and pumps. However, this is a task of increased complexity that requires several modeling assumptions and simplifications to be incorporated. Therefore it is essential that the balance between numerical accuracy and computational cost is identified. In addition, to increase confidence and permit their wide use in engineering applications in the industry, validation of these tools against experimental data is required. Many studies have focused on the development and refinement of such tools and have been successful in reproducing the flow field in hydraulic machines under cavitating conditions.

The various numerical tools that have been developed and applied to study cavitation in centrifugal pumps are discussed in Chapter 5. For the study of cavitation in hydraulic turbines the same models and methodology may be employed. In this section the application of the various cavitation models in the study of cavitation in hydraulic turbines, based on the Rayleigh-Plesset equation, are reviewed. The vast majority of numerical investigations concern the use of a transport equation–based model (TEM) to account for the mass transfer rate during evaporation and condensation. Although the effects of cavitation are as important in hydraulic turbines as in centrifugal pumps, few numerical investigations for the prediction of cavitation inception and development in turbines can be found. The associated numerical difficulty and the requirement to simulate a great number of operating conditions to fully map cavitation performance mean that such investigations are still not practical from an industrial viewpoint. On the other hand, owing to the increasing importance of off-design operation of hydraulic turbines, numerous investigations have been carried out to model and analyze the flow behavior associated with secondary flows and instabilities developed in such conditions. For this reason, particular emphasis is given to modeling applications at off-design operation.

6.4.1 Numerical Investigations of Cavitation in Hydraulic Turbines

The potentially severe implications of cavitation in hydraulic turbines require careful consideration in the early stages of the turbine design process. To avoid the development of excessively low pressure regions within the flow path, CFD may be used to guide the design process and avoid the development of erosive conditions. To this end, there have been several studies in which an acceptable agreement between numerical results and experimental data could be achieved. In particular, in reaction turbines, Muntean et al. [68] applied a homogenous multiphase model based on the Rayleigh-Plesset equation to model the inlet edge cavitation in a Francis turbine. The developed model was able to predict the extent and pressure distribution on the runner blades with sufficient accuracy. It was highlighted that in conditions in which cavitation is expected, including a cavitation model allows for an accurate prediction of the pressure field on the runner blades. Liu et al. [69] carried out a two-phase flow examination in a Francis turbine using the Singhal et al. model that accounts for surface tension effects and noncondensable gases in the mass transfer equations. Although no experimental data were used for validation, satisfactory results were achieved from a qualitative viewpoint. Similarly, Zhou et al. [70] studied the development of a two-phase flow area in the runner of a Francis turbine and validated their result with experimental data. By reducing the Thoma number, a vapor cavity begun to develop on the nose cone of the runner, which grew toward the suction side of the blades with further reductions in the pressure level. As the cavity grew and approached the critical conditions, an increase in efficiency was observed, followed by an abrupt drop as the cavitation extended in a larger area of the blade. This local increase in efficiency is attributed to the interaction of the vapor with the blade, which locally reduces friction losses. Liu et al. [71] also studied the development of the trailing edge cavity in a Francis turbine by successfully applying a similar approach, while Yao et al. [72] carried out a numerical campaign to improve the cavitation performance across numerous operating conditions by redesigning the runner.

Fig. 6.12 shows simulation results of cavitation modeling in a Francis turbine at two different pressure levels, where the increase in the vapor volume can be observed with reducing Thoma number.

The homogenous multiphase approach has been also successfully applied to other reaction turbines. Necker and Aschenbrenner, Schffer et al., and Jost et al. [73–75], carried out two-phase flow analyses in Kaplan turbines and were able to accurately predict the critical Thoma number as well as the locations most prone to cavitation inception. With reducing Thoma number, the first bubbles were initially observed at the tip near the trailing edge, while another low-pressure region originating at the hub and extending to the outer radius was predicted. The importance of transient analyses to accurately predict performance and two-phase flow in low-head machines was highlighted by Jost et al. [75] and Morgut et al. [76], as steady-state simulations were seen to greatly overpredict the losses in the draft tube, leading to larger pressure levels in the runner. Escaler et al. [77] investigated the sensitivity of the predicted cavities to variations in the cavitation model parameters.

The numerical prediction of cavitation in impulse turbines has only very recently been tackled. In an effort to replicate the observed erosion patterns in a

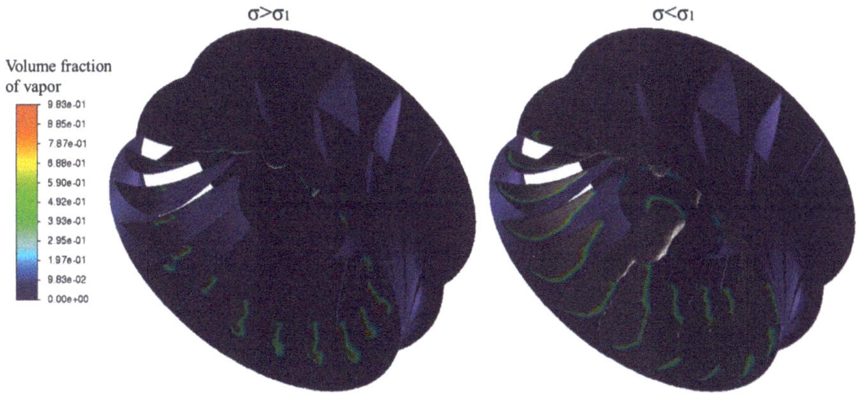

Figure 6.12 Example of predicted vapor volume fraction in a midspecific speed Francis turbine at two different pressure levels. Left: σ is higher than σ_1 where efficiency of the turbine has dropped by 1%. Right: σ is lower than the σ_1.

high-head Pelton turbine, Rosetti et al. [24] applied the homogenous TEM approach, assuming that the fluid was a mixture of water, vapor, and air, in conjunction with the Kunz et al. model. By monitoring the flow evolution and the evolution of vapor volume at the various phases of the bucket interaction with the jet, they observed the development of sheet cavitation. As the bucket began to interact with the jet, a mismatch in relative flow angle in the vicinity of the cutout and the blade angle was observed, which promoted the development of a stable sheet cavity. Furthermore, at the back of the buckets, cavitation was observed near the splitter, where high velocity was observed. The cavity disappeared as the water moved past the bucket and air was introduced, displacing the vapor. Similarly, Zeng et al. employed the same approach [25] and were able to accurately predict the loss in turbine efficiency due to the development of cavitation.

The latest application of cavitation modeling was presented by Jošt et al. [26], who demonstrated that the inclusion of cavitation permitted improved efficiency prediction, since previously regions on the buckets, where unphysically low pressure developed and gave incorrect results, were now treated correctly, resulting in a different pressure field. In their analysis, cavitation was predicted in similar locations.

6.4.2 Numerical Investigations at Off-Design Operation

Operation at off-deign conditions gives rise to several flow phenomena within hydraulic turbines that are characterized by strong unsteadiness, leading to cavitation erosion, excessive vibration, and fatigue of the mechanical components. From the various cavitation types described in Section 2 the most important instabilities that are encountered in hydraulic machines are the draft tube vortex (DTV) and the interblade vortex. To better understand the characteristics of such flows, apart from experimental observations, numerical analyses are typically employed.

In the case of the DTV (Fig. 6.13A), to deepen the understanding of the flow mechanisms that are responsible for its development, several recent studies found an acceptable agreement between numerical and experimental data. Depending on the modeling approach and numerical setup, the observed pressure fluctuation amplitudes and frequencies can generally be well reproduced.

To simplify the numerical setup and limit the computational time required to model the evolution of the DTV for sufficient time, the numerical model may be simplified by excluding upstream components. In this case, only the draft tube may be considered, but to properly reproduce the phenomenon, accurate determination of the inlet boundary conditions is required. This may be determined from experimental measurements or by numerical simulations in which the computational domain extends to all turbine components [78–80].

Alternatively, a simplified simulation can include a portion of the wicket gate and the runner with flow variables imposed at the interfaces of components by circumferentially averaging the flow field [81,82]. Even with such a simplified approach, it is possible to accurately reproduce the dynamics of the part-load vortex rope, but it fails to capture the pressure amplitude on the blades, as rotor stator interaction effects are neglected.

The selected turbulence model and mesh configuration have also been shown to significantly affect the predicted results. Jost and Lipej [83] and Xiao et al. [84]

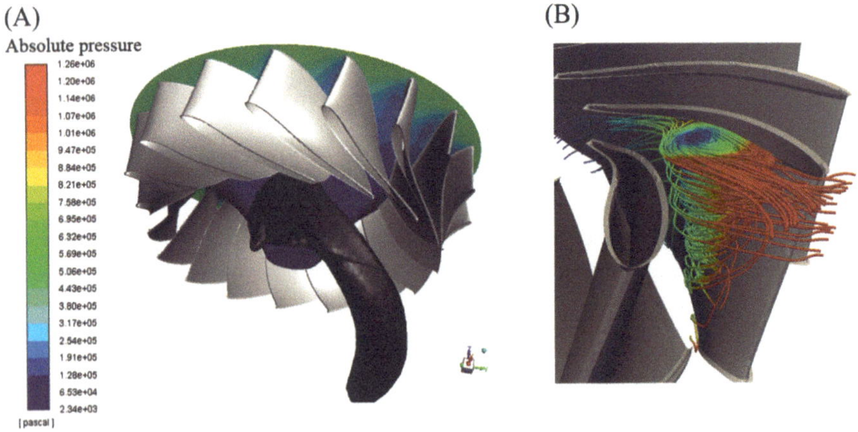

Figure 6.13 Example of predicted part-load vortex downstream of a Francis turbine. (A) The vapor volume is represented as an isosurface of vapor pressure, (B) Interblade vortex developed in a Francis runner due to inflow with high incidence at a high-head operation.

both reported good agreement between the predicted pulsation frequencies, but only when the mesh was sufficiently refined and a more accurate turbulence model was used did the prediction of the pulsation amplitude improve [83]. A coarse mesh in the draft tube has been associated with a faster dissipation of the vortical structures [85], leading to the prediction of smaller-amplitude pulsations. In a similar manner the selected turbulence model may significantly affect the prediction results, and more accurate turbulence models such as hybrid large eddy simulation (LES)/ Reynolds-averaged Navier-Stokes (RANS) have been used [86–89].

Similarly, to elucidate the mechanism for the development of interblade channel vortices (Fig. 6.13B), several studies examine the application of multiphase models in Francis turbines. To accurate calculate the low-pressure zone at the core of the vortex, sufficiently refined meshes are typically needed [90,91]. On the basis of the operating conditions, the location of the interblade vortex may shift toward the outlet of the runner or toward the shroud [92]. Furthermore, development of the vortex may be associated with different secondary flow patterns, due either to inlet separation resulting from flow incidence or to severe backflow conditions that extend from the draft tube and enter the blade channels [21,93,94]. As in the case of the DTV, the importance of the selected turbulence model was highlighted by the work of Wack et al. [91], in which the unsteady characteristics were poorly predicted by using a two-equation turbulence model, while improved results were achieved by using a hybrid LES/RANS model.

6.5 Mitigation of Hydraulic Instabilities

The need for flexible turbine operation requires a thorough analysis of the flow features and understanding of the physical process that govern the dynamic characteristics of the flow at different operating conditions. From the various cavitation types in hydraulic turbines the most frequently encountered problems arise from the DTV cavitation. The previous sections revealed that for this reason, identification of the various flow instabilities and the induced vibration and response of the hydraulic system have been the focus of numerous studies. The experimental and numerical techniques that have been applied have provided useful insights into the characteristics of such flows, and the knowledge that has been gained can be used to determine appropriate measures to deal with the undesired characteristics of off-design operation. By using this information, it is possible to identify mitigation techniques to actively and passively alleviate the detrimental effects of cavitation and of cavitating instabilities that appear during the operation of hydraulic turbines. The available measures typically incorporated involve the use of various inserts, baffle plates, and blades; air injection; water injection; and extensions of the suction cones.

Various fins can be attached on the draft tube wall parallel to the shaft centerline (Fig. 6.14A) or downstream of the nose cone [95] or an orifice plate employed at the end of the suction cone [96]. The function of these inserts is to reduce the swirl downstream of the runner by modifying the flux of the moment of momentum [97]. The inclusion of such inserts will naturally affect the hydraulic efficiency of the turbine, owing to a substantial increase in friction losses, particularly in high specific speed machines [98]. The main drawback of using inserts is the difficulty of broadly determining the positive influence of the chosen solution in a wide range of conditions and turbine designs. This can be determined only through model testing or numerical simulations of the complete turbine. The design and arrangement of fins on the draft tube wall have an impact on the rotating frequency of the DTV and can dampen pressure fluctuations [99–101]. However, they influence only the asynchronous components and may even enhance the instability if they are not properly chosen. Other inserts include stationary cylindrical components extending across the length of the suction cone or smaller ones supported by fins. Some success in dampening the instabilities has been achieved by proper extension of the runner cone toward the draft tube [102]. Inserts are typically successful in damping pressure pulsations but are associated with large efficiency losses and may be susceptible to cavitation erosion.

Air admission is another technique that is typically employed to combat the pressure pulsations in reaction turbines (Fig. 6.14B). Different air admission strategies have been employed, whereby air is introduced in the draft tube center through the shaft or, where this is not technically possible, through the side wall

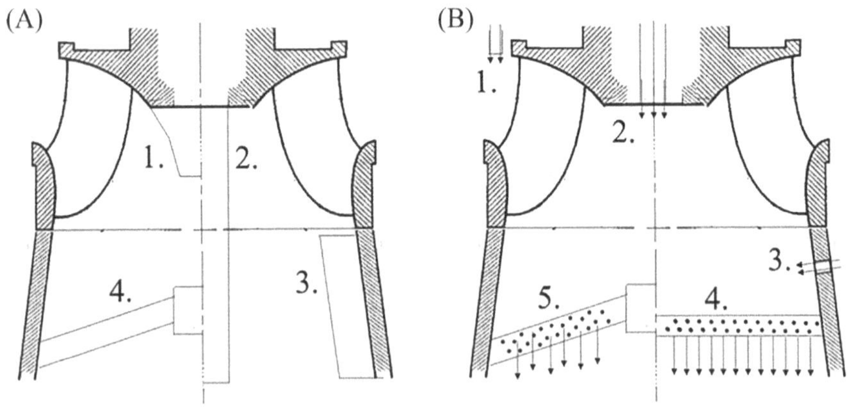

Figure 6.14 Draft tube vortex cavitation mitigation techniques. (A) The use of various inserts (1) using hub/nose cone extensions, (2) using a cylinder extension along the suction cone, (3) using fins on the suction cone wall, (4) using various vaned tripods. (B) The use of air/water injection (1) upstream of the runner, (2) through the shaft, (3) from the suction cone wall, (4) through a transversing pipe, and (5) through a pipe tripod.

of the suction cone [103]. Air admission can be carried out passively by taking advantage of the low-pressure conditions that are present in the draft tube or actively with an external compressor. Water jet pumps have also been used, allowing the introduction of a mixture of air and water [104]. Other air admission locations are in the wake of guide vanes or through transversal pipes and tripods installed in the suction cone [105]. The introduced air stabilizes the cavity beneath the runner by replacing water vapor while at the same time increasing the compressibility in the flow and damping the shock waves and noise generated by the implosion of the cavities. Air injection has been also incorporated to alleviate the effects of tip leakage cavitation, accomplishing up to a 50% reduction in vibration level [106]. Air injection is a promising solution for dampening pressure pulsations, however, as the injected air volume flow rate is increased, the turbine may suffer from higher efficiecny losses, while the system may become uneconomic due to the additional components and complexity required. Therefore, a careful consideration of the turbine's operating parameters, required operating envelope and operational characteristics is needed for an air injection system to be succesfully implemented.

An alternative to air injection is the injection of water through the nose cone. In a manner similar to air injection, the injected water aims to reduce the swirl downstream of the runner, disrupt the generated vortex, and dampen out the induced instabilities. In contrast to air injection, water injection aims to control the flux of axial momentum by utilizing a portion of the flow injected through the nose cone directly in the core of the vortex. This method is successful in significantly reducing the pulsation amplitudes that are associated with the asynchronous component of pressure fluctuations, but it fails to deal with and even magnifies the synchronous component [97,107].

6.6 Summary

In this chapter the various cavitation types that may appear in hydraulic turbines and their associated detrimental effects are presented from both experimental and numerical standpoints. Of particular importance is the study of the observed dynamic instabilities at off- design conditions as a result of the increasing need for hydraulic turbines to operate at an extended envelope. The available techniques that are used to study and detect the development of such phenomena are also presented. In hydraulic turbines, where a multitude of different operating conditions with different hydraulic characteristics may be present, the challenge lies in relating the acquired signals and observed system response to different cavitation types. Furthermore, the discussion revealed that the connection between the measured signals and the erosive potential of different cavitating types is still an open question.

This chapter also deals with the numerical application of various models to predict cavitation and identify the dynamic characteristics of various unsteady flow phenomena. Particular emphasis has been given to applications of CFD techniques to model off-design conditions, where separation and secondary flows create flow instabilities and may lead to excessive vibration and fatigue. Finally, the available techniques to mitigate the effects of such instabilities are reviewed.

Abbreviations

BPF blade-passing frequency
CCD charge-coupled device
CFD computational fluid dynamics
DTV draft tube vortex cavitation
LDV laser Doppler velocimetry
LEC leading edge cavitation
LES large eddy simulation
PIV particle image velocimetry
RANS Reynolds-averaged Navier-Stokes
TEC trailing edge cavitation
TEM transport equation-based models
TLC tip leakage vortex
TVC tip cortex cavitation

Nomenclature

σ Thoma's cavitation number (-)
σ_1 Thoma's cavitation number 1 (-)
σ_o Thoma's cavitation number 0 (-)
σ_i incipient Thoma's cavitation number (-)
g gravity acceleration (m/s^2)
Q flow rate (m^3/s)
ρ fluid density (kg/m^3)
n rotational velocity (rpm)
p_{sat} vapor pressure of fluid (Pa)
p pressure (Pa)
ω angular velocity (r/s)
H_B pressure at the surface of the suction tank (Pa)
H turbine head (m)
E specific hydraulic energy (J/kg)
T shaft torque (Nm)
c_t tangential velocity components (m/s)
c_a axial velocity components (m/s)
S swirl number (-)
P power output (W)

NPSE net positive specific energy
n_q specific speed (-)
n_{ED} speed factor
Q_{ED} discharge factor
P_{ED} power factor

References

[1] J. Obrovsky, J. Zouhar, M. Abraham, A. Skotak, Development of wide operating range runner for Francis turbine upgrading, in: IOP Conference Series: Earth and Environmental Science 240 (2019) 022014. Available from: https://doi.org/10.1088/1755−1315/240/2/022014.

[2] C. Trivedi, I. Iliev, O.G. Dahlhaug, Z. Markov, F. Engstrom, H. Lysaker, Investigation of a Francis turbine during speed variation: inception of cavitation, Renewable Energy 166 (2020) 147−162. Available from: https://doi.org/10.1016/j.renene.2020.11.108.

[3] S.C. Li, Cavitation of Hydraulic Machinery, vol. 1, Imperial College Press, London, 2000.

[4] C.E. Brennen, Cavitation and Bubble Dynamics, Cambridge University Press, New York, 2014.

[5] R.E.A. Arndt, Cavitation in fluid machinery and hydraulic structures, Annual Review of Fluid Mechanics 13 (1981) 273−326. Available from: https://doi.org/10.1146/annurev.fl.13.010181.001421.

[6] F. Avellan, Introduction to cavitation in hydraulic machinery, in: Scientific Bulletin of the Politehnica University of Timisoara: Transactions on Mechanics, 2004.

[7] C. Arn, P. Dupont, F. Avellan, Efficiency alteration of francis turbines by travelling bubble cavitation, in: E. Cabrera, V. Espert, F. Martínez (Eds.), Hydraulic Machinery and Cavitation, Springer, Dordrecht, 1996. Available from: https://doi.org/10.1007/978-94-010-9385-9_53.

[8] X. Escaler, E. Egusquiza, M. Farhat, F. Avellan, M. Coussirat, Detection of cavitation in hydraulic turbines, Mechanical Systems and Signal Processing 20 (4) (2006) 983−1007. Available from: https://doi.org/10.1016/j.ymssp.2004.08.006.

[9] P. Bourdon, R. Simoneau, F. Avellan, Erosion vibratory fingerprint of leading edge cavitation of a NACA profile and of a Francis model and prototype hydroturbine, in: Proceedings of Bubble Noise and Cavitation Erosion in Fluid Systems, ASME Winter Annual Meeting, New Orleans, USA, November, 1993.

[10] X.-wu Luo, B. Ji, Y. Tsujimoto, A review of cavitation in hydraulic machinery, Journal of Hydrodynamics 28 (3) (2016) 335−358. Available from: https://doi.org/10.1016/S1001-6058(16)60638-8.

[11] A. Favrel, A. Müller, C. Landry, K. Yamamoto, F. Avellan, LDV survey of cavitation and resonance effect on the precessing vortex rope dynamics in the draft tube of Francis turbines, Experiments in Fluids 57 (11) (2016). Available from: https://doi.org/10.1007/s00348-016-2257-y.

[12] M. Nishi, S. Liu, An outlook on the draft-tube-surge study, International Journal of Fluid Machinery and Systems 6 (1) (2013) 33−48.

[13] A. Favrel, J.G.P. Junior, C. Landry, A. Müller, C. Nicolet, F. Avellan, New insight in Francis turbine cavitation vortex rope: role of the runner outlet flow swirl number, Journal of Hydraulic Research 56 (3) (2018) 367−379. Available from: https://doi.org/10.1080/00221686.2017.1356758.

[14] Y. Hosoi, Contribution to model tests of draft tube surges of Francis turbines, in: Proc. Joint Symp. on Design and Operation of Fluid Machinery, Fort Collins, CO, vol. 1, 1978, pp. 141−150.

[15] M. Nishi, S. Matsunaga, T. Kubota, Y. Senoo, Effect of draft tube shape on the characteristics of pressure surge and swirl flow, in: IAHR 13th IAHR Symposium, Montreal, Canada, vol. 1, 7/1–13.

[16] D. Chirkov, P. Scherbakov, S. Cherny, A. Zakharov, V. Skorospelov, Mitigation of self-excited oscillations at full load: CFD analysis of air admission and effects of runner design, IOP Conference Series: Earth and Environmental Science 49 (2016).

[17] I. Kassanos, J. Anagnostopoulos, D. Papantonis, Numerical investigation of draft tube pressure pulsations in a Francis turbine with splitter blades, IOP Journal of Physics: Conference Series 813 (2017).

[18] A. Müller, M. Dreyer, N. Andreini, et al., Draft tube discharge fluctuation during self-sustained pressure surge: fluorescent particle image velocimetry in two-phase flow, Experiments in Fluids 54 (2013) 1514. Available from: https://doi.org/10.1007/s00348-013-1514-6.

[19] K. Yamamoto, A. Müller, A. Favrel, et al., Experimental evidence of inter-blade cavitation vortex development in Francis turbines at deep part load condition, Experiments in Fluids 58 (2017) 142. Available from: https://doi.org/10.1007/s00348-017-2421-z.

[20] F. Doussot, G. Balarac, J. Brammer, Y. Laurant, O. Métais, RANS and LES simulations at partial load in Francis turbines: three-dimensional topology and dynamic behaviour of inter-blade vortices, International Journal of Fluid Machinery and Systems, Korean Society for Fluid Machinery 13 (2020) 12–20. Available from: https://doi.org/10.5293/IJFMS.2019.13.1.012.

[21] K. Yamamoto, A. Müller, A. Favrel, F. Avellan, Physical mechanism of interblade vortex development at deep part load operation of a Francis turbine, ASME Journal of Fluids Engineering 141 (11) (2019) 111113. Available from: https://doi.org/10.1115/1.4043989.

[22] A. Zobeiri, F. Avellan, M. Farhat, Physics behind vortex-induced vibration reduction using an oblique trailing edge hydrofoil, 2010.

[23] P. Ausoni, M. Farhat, X. Escaler, E. Egusquiza, F. Avellan, Cavitation influence on Kármán vortex shedding and induced hydrofoil vibrations, Journal of Fluids Engineering 129 (8) (2007) 966–973. Available from: https://doi.org/10.1115/1.2746907.

[24] A. Rossetti, G. Pavesi, G. Ardizzon, A. Santolin, Numerical analyses of cavitating flow in a pelton turbine, ASME Journal of Fluids Engineering. 136 (8) (2014) 081304. Available from: https://doi.org/10.1115/1.4027139.

[25] C.J. Zeng, Y.X. Xiao, W. Zhu, Y.Y. Yao, Z.W. Wang, Numerical simulation of cavitation flow characteristic on Pelton turbine bucket surface, IOP Conference Series: Materials Science and Engineering 72 (2015) 4.

[26] D. Jošt, A. Škerlavaj, V. Pirnat, M. Morgut, E. Nobile, Detailed analysis of flow in two pelton turbines with efficiency and cavitation prediction, International Journal of Fluid Machinery and Systems 12 (2019) 388–399. 10.5293/IJFMS.2019.12.4.388.

[27] IEC Standards, 60193: hydraulic turbines, storage pumps and pump-turbines–model acceptance tests, in: International Electrotechnic Commission, third ed., 2018.

[28] B. Carola, A. François, The IMHEF system for cavitation nuclei injection, in: Proceedings of the 16th Symposium of the IAHR, Section on Hydraulic Machinery and Cavitation: Large and Medium Size Hydraulic Machinery and Installations from Planning to Operation, Brazil, 1992.

[29] F. Avellan, Cavitation test of hydraulic machines: procedure and instrumentation, in: ASME Winter Annual Meeting, New Orleans, Louisiana, November 28–December 3, 1993.

[30] F. Avellan, S. Etter, J. Gummer, U. Seidel, Dynamic pressure measurements on a model turbine runner and their use in preventing runner fatigue failure, in: Proceedings of the 20th IAHR Symposium, Charlotte, NC, USA, 2000.

[31] H. Bjørndal, T. Moltubakk, H. Aunemo, Flow induced stresses in a medium head Francis runner – strain gauge measurements in an operating plant and comparison with finite element analysis, in: 10th international Meeting of the Work Group on The Behaviour of Hydraulic Machinery Under Steady Oscillatory Conditions, 2001.

[32] M. Farhat, S. Natal, F. Avellan, F. Paquet, P.-Y. Lowys, M. Couston, Onboard measurements of pressure and strain fluctuations in a model of low head Francis turbine. Part 1: instrumentation, in: Proceedings of the 21st IAHR Symposium on Hydraulic Machinery and Systems, Lausanne, Switzerland, 2002.

[33] X. Escaler, E. Egusquiza, M. Farhat, F. Avellan, Cavitation erosion prediction in hydro turbines from onboard vibrations, in: Proceedings of the 22nd IAHR Symposium on Hydraulic Machinery and Systems, Stockholm, Sweden, 2004.

[34] T. Rus, M. Dular, B. Širok, M. Hočevar, I. Kern, An investigation of the relationship between acoustic emission, vibration, noise, and cavitation structures on a Kaplan turbine, ASME Journal of Fluids Engineering 129 (9) (2007) 1112−1122. Available from: https://doi.org/10.1115/1.2754313.

[35] K. Yamamoto, A. Müller, A. Favrel, C. Landry, F. Avellan Guide vanes embedded visualization technique for investigating Francis runner inter-blade vortices at deep part load operation, 2015.

[36] A. Favrel, A. Müller, C. Landry, K. Yamamoto, F. Avellan, Study of the vortex-induced pressure excitation source in a Francis turbine draft tube by particle image velocimetry, Experiments in Fluids 56 (2015). Available from: https://doi.org/10.1007/s00348-015-2085-5.

[37] M. Ocepek, Z. Peršin, I. Kern, V. Djelić, S. Muhic, A. Lipej, Experimental and numerical analysis of cavitation and pressure fluctuations in large high head propeller turbine, IOP Conference Series: Earth and Environmental Science 405 (2019) 012032. Available from: https://doi.org/10.1088/1755-1315/405/1/012032.

[38] D. Rodriguez, A. Rivetti, M. Angulo, C. Lucino, S. Liscia, Francis turbine high load instabilities − model test and CFD simulation, IOP Conference Series: Earth and Environmental Science 240 (2019) 022053. Available from: https://doi.org/10.1088/1755-1315/240/2/022053.

[39] A. Müller, A. Favrel, C. Landry, K. Yamamoto, F. Avellan, On the physical mechanisms governing self-excited pressure surge in Francis turbines, IOP Conference Series: Earth and Environmental Science 22 (2014) 032034. Available from: https://doi.org/10.1088/1755-1315/22/3/032034.

[40] K. Yamamoto, A. Müller, A. Favrel, C. Landry, F. Avellan, Pressure measurements and high speed visualizations of the cavitation phenomena at deep part load condition in a Francis turbine, IOP Conference Series: Earth and Environmental Science 22 (2014) 022011. Available from: https://doi.org/10.1088/1755-1315/22/2/022011.

[41] M. Iliescu, S. Houde, C. Deschênes, S. Lemay, Investigations of the cavitational behavior of an axial hydraulic turbine operating at partial discharge by 3D-PIV, 2011.

[42] A. Podnar, M. Dular, B. Širok, M. Hočevar, Experimental analysis of cavitation phenomena on kaplan turbine blades using flow visualization, ASME Journal of Fluids Engineering 141 (7) (2019) 071101. Available from: https://doi.org/10.1115/1.4041985.

[43] F. Durst, A. Melling, J.H. Whitelaw, Principles and Practice of Laser-Doppler Anemometry, Academic Press, London, 1981.

[44] G. Ciocan, F. Avellan, J.-L. Kueny, Optical measurement techniques for experimental analysis of hydraulic turbines rotor − stator interaction, in: Proceedings of the ASME Fludis Engineering Division Summer Meeting, Boston, MA, USA, 2000.

[45] G. Ciocan, F. Avellan, Flow investigations in a francis draft tube: advanced experimental methods, in: Proceedings of the 3rd Conference of Romanian Hydropower Engineers, 2004.

[46] R. Beauvais, Laser-Doppler velocimetry (LDV), in: F. Mayinger (Ed.), Optical Measurements, Springer, Berlin, Heidelberg, 1994, pp. 179−193. Available from: https://doi.org/10.1007/978-3-662-02967-1_10.

[47] H.E. Albrecht, M. Borys, N. Damaschke, C. Tropea, Basic measurement principles, Laser Doppler and Phase Doppler Measurement Techniques. Experimental Fluid Mechanics, Springer, Berlin, Heidelberg, 2003, pp. 9−26. Available from: https://doi.org/10.1007/978-3-662-05165-8_2.

[48] S. Skripkin, M. Tsoy, P. Kuibin, S. Shtork, Vortex rope patterns at different load of hydro turbine model, MATEC Web of Conferences 115 (2017) 06004. Available from: https://doi.org/10.1051/matecconf/201711506004.

[49] M.D. Atkins, Chapter 5 – Velocity field measurement using particle image velocimetry (PIV), in: T. Kim, T.J. Lu, S.J. Song (Eds.), Application of Thermo-Fluidic Measurement Techniques, Butterworth-Heinemann, 2016, pp. 125–166. Available from: https://doi.org/10.1016/B978-0-12-809731-1.00005-8.

[50] M.S. Iliescu, G.D. Ciocan, F. Avellan, Two phase PIV measurements at the runner outlet in a Francis turbine, in: Proceedings of the ASME/JSME 2003 4th Joint Fluids Summer Engineering Conference. Volume 2: Symposia, Parts A, B, and C. Honolulu, Hawaii, USA. July 6–10, 2003, pp. 2175–2181. ASME. https://doi.org/10.1115/FEDSM2003-45756.

[51] M.S. Liescu, G.D. Ciocan, F. Avellan, Analysis of the cavitating draft tube vortex in a Francis turbine using particle image velocimetry measurements in two-phase flow, ASME Journal of Fluids Engineering. 130 (2) (2008) 021105. Available from: https://doi.org/10.1115/1.2813052.

[52] S. Tridon, S. Barre, G.D. Ciocan, L. Tomas, Experimental analysis of the swirling flow in a Francis turbine draft tube: focus on radial velocity component determination, European Journal of Mechanics – B/Fluids 29 (4) (2010) 321–335. Available from: https://doi.org/10.1016/j.euromechflu.2010.02.004.

[53] A. Favrel, A. Müller, C. Landry, K. Yamamoto, F. Avellan, Space and time reconstruction of the precessing vortex core in Francis turbine draft tube by 2D-PIV, IOP Conference Series: Earth and Environmental Science 49 (2016) 082011. Available from: https://doi.org/10.1088/1755-1315/49/8/082011.

[54] R. Goyal, M.J. Cervantes, B.K. Gandhi, Vortex rope formation in a high head model Francis turbine, ASME Journal of Fluids Engineering. 139 (4) (2017) 041102. Available from: https://doi.org/10.1115/1.4035224.

[55] B. Branko, Methods for vibro-acoustic diagnostics of turbine cavitation, Journal of Hydraulic Research 41 (1) (2003) 87–96. Available from: https://doi.org/10.1080/00221680309499932.

[56] D. Valentín, A. Presas, M. Egusquiza, C. Valero, E. Egusquiza, Transmission of high frequency vibrations in rotating systems. application to cavitation detection in hydraulic turbines, Applied Science 8 (2018) 451. Available from: https://doi.org/10.3390/app8030451.

[57] X. Escaler, E. Egusquiza, T. Mebarki, F. Avellan, M. Farhat, Field assessment of cavitation detection methods in hydropower plants, in: Proceedings of the 21st IAHR Symposium on Hydraulic Machinery and Systems, Lausanne, Switzerland, 2002.

[58] P. Bourdon, R. Simoneau, F. Avellan, Erosion vibratory fingerprint of leading edge cavitation of a NACA profile and of a Francis model and prototype hydroturbine, in: ASME Winter Annual Meeting, New Orleans, Louisiana, USA, November 28-December 3, 1993.

[59] B. Bajić, A. Keller, Spectrum normalization method in vibro-acoustical diagnostic measurements of hydroturbine cavitation, ASME Journal of Fluids Engineering. 118 (4) (1996) 756–761. Available from: https://doi.org/10.1115/1.2835506.

[60] X. Escaler, F. Avellan, E. Egusquiza, Cavitation erosion prediction from inferred forces using material resistance data, In: CAV 2001: Fourth International Symposium on Cavitation, June 20–23, 2001, California Institute of Technology, Pasadena, CA USA, 2001.

[61] X. Escaler, E. Egusquiza, M. Farhat, Vibration cavitation detection using onboard measurements, in: The Fifth International Symposium on Cavitation, Osaka, Japan, 1–4 November, paper no. Cav03-OS-6–007, 2003. Available from: http://flow.me.es.osaka-u.ac.jp/cav2003/Papers/Cav03-OS-6-007.pdf.

[62] X. Escaler, M. Farhat, P. Ausoni, E. Egusquiza, F. Avellan, Cavitation monitoring of hydroturbines: tests in a Francis turbine model, in: Sixth International Symposium on Cavitation (CAV2006), Wageningen, The Netherlands, 2006.

[63] M.T.C. Faria, F.R. Queiroz, E.B. Medeiros, C.B. Martinez, Acoustic emission tests in the monitoring of cavitation erosion in hydraulic turbines, in: Proceedings of the ASME 2013 International Mechanical Engineering Congress and Exposition. Volume 14: Vibration, Acoustics and Wave Propagation, San Diego, California, USA, November 15–21, 2013. V014T15A016. ASME. Available from: https://doi.org/10.1115/IMECE2013-62960.

[64] H. Schmidt, O. Kirschner, S. Riedelbauch, J. Necker, E. Kopf, M. Rieg, et al., Influence of the vibro-acoustic sensor position on cavitation detection in a Kaplan turbine, IOP Conference Series: Earth and Environmental Science 22 (2014). Available from: https://doi.org/10.1088/1755-1315/22/5/052006.

[65] X. Escaler, J.V. Ekanger, H.H. Francke, M. Kjeldsen, T.K. Nielsen, Detection of draft tube surge and erosive blade cavitation in a full-scale Francis turbine, ASME Journal of Fluids Engineering 137 (1) (2014) 011103. Available from: https://doi.org/10.1115/1.4027541.

[66] V. Hasmatuchi, J. Decaix, M. Titzschkau, O. Pacot, C. Münch, Detection of Harsh Operating Conditions on a Francis Prototype Based on In-situ Non-Intrusive Measurements, Hydro 2019, Porto, 2019.

[67] A. Presas, Y. Luo, D. Valentin, M. Egusquiza Montagut, Z. Wang, C. Valero, et al., Detection of erosive cavitation on hydraulic turbines through demodulation analysis, IOP Conference Series: Earth and Environmental Science 240 (2019) 062048. Available from: https://doi.org/10.1088/1755-1315/240/6/062048.

[68] S. Muntean, S. Bernad, R. Resiga, I. Anton, Three-dimensional cavitating flow in hydraulic Francis turbines, in: Workshop on Numerical Methods in Fluid Mechanics and Fluent, Timisoara, Romania, 2003.

[69] S. Liu, Y. Wu, X. Luo, "Numerical Simulation of 3D Cavitating Turbulent Flow in Francis Turbine." Proceedings of the ASME 2005 Fluids Engineering Division Summer Meeting. Volume 1: Symposia, Parts A and B. Houston, Texas, USA. June 19–23, 2005. pp. 1551-1556. ASME. Available from: https://doi.org/10.1115/FEDSM2005-77017.

[70] Zhou, L, Wang, Z, Luo, Y, & Peng, G. "Numerical Simulation of Cavitating Flow in a Francis Turbine." Proceedings of the ASME 2008 Fluids Engineering Division Summer Meeting collocated with the Heat Transfer, Energy Sustainability, and 3rd Energy Nanotechnology Conferences. Volume 1: Symposia, Parts A and B. Jacksonville, Florida, USA. August 10–14, 2008. pp. 1193-1197. ASME. Available from: https://doi.org/10.1115/FEDSM2008-55242.

[71] S. Liu, L. Zhang, M. Nishi, Y. Wu, Cavitating turbulent flow simulation in a Francis turbine based on mixture model, ASME Journal of Fluids Engineering. 131 (5) (2009) 051302. Available from: https://doi.org/10.1115/1.3112382.

[72] Z. Yao, R. Xiao, F. Wang, W. Yang, Numerical investigation of cavitation improvement for a Francis turbine, Journal of Physics: Conference Series 656 (2015).

[73] J. Necker, T. Aschenbrenner, Model test and CFD calculation of a cavitating bulb turbine, IOP Conference Series: Earth and Environmental Science 12 (2010).

[74] J. Schiffer, H. Benigni, H. Jaberg, Cavitation in a high specific speed Kaplan pit-type turbine-two-phase CFD-simulations and experimental verification, in: Conference on Modelling Fluid Flow (CMFF), Budapest, 2018.

[75] D. Jošt, M. Morgut, A. Škerlavaj, E. Nobile, Cavitation prediction in a Kaplan turbine using standard and optimized model parameters, in: 6th IAHR International Meeting of the Workgroup on Cavitation and Dynamic Problems in Hydraulic Machinery and Systems, Ljubljana, Slovenia, 2015.

[76] M. Morgut, E. Nobile, I. Biluš, Comparison of mass transfer models for the numerical prediction of sheet cavitation around a hydrofoil, International Journal of Multiphase Flow 37 (6) (2010) 620–626.

[77] X. Escaler, R. Roig, V. Hidalgo, Sensitivity analysis of Zwart-Gerber-Belamri model parameters on the numerical simulation of Francis runner cavitation, in: J. Katz (Ed.), Proceedings of the 10th International Symposium on Cavitation (CAV2018), ASME Press, New York, 2018, pp. 911–914.

[78] R. Susan-Resiga, G. Ciocan, I. Anton, F. Avellan, Analysis of the swirling flow downstream of a Francis turbine runner, Journal of Fluids Engineering. 128 (2006) 177–189.

[79] L. Zhou, Z. Qnag, Y. Tian, Numerical simulation of vortex cavitation in a draf tube, in: Proc. of the 23th IAHR Symposium on Hydraulic Machinery and Systems, (Yokohama, Japan), 2003.

[80] P. Guo, X. Luo, Z. Wang, P. Lu, Numerical investigation of steady and unsteady flow fields in an elbow draft tube, in: Proceedings of the 23th IAHR Symposium on Hydraulic machinery and systems, (Yokohama, Japan), 2003.

[81] D.J. Leonard, J.W. Lindau, Multiphase computation of cavitation breakdown in model and prototype scale Francis turbines, in: Proceedings of the ASME/JSME/KSME 2015 Joint Fluids Engineering Conference. Volume 2: Fora. Seoul, South Korea. July 26–31, 2015. V002T05A001. ASME. https://doi.org/10.1115/AJKFluids2015-5123.

[82] J. Brammer, C. Ségoufin, F. Duparchy, P.-Y. Lowys, A. Favrel, F. Avellan, Unsteady hydraulic simulation of the cavitating part load vortex rope in Francis turbines, Journal of Physics: Conference Series 813 (2017) 012020. Available from: https://doi.org/10.1088/1742-6596/813/1/012020.

[83] D. Jost, A. Lipej, Numerical prediction of the vortex rope in the draft tube, in: Proceedings of the 3rd IAHR International Meeting of the Work Group on Cavitation and Dynamic Problems in Hydr. Machinery and Systems (Brno, Czech Republic), 2009.

[84] Y. Xiao, Z. Wang, Z. Yan, J. Zhang, Experimental and numerical analysis of pressure pulses characteristics in a Francis turbine with partial load, in: Proceedings of the 25th IAHR Symposium on Hydraulic Machinery and Systems, (Timisoara, Romania), 2010.

[85] T. Krappel, A. Ruprecht, S. Riedelbauch, Flow simulation of a Francis turbine using the SAS turbulence model, in: W. Nagel, D. Kröner, M. Resch (Eds.), High Perform. Comput. Sci.Eng. '13, Springer, Cham, 2013, pp. 455–463. Available from: https://doi.org/10.1007/978-3-319-02165-2_31.

[86] A.V. Minakov, D.V. Platonov, A.A. Dekterev, A.V. Sentyabov, A.V. Zakharov, The numerical simulation of low frequency pressure pulsations in the high-head Francis turbine, Computers & Fluids 111 (2015) 197–205. Available from: https://doi.org/10.1016/j.compfluid.2015.01.007.

[87] J. Wack, S. Riedelbauch, Numerical simulation of a cavitating draft tube vortex rope in a Francis turbine at part load conditions for different σ-levels, IOP Journal of Physics: Conference Series 813 (2017).

[88] D. Jošt, A. Škerlavaj, M. Morgut, E. Nobile, Numerical prediction of cavitating vortex rope in a draft tube of a Francis turbine with standard and calibrated cavitation model, Journal of Physics: Conference Series 813 (2017). Available from: https://doi.org/10.1088/1742-6596/813/1/012045.

[89] J. Wack, S. Riedelbauch, Cavitation simulations of a tip leakage vortex for a NACA0009 hydrofoil and a Francis turbine at stable full load operating point, in: W. Nagel, D. Kröner, M. Resch (Eds.), High Performance Computing in Science and Engineering 18, Springer, Cham, 2019, pp. 351–368. Available from: https://doi.org/10.1007/978-3-030-13325-2_22.

[90] J. Wack, S. Riedelbauch, Numerical simulations of the cavitation phenomena in a Francis turbine at deep part load conditions, Journal of Physics: Conference Series 656 (2015) 012074. Available from: https://doi.org/10.1088/1742-6596/656/1/012074.

[91] J. Wack, S. Riedelbauch, K. Yamamoto, F. Avellan, Two-phase flow simulations of the interblade vortices in a Francis turbine, in: 9th International Conference on Multiphase Flow, Florence, Italy, 2016.

[92] M. Liu, L. Zhou, Z. Wang, D. Liu, Y. Zhao, Investigation of channel vortices in francis turbines, in: IOP Conference Series, Earth and Environmental Science 49 (2016) 082003. Available from: https://doi.org/10.1088/1755-1315/49/8/082003.

[93] X. Liu, Z. Wang, L. Zhou, X. Escaler, Y. Luo, Failure investigation of a Francis turbine under the cavitation conditions, in: 16th International Symposium on Transport Phenomena and Dynamics of Rotating Machinery, Apr, Honolulu, United States, 2016.

[94] L. Zhou, M. Liu, Z. Wang, D. Liu, Y. Zhao, Numerical simulation of the blade channel vortices in a Francis turbine runner, Engineering Computations 34 (2) (2017) 364–376. Available from: https://doi.org/10.1108/EC-10-2015-0302.

[95] R. Resiga, A. Stuparu, S. Muntean, Francis turbine with tandem runners: a proof of concept, IOP Conference Series: Earth and Environmental Science 240 (2019) 022012. Available from: https://doi.org/10.1088/1755-1315/240/2/022012.

[96] C. Tănasă, A. Bosioc, S. Muntean, R. Susan-Resiga, A novel passive method to control the swirling flow with vortex rope from the conical diffuser of hydraulic turbines with fixed blades, Applied Science. 9 (2019) 4910. Available from: https://doi.org/10.3390/app9224910.

[97] R. Susan-Resiga, S. Muntean, V. Hasmatuchi, I. Anton, F. Avellan, Analysis and prevention of vortex breakdown in the simplified discharge cone of a Francis turbine, ASME Journal of Fluids Engineering. 132 (5) (2010) 051102. Available from: https://doi.org/10.1115/1.4001486. May 2010.

[98] H. Grein, Vibration Phenomena in Francis Turbines: Their Causes and Prevention, 54, Escher Wyss News, Switzerland, 1981, pp. 905–914.

[99] An experimental study on fins, their role in control of the draft tube surging, in: M. Nishi, X. M. Wang, K. Yoshida, T. Takahashi, T. Tsukamoto, E. Cabrera, V. Espert, et al. (Eds.), Hydraulic Machinery and Cavitation, Springer, Dordrecht, 1996, pp. 905–914. Available from: https://doi.org/10.1007/978-94-010-9385-9_92.

[100] J. Feng, W. Li, H. Wu, J. Lu, W. Liao, X. Luo, Investigation on pressure fluctuation in a Francis turbine with improvement measures, IOP Conference Series: Earth and Environmental Science 22 (2014) 032006. Available from: https://doi.org/10.1088/1755-1315/22/3/032006.

[101] D. Platonov, A. Minakov, D. Dekterev, The study of the influence of stabilizing devices on the pressure pulsations at the free discharge of water through the turbine, MATEC Web of Conferences 115 (2017) 06002. Available from: https://doi.org/10.1051/matecconf/201711506002.

[102] R.H. Thicke, Practical solutions for draft tube instability, Water power and Dam Constructions 33 (2) (1981) 31–37.

[103] B. Papillon, M. Sabourin, M. Couston, C. Deschenes, Methods for AI admission in hydroturbines, in: Proceedings of the 21st IAHR Symposium on Hydraulic Machinery and Systems, Lausanne, Sept., 2002, pp. 81–91.

[104] K. Nakanishi, T. Ueda, Air supply into draft tube of Francis turbine, Fuji Electric Review 10 (3) (1964) 81–91.

[105] P. Doerfler, Design criteria for air admission systems in Francis turbines, in: Proceedings of the 13th IAHR Symposium of Hydraulic Machinery and systems, Montreal, 1986.

[106] A. Rivetti, M. Angulo, C. Lucino, S. Liscia, Pressurized air injection in an axial hydro-turbine model for the mitigation of tip leakage cavitation, Journal of Physics Conference Series 656 (1) (2015). Available from: https://doi.org/10.1088/1742-6596/656/1/012069.

[107] A.I. Bosioc, R. Susan-Resiga, S. Muntean, C. Tanasa, Unsteady pressure analysis of a swirling flow with vortex rope and axial water injection in a discharge cone, ASME Journal of Fluids Engineering. 134 (8) (2012) 081104. Available from: https://doi.org/10.1115/1.4007074.

Chapter 7

X-ray Measurement Techniques Used for Wall-Bounded Cavitating Flows

Milad Heidari-Koochi[1], Ioannis K. Karathanassis[2] and Manolis Gavaises[2]

[1]*School of Mathematics, Computer Science and Engineering, CITY, University of London, London, United Kingdom,* [2]*Department of Mechanical Engineering and Aeronautics, School of Mathematics, Computer Science and Engineering, CITY, University of London, London, United Kingdom*

7.1 Introduction

The emergence and collapse of cavitation bubbles can have a crucial impact on a multitude of flow processes that evolve at highly diverse length and time scales. Onset of cavitation is accompanied by flow phenomena, such as formation of shock waves and high-velocity liquid jets, that, if properly controlled, can be beneficial to various industrial and biomedical processes. For instance, cavitation may be used to induce soft or solid matter fragmentation, particle/deposits deagglomeration, and spray atomization [1–7]. Nevertheless the uncontrollable formation of vapor bubbles is associated with undesirable aftereffects, including vibrations, noise, hydraulic efficiency deterioration, and material erosion [8–17].

The occurrence of inertial cavitation is triggered by the decrease of the local pressure below the liquid saturation pressure, which causes bubbles to form, owing to an abrupt flow acceleration or recirculating motion. Cloud cavitation and vortical cavitation are two distinct cavitation regimes in high-speed turbulent flows, setting in as vapor pockets attached to or in the vicinity of solid boundaries or as coherent

Cavitation and Bubble Dynamics. DOI: https://doi.org/10.1016/B978-0-12-823388-7.00009-6

elongated cavities in the flow core, respectively. The presence of adverse pressure gradients leading to flow separation and/or coherent vortex roll-up constitutes the underlying cause that provides each cavitation regime its characteristic morphology. The dynamic behavior of transient cavities has been illustrated by employing high-speed imaging in fundamental internal and external flow layouts such as hydrofoils, extruded geometries with a constriction [18–27], axisymmetric geometries [28,29], and bluff bodies [30–32], as well as several types of hydraulic devices, including marine propellers, hydroturbines, high-pressure pumps, fuel injection equipment, valves and piping components [33–38], and even mechanical heart valves [39].

Optical imaging has been widely employed to elucidate either wall-bounded or unconfined cavitating flows, as it offers insight into the extent and overall topology of vapor structures as well as the dynamic evolution of the two-phase field. However, the nature of visible light imposes certain crucial limitations on such types of flow visualization. A principal drawback, especially in regard to wall-bounded flows, is the necessity of optical access, which requires the fabrication of transparent replicas that are, in principle, capable of operating at conditions similar to those of the actual flow devices, which in high-pressure and/or high-temperature applications is inherently impossible. Besides, the presence of minor amounts of vapor leads to extensive light scattering, a shortcoming that is accentuated in complex two-phase fields that exhibit extensive three-dimensional (3D) structures, which can render the entire visualization region opaque. However, it could still be said that the inability of optical imaging to offer any kind of quantification with reference to the actual amount of vapor formation constitutes its major flaw as an evaluation technique for cavitating flows.

X-ray imaging has gained momentum in the last two decades as an attractive alternative to visible light imaging that, albeit more complex to implement, is capable of alleviating all the referenced technical concerns. This chapter discusses in detail the relevant X-ray imaging methods that have been proven suitable for the visualization of complex two-phase flows. It also outlines the experimental infrastructure that is required for their realization and the amount and nature of physical information that can be extracted through their use.

7.2 Fundamentals of X-rays

X-rays are generated through physical mechanisms stemming from interactions of high-energy electrons with materials atoms. Two types of interactions lead to the production of X-rays with either a characteristic spectrum or a continuous spectrum; the latter type of X-ray radiation is commonly referred to as bremsstrahlung, a term that is derived from *bremsen*, the German word for braking. So-called characteristic radiation is produced as a result of the collision of a high-energy electron with an electron in the inner shell of an atom, which obtains energy higher than its binding energy and is therefore ejected by the atom.

Subsequently, an outer shell electron fills the hole in the inner layer, and the relevant loss of energy is emitted as an X-ray photon. When an electron has a trajectory close enough to an atomic nucleus, so that electrostatic forces come into play, the particle is both decelerated and redirected. The relevant energy loss is emitted as a bremsstrahlung X-ray photon [40]. X-ray energies vary from 0.1 keV to more than 100 KeV, the latter being referred to as "hard" X-rays characterized by short wavelengths that are able to penetrate dense objects. In contrast, the higher-wavelength and thus lower-energy "soft" X-rays are useful for enhancing the image contrast in obtained radiographs. Most commercial, laboratory-scale X-ray sources are polychromatic and produce a spectrum of energy encompassing both soft and hard X-rays [41].

Because X-rays constitute electromagnetic waves, refraction occurs at interfaces of media with different densities. Nevertheless, the refractive index concerning X-rays is expressed as a complex number as follows:

$$n = 1 - \delta - i\beta \tag{7.1}$$

where the real part δ designates the phase shift of the wave and the imaginary part β describes the amplitude attenuation [42,43]; δ is of the order of 10^{-5} in solids and about 10^{-8} in air, while β obtains even smaller values [40]. In general, the visible light refractive index varies in the range 1.3–4.0 for liquid and solid materials, whereas for X-rays the difference from unity is infinitesimal.

The potential interactions of X-ray photons with matter comprise absorption, coherent (Rayleigh) scattering, and incoherent (Compton) scattering [42]. Various aspects of X-ray imaging are summarized in Fig. 7.1. For low-energy X-rays, the favored interactions are absorption and Rayleigh scattering, while Compton scattering prevails for high energies in comparison to the atomic binding energy [44]. The overall atomic interaction cross section σ_{tot} (cm^2/atom) can be defined as the sum of individual cross sections as follows:

$$\sigma_{\text{tot}} = \sigma_R + \sigma_C + \tau_{\text{PE}} \tag{7.2}$$

where σ_{R} and σ_{c} are the cross sections for Rayleigh scattering and Compton scattering, respectively, and τ_{PE} is the cross section for photoelectric absorption. Additionally, the atomic cross section of each interaction is correlated with the respective cross section per unit mass μ/ρ, or mass attenuation coefficient, as follows:

$$\tau_{\text{PE}} = \frac{A}{N_A}\frac{\mu_a}{\rho} \tag{7.3}$$

$$\sigma_C = \frac{A}{N_A}\frac{\mu_C}{\rho} \tag{7.4}$$

$$\sigma_R = \frac{A}{N_A}\frac{\mu_R}{\rho} \tag{7.5}$$

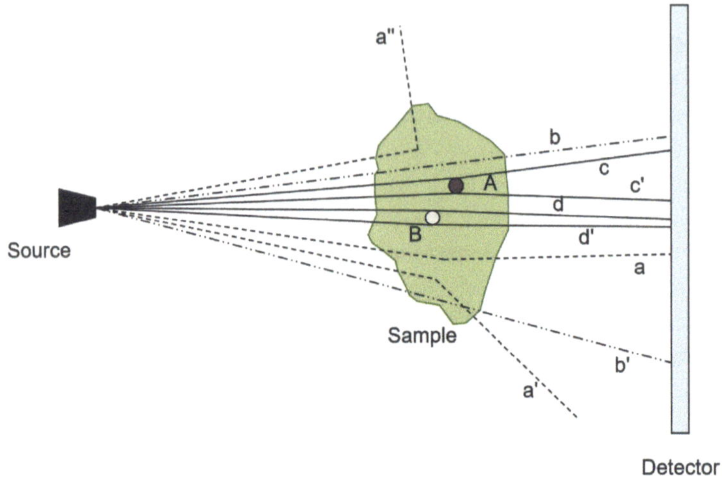

Figure 7.1 Schematic illustration of X-ray planar imaging. X-rays diverge from the cone-beam source, and after they interact with matter, their trajectory either deviates by scattering (a, a′, and a″) or remains unaltered (b and b′) by transmission through the sample. Additionally, X-ray refraction can be observed as rays at the boundaries of comparatively denser objects diverge (c and c′), while the opposite occurs for lighter objects (d and d′). The transmitted X-ray intensity recorded by the detector is dependent on the local properties of absorption, scattering, and refraction in the sample. *Redrawn from P. Russo, Physical basis of X-ray imaging, in: Comprehensive Biomedical Physics, Elsevier, 2014, pp. 1–48, Copyright © 2014 Elsevier B.V.*

where $\frac{\mu_a}{\rho}$, $\frac{\mu_C}{\rho}$, and $\frac{\mu_R}{\rho}$ (all units in cm^2/g) are the mass attenuation coefficients corresponding to photoelectric effect, Compton scattering, and Rayleigh scattering, respectively, N_A is Avogadro's number, and A is the atomic mass [45]. As a result, the total mass attenuation coefficient $\frac{\mu}{\rho}$ (cm^2/g) of a specific material can be obtained as follows:

$$\frac{\mu}{\rho} = \frac{\mu_R}{\rho} + \frac{\mu_C}{\rho} + \frac{\mu_a}{\rho} \tag{7.6}$$

7.2.1 Principles of X-ray

An electromagnetic wave of amplitude E_0 propagating along the z direction in free space is characterized by a wave equation of the complex form:

$$E = E_0 \times e^{i\left(\omega t - \frac{\omega z}{c}\right)} \tag{7.7}$$

If the wave propagation takes place in a homogeneous medium with a refractive index n, then incorporating Eq. (7.1) into Eq. (7.7) gives

$$E' = E_0 \times e^{i\left(\omega t - \frac{\omega n z}{c}\right)} = E_0 \times e^{i\left(\omega t - \frac{\omega z}{c}\right)} \times e^{i\left(\frac{\delta \omega z}{c}\right)} \times e^{\left(\frac{-\beta \omega z}{c}\right)} \tag{7.8}$$

Therefore the change in wave amplitude compared to propagation in free space would be

$$\frac{E'}{E} = e^{i\left(\frac{\delta \omega z}{c}\right)} \times e^{\left(\frac{-\beta \omega z}{c}\right)} \tag{7.9}$$

where $\omega = 2\pi c/\lambda$ is the angular velocity (c is the speed of light in vacuum, and λ is the X-ray wavelength). Hence the terms on the right-hand side of Eq. (7.9) can be rewritten as follows:

$$\frac{E'}{E} = e^{i\frac{2\pi}{\lambda}\delta z} \times e^{\frac{-2\pi}{\lambda}\beta z} \tag{7.10}$$

When Eq. (7.10) is compared with Eq. (7.7), it becomes evident that the change in wave phase (φ) is correlated with the real part δ of the refractive index, while the change in amplitude (E) is correlated with the imaginary part β:

$$\Delta\varphi = \frac{2\pi}{\lambda}\delta \Delta z \tag{7.11}$$

$$\frac{E'}{E} = e^{-\frac{2\pi}{\lambda}\beta \Delta z} \tag{7.12}$$

By defining the attenuation coefficient as $\mu = \frac{4\pi}{\lambda}\beta$ [46], Eq. (7.12) can be rewritten as follows:

$$\frac{E'}{E} = e^{-\frac{\mu}{2}\Delta z} \tag{7.13}$$

The attenuation coefficient μ is designated by photon energy, material density, and elemental atomic number [47,48]. Moreover, since the intensity (I) of the wave field is proportional to the square of the amplitude, Eq. (7.13) takes the following form:

$$\frac{I}{I_0} = e^{-\mu \Delta z} \tag{7.14}$$

Eq. (7.14) is known as the Beer-Lambert law [49] describing the intensity attenuation of an electromagnetic wave (X-rays in this chapter) interacting with matter. As was described earlier, this decrease in intensity occurs as a result of the mechanisms of photoelectric effect, coherent and incoherent scattering, and the overall attenuation coefficient results as the sum of the respective values corresponding to each process [50].

7.2.2 X-RAY SYSTEMS

The main constituents of an X-ray system are sources, detectors, optical lenses, and choppers, as described in the following sections.

7.2.2.1 X-ray Sources

X-ray sources are typically categorized in two classes of laboratory-scale and large-scale, synchrotron facilities. In the former, the electron beam generated in a cathode-ray tube impinges onto a metal anode (usually tungsten) to generate X-rays. In the latter, a high-energy beam of particles (electrons or positrons) is constantly shifting its direction in the synchrotron storage ring, leading to the constant emission of X-ray radiation [51]. To select the appropriate X-ray source for a specific imaging application, considerations regarding the size of the beam focal spot, energy spectrum of generated X-rays, and photon flux are crucial. The focal spot size determines the potential spatial resolution of the experiment, the energy spectrum defines the penetrability of the beam, where higher-energy rays penetrate more at a cost of losing sensitivity to changes in material density, and the flux marks the signal-to-noise ratio, that is, the clarity of the acquired radiograph [52].

7.2.2.1.1 X-ray Tube

In principle, a simplified X-ray tube employs a glowing filament to produce electrons, which, through application of high voltage, are accelerated toward a water-cooled metal anode, as demonstrated in Fig. 7.2A. The anode produces X-rays mainly as bremsstrahlung radiation, as described earlier. Both cathode and anode are accommodated in a vacuum envelope made of metal, ceramic, or glass that is able to withstand the amount of heat generated during the production of X-rays. A beryllium window, owing to its low atomic number, allows X-rays to be emitted from the source. In the 1960s a rotating anode generator was introduced to the setup to increase the effective area exposed to the beam and facilitate higher heat dissipation [40]. During the interaction of fast electrons with the anode matter, the continuous bremsstrahlung X-ray component increases with electron energy and elemental atomic number. This process is used to adjust the energy spectrum of the emitted beam by passage of X-rays through absorber materials, such as molybdenum or gadolinium oxide, which cut off the lower-energy photons [53,54]. It is worth mentioning that the emitted energy spectrum is designated by a series of factors, such as the potential difference between the anode and cathode, target material, electron incidence angle on the target, and X-ray emission angle [41]. A prototype X-ray tube source that was developed at the University of Bergamo is shown in Fig. 7.2B. The source can operate at 160 kV/400 μA and has

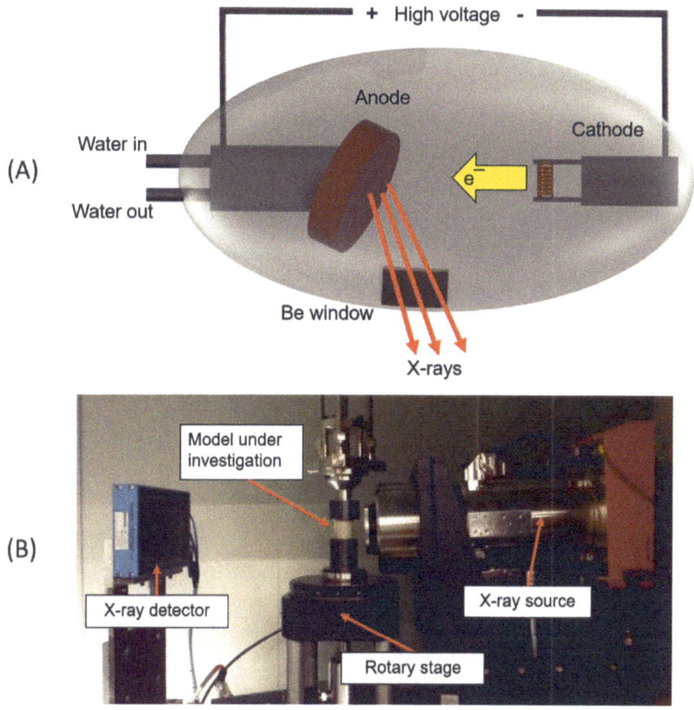

Figure 7.2 (A) Schematic representation of standard X-ray tube operation. (B) Prototype X-ray tube source available at the University of Bergamo. The facility also includes a rotating stage, evident in the figure, which enables the conduction of computed tomography investigations (CT). *Image in (B) is reprinted from M. Lorenzi, N. Mitroglou, M. Santini, M. Gavaises, Novel experimental technique for 3D investigation of high-speed cavitating diesel fuel flows by X-ray micro computed tomography, Review of Scientific Instruments 88 (3) (2017) 033706, with the permission of AIP Publishing.*

been proven to be suitable for the characterization of different types of two-phase flows [55–57].

7.2.2.1.2 Synchrotron Radiation

Synchrotron storage rings are used to produce radiation due to the acceleration of charged particles under the influence of magnets positioned around the periphery of the circular layout. Two types of X-ray sources are employed in modern synchrotron facilities: bending magnets and undulators. Bending magnets are required to keep electrons in a closed orbit or insertion devices, such as undulators, located in a straight section of the storage ring, which force the electrons to follow an oscillating path by means of an alternating magnetic field [51]. Bending magnets produce broadband-emitted radiation as a horizontally broad and vertically narrow fan, as illustrated in Fig. 7.3A. Undulators, comprising a sequence of short bending magnets (Fig. 7.3B), produce a repeating magnetic field, which eventually leads to a highly collimated X-ray beam. When a beam of high-energy electrons is bent, a small fraction of its energy is radiated away and extracted in a direction tangential to the orbit. The radiation appears as an intense and highly directional beam of photons [40,54] in the so-called beamline with energy ranges from a few electron volts to $\sim 10^5$ eV. Storage rings store electrons in bunches, and depending on the application, a single bunch or a sequence of bunches can be utilized. For instance, the storage ring in the Advanced Photon Source (APS) of the Argonne National Laboratory has an 1104-m circumference, and its standard pulsation mode consists of 24 singlets (single bunches) with a nominal current of 4.25 mA for each bunch. The duration of each pulse is 33.5 ps with a gap of 153 ns between pulses, as presented in Fig. 7.4. The pulsating nature of the synchrotrons is utilized in various operating modes of the storage rings to facilitate the acquisition of time-resolved measurements.

The spectral distribution of bending magnet radiation (Fig. 7.5A) is a function of photon energy and therefore results in a continuous flux

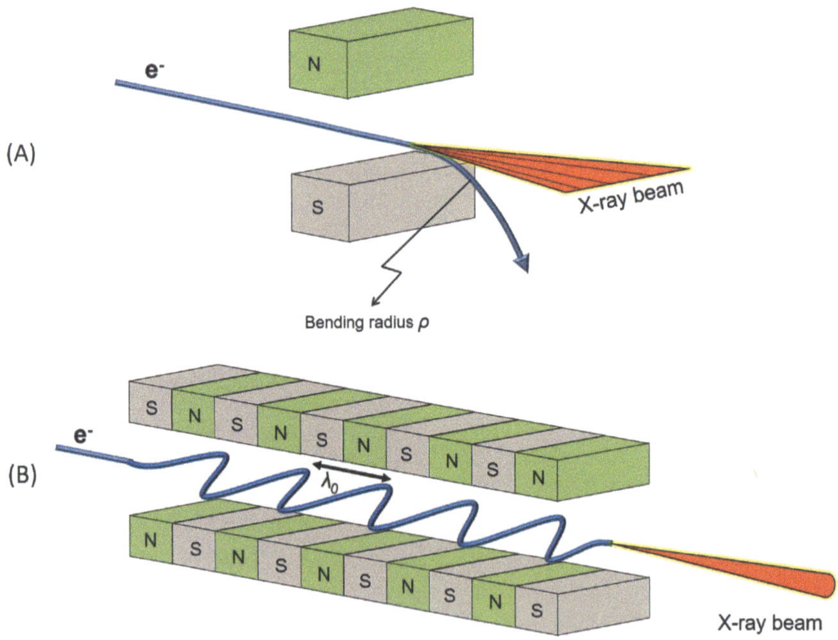

Figure 7.3 (A) Schematic representation of a bending magnet in which a horizontally broad beam is emitted, as the electrons direction changes. (B) Schematic representation of an undulator, in which electrons are accelerated through altering magnetic fields and a focused X-ray beam is emitted.

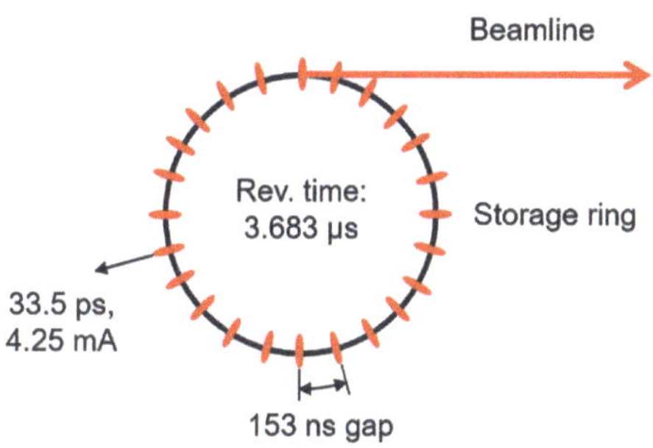

Figure 7.4 Standard pulsation mode of an APS storage ring.

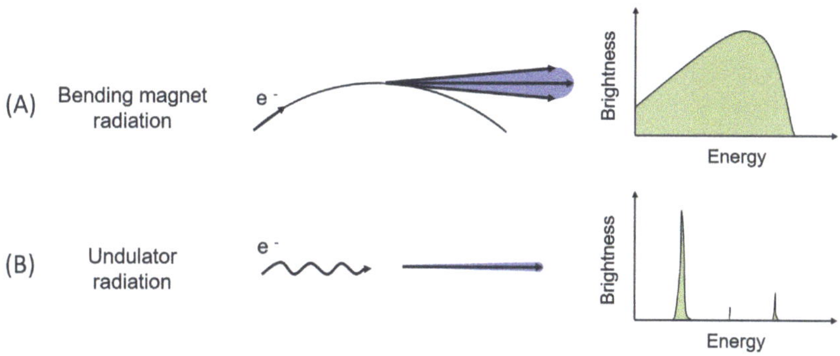

Figure 7.5 Spectral bandwidth of (A) bending magnet radiation exhibiting a continuous range of X-ray energies known as white beam and (B) undulator radiation exhibiting peaks at specific harmonics.

distribution over a range of X-ray energies [40]. On the other hand, in undulators a given electron at one trajectory oscillation emits radiation, which is in phase with the subsequent oscillation event. Therefore as the frequency remains the same, only the wave amplitudes are added, and subsequently, the sum is squared to acquire the resultant intensity. This coherent addition of amplitude implies a quasimonochromatic spectrum with harmonics, as illustrated in Fig. 7.5B.

With reference to bending magnets it should be pointed out that the design parameters designating the beam spectrum are the ratio of electron energy to rest mass energy, the cyclic frequency, and the bending radius. For undulators the key parameters are the spatial period (λ_0 in Fig. 7.3B) and the dimensionless parameter K, known as deflection parameter, which defines the maximum angular deviation of the beam from the undulator axis and can be calculated from the maximum magnetic field B_0 and undulator spatial period λ_0 as follows (for a planar undulator):

$$K = 0.0934 \times B_0 \times \lambda_0 \tag{7.15}$$

The spectrum distribution can be tuned through this parameter, as small values of K (≤ 1) result in a dominant peak at the fundamental harmonic. On the contrary, large K results in a series of narrow spectral peaks (high harmonics) [40], as illustrated in Fig. 7.6.

7.2.2.2 X-ray Detectors

Ionization due to X-ray absorption is the working principle of X-ray detectors. The main classifications between them refers to photon-counting over charge-integrating detectors. Photon-counting detectors have a moderate ability to distinguish between different photon energies at a cost of maximum input flux, while charge-integrating sensors accept higher fluxes but exhibit reduced sensitivity at low fluxes [51]. An ionization chamber and a silicon PIN diode constitute the main components of a charge-integrating detector, while a photon-counting detector comprises scintillators, silicon drift diodes, and avalanche photodiodes. In visualization experiments in which spatially resolved information must be extracted from the irradiated region, scintillator crystals are used to convert X-rays to visible light [51]. Image quality is affected by the size, quantity, and efficiency of the detectors that are used. For instance, in CT a single detector averages a number of X-ray interactions through a portion of the irradiated object and turns it into a single intensity reading, while the information that can be acquired at each slice is determined by the total number of detectors [52]. A limiting parameter regarding the detector performance is the so-called dead time of the counting system corresponding to the minimum time period between photon impingement events for which the events can be differentiated. Therefore in synchrotron-based experiments in which photons reach the detector at infinitesimally small time intervals, dead time corrections are used to account for the detector response [58]. A comprehensive review of X-ray detectors can be found in [54].

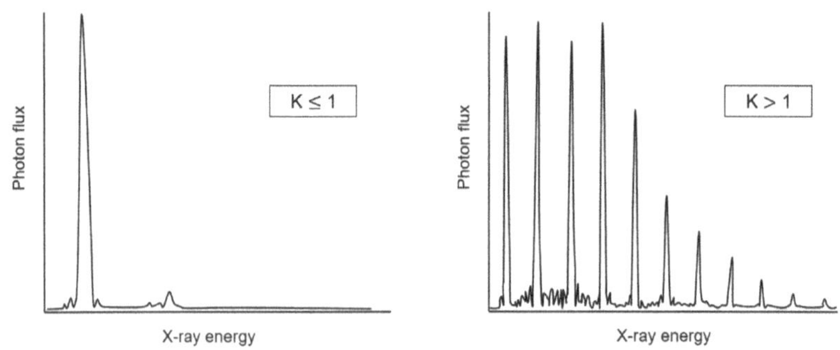

Figure 7.6 Influence of deflection parameter K on the radiation spectrum: Left plot represents a spectrum dominated by a single peak at the fundamental harmonic for small K, while right plot exhibits a series of narrow peaks for "large" K.

7.2.2.3 X-ray Optics

Refraction of X-rays takes place at an interface of media with different densities. However, as was described earlier in this chapter, owing to the fact that the relevant refractive index is slightly less than unity, the shape of converging X-ray lens would be similar to a diverging optical lens [40]. Moreover, also owing to the weak refraction of X-rays, the focal length of a single lens would be of the order of hundreds of meters, making it impractical. Therefore a compound form of lenses is used to alleviate this limitation. In a simplified form, drilling N holes into a solid block of material creates an N-compound lens. As reported by Snigirev et al. [59], perforating an aluminum block with 30 holes, each having a diameter of 0.6 mm with a spacing of 25 μm between them, led to a reduction of the focal length from 54 to 1.8 m and enabled the focusing of a 14-keV X-ray beam onto an 8-μm spot. Alternatively, Fresnel plates made of corrugated concentric rings can be used to focus X-rays to submicron focal spot sizes [60]. Moreover, curved mirrors with surfaces coated with a heavy material, such as gold or platinum, to achieve a rather large electron density [40] constitute another category of focusing optics. Similar to visible light interaction with matter, X-rays undergo total external reflection, yet at very small incidence angles, less than 1 degree [61]. Finally, perfect crystals made of silicon or germanium are capable of filtering broadband X-rays, in essence operating as monochromators [62,63], as their lattice planes diffract the rays. They can operate at larger incidence angles compared to mirror-based optics; however, they accept a narrow range of X-ray energies.

7.2.2.4 X-ray Choppers

As was mentioned earlier, modern synchrotrons are capable of producing pulses with a duration of around 100 ps. However, temporal resolution of detectors is orders of magnitude slower, hence posing a technical limitation in capturing transient phenomena in a time-resolved manner. Therefore X-ray choppers have been designed to manipulate the beam pulsation frequency so that the state-of-the-art detectors can fully resolve the incoming photon flux [64]. Two main approaches have been proposed in the literature. The first one includes choppers based on either rotating crystals, such as Si(111), or X-ray mirrors. The rationale behind it is based on the small acceptance angle of the diffracting crystal or reflection angle of the mirror, respectively [65]. For instance, McPherson et al. [66] managed to increase the time period between pulses from 2.84 to 10 ns in a 10-keV beam using a cubic silicon crystal rotating at 32,586 rpm. The second approach involves rotating apertures, which block X-rays at a specified timing designated by their geometrical layout, as depicted in Fig. 7.7. Distinct chopper types

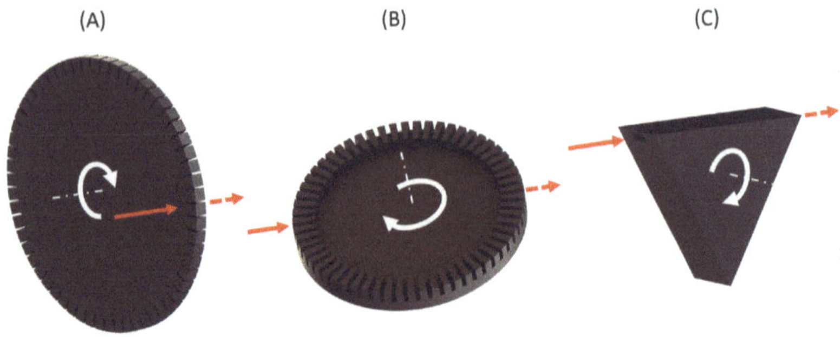

Figure 7.7 Rotating aperture X-ray choppers. (A) Slitted disk having high repetition rates. (B) Grooved disk having both high repetition rates and short opening times. (C) Prismatic aperture having short opening times. The opening time and repetition rate are designated by the slot size and the number of slots and rotation speed of the chopper, respectively. *Adapted from H. Osawa, T. Kudo, S. Kimura, Development of high-repetition-rate X-ray chopper system for time-resolved measurements with synchrotron radiation, Japanese Journal of Applied Physics 56 (4) (2017) 048001, licensed under a Creative Commons Attribution 4.0 international (CC BY 4.0), https://creativecommons.org/licenses/by/4.0/.*

constitute disks rotating about an axis parallel to the beam with slits realized at their perimeter (Fig. 7.7A) or rotating perpendicular to the beam and adjusting the pulsation through radial grooves on their surface, (Fig. 7.7B), as well as prismatic apertures rotating perpendicular to the beam with a channel carved into one of their sides (Fig. 7.7C) [67].

7.3 X-ray Measurement Techniques

The principal X-ray measurement techniques with relevance to wall-bounded cavitating flows are discussed in this section. These are X-ray absorption imaging, X-ray CT, X-ray phase contrast imaging, and X-ray fluorescence spectroscopy.

7.3.1 X-ray Absorption Imaging

A simplified schematic representation demonstrating the basic constituents of an imaging setup suitable for obtaining quantitative void fraction measurements of a gas-liquid mixture inside a section of an arbitrary duct is shown in Fig. 7.8 [68]. The intensity of the attenuated radiation beam I on the detector, positioned on the opposite side of the X-ray source, after passing through the gas-liquid medium of path length L is as follows:

$$I = I_0 \exp[-(2\Sigma_W T + \Sigma_T L)] \tag{7.16}$$

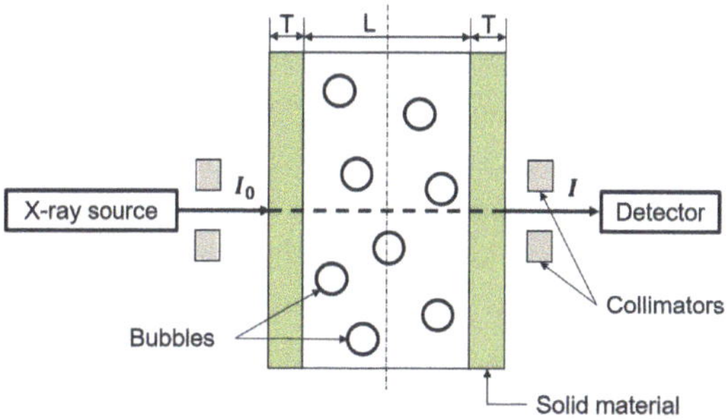

Figure 7.8 Simplified arrangement of a void fraction–measuring X-ray setup. Collimators are used to focus the beam onto the detector by filtering rays propagating at random trajectories. *Redrawn from A.A. Kendoush, A comparative study of the various nuclear radiations used for void fraction measurements, Nuclear Engineering and Design 137 (2) (1992) 249–257, Copyright © 1992 Elsevier Science Publishers B.V.*

where I_0 is the intensity of beam radiation emitted from the X-ray source ($cm^{-2}\ s^{-1}$), Σ_W is the macroscopic cross section of the solid material (cm^{-1}) with a thickness of T, and Σ_T is the overall macroscopic cross section of the gas-liquid mixture. The macroscopic cross section of the mixture can be further analyzed in terms of void fraction α, as follows:

$$\Sigma_T = \alpha\Sigma_g + (1-\alpha)\Sigma_l \tag{7.17}$$

where Σ_g, and Σ_l are the cross sections of the liquid and gas phases, respectively. Using Eqs. (7.16) and (7.17), the void fraction of the two-phase mixture can be derived as follows:

$$\alpha = \frac{\ln(I/I_l)}{\ln(I_g/I_l)} \tag{7.18}$$

where I_g and I_l are the attenuated intensities of an X-ray beam with initial intensity I_0 after passing through the entire duct thickness ($L+2T$) when it is filled only with the gas phase or the liquid phase, respectively.

7.3.1.1 X-ray Absorption Examples in Wall-Bounded Cavitating Flows

Pike et al. [69] were among the first researchers to use X-ray attenuation for the measurement of void fraction in a two-phase steam-water system. They used a Picker PX-3B X-ray tube with a tungsten target operated at a peak voltage of 45 kV. A mockup technique was implemented to simulate the steam-water flow by realizing a Lucite-air system with the same absorption characteristics. Seven samples with different dimensions were blended with known void fraction compositions, calculated by measuring the average system density, and were subsequently compared against the X-ray results. The detected intensities were observed to be essentially zero for X-ray energies below 30 keV. Therefore energies in the range 33−42 keV were selected, for which the attenuation coefficient was found to be essentially constant and the spectrum of radiation could be considered to be monochromatic for the specific conditions of the experiment. To collimate the beam, a hole with a diameter of 2.4 mm in a 6.4-mm-thick lead sheet was placed between the tube source and test section. Both X-ray source and collimator were mounted on a carriage that was capable of translating in a horizontal line, perpendicular to the test cross section, to measure the vapor fraction at different chords. The authors concluded that an X-ray tube could be considered to be an effective source for measuring void fraction if proper tube operating conditions and collimation are selected.

Jones and Zuber [70] used an X-ray system to generate the probability density function (PDF) of the fluctuations in void fraction at the centerline of a two-dimensional (2D) rectangular duct in order to distinguish between three dominant patterns of bubbly, slug, and annular flow. A transparent flow passage was realized, allowing the injection of both air and water through a regulated supply to form a two-phase mixture. X-ray void fraction was measured by using a commercial 700-W tube with a tungsten target, working at 35 kV and 20 mA. The beam size was approximately $1.27 \times 3.8\ mm^2$ at the test section. The authors demonstrated that the bubbly flow exhibited a single-peaked PDF at low void fraction, while annular flow presented a single-peaked PDF at high void fraction. By contrast, slug flow was characterized by a twin-peaked PDF. Vince et al. [71] performed an experiment similar to that of Jones and Zuber [70], with the X-ray system irradiating a vertical Plexiglas tube with a diameter of 2.54 cm. They made a series of chordal-averaged void fraction measurements to generate the respective PDFs and analyzed the first four moments associated with the distributions. Six chordal measurements, including one along the diameter and one in the vicinity of the tube wall, were taken on half of the tube cross section, since the flow was found to be axisymmetric. The researchers found that the PDF variance, which is the second moment about the mean, was the best indicator of flow regime. Kendoush [68] comparatively assessed the capability of different types of nuclear radiation (gamma, beta, and neutrons) and X-rays for void fraction quantitative measurements. A flow configuration with well-controlled void fraction values was realized by inserting empty plastic tubes into a water-filled acrylic duct to illustrate the accuracy of each imaging technique. Regarding X-rays, a typical cone-beam source was used, operated at 70 kV. It was observed that for all types of radiation the statistical error was increased as the measured void fraction decreased. Beta radiation exhibited the smallest statistical error in void fraction measurements, followed by X-ray, neutron, and gamma radiation. It was also concluded that errors increased for increasing the radiation energy, regardless of the radiation type employed.

Although the early studies that have been discussed so far are of value, since they demonstrated the applicability of X-ray imaging to the visualization of two-phase flows, only idealized configurations were considered in the absence of complex flow phenomena and phase change. Stutz and Legoupil [26] were probably the first in the open literature to perform instantaneous volume fraction measurements of a two-phase cavitating flow by means of X-ray attenuation. Their flow loop consisted of a cavitation tunnel equipped with a planar converging-diverging test section having width and length dimensions of 44 × 520 mm, respectively. This flow layout was introduced in a previous work by Stutz [72], in which a double optical probe device and piezoresistive pressure transducers were employed to measure the vapor quantity of the liquid-vapor mixture arising in the region of geometrical constriction, as well as the local flow velocity. The X-ray setup of Stutz and Legoupil [26] consisted of a 160 kV/400 μA X-ray generator and 24 NaI(Ti) detectors in a linear array to detect the attenuated beam. Vapor volume fraction was measured at 500 Hz and compared with the respective spatially averaged values obtained by the dual optical probe method. It was verified that the measurements were in agreement within the uncertainty range. It was observed that by increasing the flow velocity from 6 to 10 m/s, the time-averaged vapor volume fraction increased by 50%, and the frequency of the cloud cavity shedding taking place downstream the constriction apex also increased by 65%, whereas the mean size of cavitation area remained unchanged. Coutier-Delgosha et al. [73] used X-ray attenuation to investigate the two-phase morphology and vapor content of sheet cavitation forming on the suction side of a 150-mm-long 2D hydrofoil section with sharp leading and trailing edges. An X-ray imaging setup similar to that of Stutz and Legoupil [26] was employed to measure the ray intensity from the leading to the trailing edge of the hydrofoil at three different angles of attack, which resulted in volumes with a cross section of 6 × 3 mm in the middle of the cavitation tunnel. The generator and detectors moved vertically to perform measurements in the transverse direction, and quantitative data were acquired at 1000 Hz for a total irradiation time of 30 s. A maximum time-averaged void fraction of 0.5 was measured in the irradiated section, which was approximately double the value measured by Stutz and Reboud [74] in the same flow layout using an optical probe. Furthermore, the sheet cavities were observed to exhibit an unsteady oscillatory behavior, the frequency of which was captured by using piezoelectric transducers and resulted in Strouhal numbers of around 0.25.

Aeschlimann et al. [75] performed X-ray absorption imaging of a 2D cavitating shear layer representative of the flow conditions prevailing in rocket engine turbopump inducers. A 300-mm-long rectangular test section with an 80 × 80 mm^2 inlet cross section was utilized, divided into two halves by a splitting plate. The initial liquid velocity in the upper half was 4.5 times larger than the respective in the lower half, thus creating a cavitating mixing layer. A 160-kV X-ray generator at 18.75 mA current was employed together with an image intensifier to convert and amplify the X-ray flux to 2D images visible by a high-speed camera. Due to the finite size of the source and large size of the test section, the spatial resolution was limited to 0.5 mm, which made it infeasible to resolve small vapor structures. Using a shutter speed of 20 μs at a frequency of 2000 frames per second (fps), instantaneous 2D void fraction fields were produced after intensity conversion and necessary corrections. The time-resolved quantitative data revealed that the vapor pockets forming in the shear layer stemmed from underlying vortices, which in turn were a consequence of Kelvin-Helmholtz instabilities. Also, the time-averaged void fraction field demonstrated that the cavities were contained inside the mixing layer after their growth. Finally, the Strouhal number characterizing the vortical cavity shedding was found to be approximately 0.22, a value that was reported in the literature for cavitation shedding in different flow configurations.

Makiharju et al. [76] developed a laboratory-scale 2D X-ray densitometry system to obtain quantitative measurements of relatively high-speed two-phase flows. The flow loop that was employed for the X-ray measurements realized a backward-facing step, where air was introduced into the flow at the contraction location, forming an attached cavity at the separation region. The X-ray source was a 150-kV rotating anode tube capable of operating at 433 mA. The accuracy of the system was investigated and compared against high-speed videos and vapor fraction measurements obtained by a dual-fiber optical probe and an optoelectronic module. The signal produced by the optical system was proportional to the amount of light reflected back to the optoelectronic module, which varied depending on the probe being immersed in either gas or liquid, owing to the different refractive indices. Since the X-ray system employed a polychromatic beam, a carbon steel plate was used as a filter to "preharden" it, in other words, to cut off low-energy photons, as will be described later in this chapter. Placement of the filter led to an error reduction in the quantitative measurements from 2.77% to 0.24% at a void fraction of 0.5. High-speed radiographs recorded at a rate of 1 kHz with a spatial resolution of 0.46 mm/pixel illustrated the topological evolution of the cavity in the flow recirculation region immediately downstream from the stepwise flow path reduction. Vapor-fraction data that were obtained by using the optical probe and X-ray imaging were found to be in good agreement for values above 0.05.

Duke et al. [77] performed X-ray radiography of the two-phase flow emerging in a polycarbonate nozzle with a diameter of 500 μm. In this experiment, the use of collimated, high-flux synchrotron X-rays in a raster-scanning method illustrated the most advantageous combination of spatial and temporal resolution over traditional flat-field X-ray imaging. The authors used a monochromatic beam tuned to a mean energy of 8 keV and focused to a spot size of $5 \times 6\ \mu m^2$ to assess the extent of vapor in the test nozzle, as schematically shown in Fig. 7.9. Since the investigation was relevant to fuel injection equipment, a gasoline surrogate (Viscor 16 BR) was selected as the working medium. The selection of beam energy was based on a tradeoff between contrast in the liquid and vapor phases and overall transmission through the polycarbonate nozzle. The attenuation coefficient of the working fluid was determined through normalization against samples filled with air and water (materials with well-documented coefficient values), respectively. Initially, a flat-field X-ray setup was used to reveal the time-averaged morphology of strongly cavitating regions, where changes in intensity were at a maximum of only 5%–6% compared to the overall image intensity. At a second step, focused-beam measurements were conducted over a raster scan grid of 100 transverse positions at 19 streamwise locations, acquiring radiographs at a sampling rate of 272 kHz at each position, which were eventually averaged to produce data of the projected void fraction. The time-averaged field illustrated the formation of a sheet cavity in the vicinity of the sharp nozzle inlet, subsequently detaching from the wall and being entrained by the core flow. Time-resolved measurements performed at specific locations, for example, close to the orifice wall and in the entrainment region, were analyzed, and the void fraction root mean square (RMS) was found to be up to 1% of the time-averaged signal, an indication of an overall steady cavitation process.

Employing the system developed at the University of Michigan [76], Ganesh et al. [18] obtained time-resolved X-ray density measurements of the transient cavity forming at the apex of a planar wedgelike flow constriction, essentially having a 2D topology with a well-defined trailing edge. Radiographs were recorded at 1000 Hz

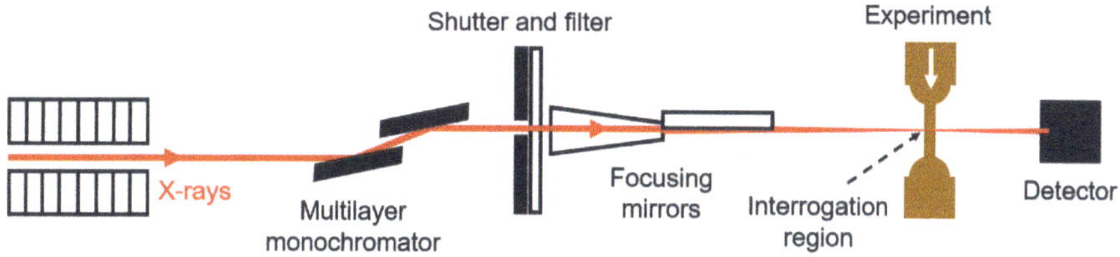

Figure 7.9 Schematic representation of the basic principles of a synchrotron radiation flow visualization setup. The X-ray beam is produced by the undulator, and subsequently, unwanted white radiation is cut off by the monochromator. Focusing X-ray mirrors are utilized to adjust the beam to the desired focal point. Eventually, the attenuated beam after interaction with the sample under investigation impinges on the detector device, producing measurable signal. The diagram is not to scale; the distance from source to experiment is of the order of meters, while the detector is located some centimeters away from the test piece. *Adapted from D.J. Duke, A.L. Kastengren, F.Z. Tilocco, A.B. Swantek, C.F. Powell, X-ray radiography measurements of cavitating nozzle flow, Atomization and Sprays 23 (9) (2013) 841–860 [77].*

with a spatial resolution of 0.25 mm/pixel. X-ray data revealed two mechanisms that govern the cavity periodic shedding sequence downstream from the constriction. For moderate values of the cavitation number σ $(=(P_0 - P_v)/(0.5 \times \rho U_0^2))$, where P_0 and P_v are the local pressure upstream of the wedge and vapor pressure, respectively, and U_0 is the flow velocity at the test section inlet, the typical reentrant jet process, that is, the boundary layer flow in the direction opposite that of the free stream, owing to the adverse pressure gradient in the separation region [78], designates the detachment frequency of coherent cavities. Nevertheless, as σ obtained lower values, that is, cavitation effects were enhanced according to the definition, pinch-off of extensive vapor clouds from the wedge apex was observed. It was verified that the cloud collapse dynamics were determined by the upstream propagation of a shock wave forming within the cavity, owing to the adjustment of the local speed of sound by the composition of the liquid-vapor mixture. Although the first shedding mechanism was also observable with high-speed optical imaging, this latter one could be properly elucidated only through X-ray densitometry measurements. Finally, the authors managed to measure the speed of sound in the mixture upstream of the shock front and the Mach number based on the observed shock-propagation speed. The bubbly shock mechanism is illustrated in Fig. 7.10, which depicts the evolution of the cavity-shedding cycle.

Similar to [77], Duke et al. [79] carried out time-resolved X-ray absorption imaging of the two-phase flow arising in a metallic nozzle with a diameter of 0.5 mm and a length of 3.05 mm. Beryllium was chosen as the test piece material, owing to its low atomic number compared to other metals, to maintain the level of beam attenuation low enough to allow the acquisition of measurable signal. Two test cases were examined, characterized by Reynolds and cavitation numbers representative of GDI injectors, that is, $Re = 15{,}000-24{,}000$ and $K = 1.01-1.11$. Cavitation number was defined as $K = (P_1 - P_v)/(P_1 - P_2)$, where P_1, P_2, and P_v were inlet, outlet, and fuel vapor pressures, respectively. A raster scan method on a 37×66 grid was once again utilized with a 9-KeV monochromatic beam collimated to a $5 \times 6\ \mu m^2$ spot. Although the acquired samples were spatially uncorrelated, as the flow had a steady behavior and the sampling period was sufficiently long in comparison to the flow time scales, time-invariant properties such as void fraction RMS and power spectrum could be mapped over the nozzle. Spectral analysis showed that the dominant frequencies at the locations where attached cavities separated from the nozzle walls were in the region of several kilohertz, presumably linked to cloud shedding. By contrast, at the nozzle exit where bubble collapse was expected, higher frequencies of the order of 10 kHz to 1 MHz were captured.

To study the effect of noncondensable gas on cavity-shedding dynamics, Makiharju et al. [20] conducted high-speed visualization and X-ray densitometry measurements on a flow setup similar to [18] but modified to allow gas injection. The flow configuration, as described earlier, realized a 2D wedge in a water

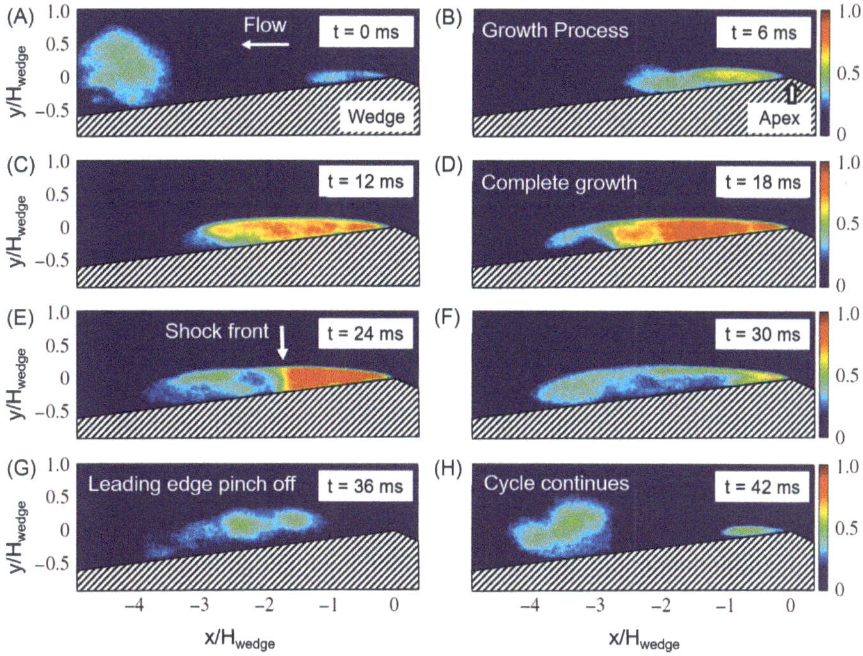

Figure 7.10 Void fraction measurements for a periodically shedding cavity over a wedge through X-ray densitometry at a cavitation number of 2.0 and an average flow velocity of 8 m/s. As the separated flow region at the wedge apex fills with vapor, the void fraction of the mixture increases and the speed of sound decreases to a point at which the Mach number becomes supersonic, which results in the formation of a well-defined shock front that propagates upstream, as highlighted in panel E. Once this front reaches the apex, a cloud of vapor is shed and entrained downstream, and the cycle repeats. *Reproduced from S.A. Mäkiharju, H. Ganesh, S.L. Ceccio, The dynamics of partial cavity formation, shedding and the influence of dissolved and injected non-condensable gas, Journal of Fluid Mechanics 829 (2017) 420–458 with the permission, from Cambridge University Press.*

tunnel. In this specific experiment, gas was injected at two spanwise locations downstream of the wedge apex through a row of 72 holes with a diameter of 0.51 mm per injector. The optical setup resulted in a spatial resolution of 1 mm/pixel at a sampling frequency of 1 kHz. Gas injection at the wedge apex actually suppressed the formation of vapor and reduced the mean vapor volume fraction, owing to an increase in the local pressure at the suction peak. Gas injection farther downstream, at the midcavity length, resulted in a moderate increase in the mean local pressure, also causing a decrease in vapor production but significantly less in comparison to apex injection. Noncondensable gas effects played a significant role in the evolution of bubbly shock waves, leading to their suppression or even elimination, owing once again to their influence on the local pressure field.

As a consequence, reentrant jet was observed as the dominant mechanism with reference to ventilated cavities.

Wu et al. [24] investigated shedding dynamics of partial cavities over a hydrofoil, also employing the X-ray facility and water tunnel developed at the University of Michigan [18]. A NACA0015 hydrofoil with a chord of 50 mm and a span of 76.8 mm was mounted to the tunnel side-wall and allowed the adjustment into two discrete angles of attack equal to 7 degrees and 10 degrees. Time-resolved X-ray densitometry similar to [22,76] was performed at 7.5 kHz, calibrated to produce absolute void fraction data through preliminary measurements of gas-water phantoms representative of the maximum expected range. The quantitative measurements around the hydrofoil revealed an extensive range of cavity dynamics based on cavitation number and angle of attack encompassing incipient cavitation, shedding from the aft portion of the shorter cavities, and reentrant jet-induced as well as shock-induced shedding, which demonstrated a complex and multimodal behavior of the hydrofoil partial cavity flow.

Barbaca et al. [80] investigated the unsteady behavior of both natural and ventilated cavities over a 2D wall-mounted obstacle. The cinematographic X-ray densitometry setup that was implemented was similar to [18,76]. Radiographs were acquired at a 1-kHz sample rate for approximately 0.8 s at a spatial resolution of 125 μm/pixel. Investigation of instantaneous X-ray radiographs found that the void fraction values at the flow region affected by the reentrant jet were remarkably lower in the naturally cavitating flow compared to the ventilated counterpart, since vapor and gas bubbles were indistinguishable in the latter case. RMS deviation of the obtained data revealed that the void fraction fluctuations were much higher in the case of natural cavitation, a behavior that was attributed to its violent nature, owing to vapor condensation and the occurrence of shock waves. Furthermore, space–time plots of void fraction distributions acquired in the location of the vapor pocket were examined and led to the proposal of two variations of the reentrant jet mechanism depending on the jet thickness compared to that of the cavity. A periodic vapor shedding takes place for a "thin" jet, whereas for a jet of thickness comparable to that of the cavity, breakup occurs in an incoherent manner, leading to a nonperiodic shedding of small-scale structures.

7.3.2 X-ray Computed Tomography

X-ray CT is used to evaluate the 3D internal structure of opaque objects in a nonintrusive manner and comprises two principal processes: the physical measuring technique and the mathematical reconstruction algorithm [47] necessary to extract physical information. During the CT process, a 3D object is projected onto a 2D plane over a wide range of angles by recording the respective radiographs, known as sinograms [40]. In modern X-ray CT setups employing a fan beam, a slice of the object is scanned, and then by rotating either the object or the source-detector setup, the process is repeated, as can be seen in Fig. 7.11. In a second step, a 3D structure of the object is generated by using reconstruction algorithms, several of which have been proposed in the literature [81]. The filtered back-projection algorithm [82,83] has been widely employed, according to which a filter is used to convolve the data and each view is successively superimposed over a square grid corresponding to its acquisition angle. During the reconstruction the raw intensity data in the sinogram files are converted to voxel brightness values, mapped linearly to the effective attenuation coefficient of the material [52]. It should be noted that for dynamic measurements such as cavitating flows, as each projection is acquired over a specific time scale, any fluctuations occurring at smaller time scales are averaged [41]. It is also noteworthy that one of the main sources of artifacts in the reconstruction phase is the so-called beam hardening. As a polychromatic beam contains a wide range of energies, the low-energy X-rays are absorbed preferentially by the sample and leave the remaining beam proportionally richer in high-energy photons [84]. This effect causes the edges of the object to appear brighter and can be mitigated by filtering low-energy X-rays through metal plates to narrow the energy span through the object [55,85].

7.3.2.1 X-ray CT Examples in Wall-Bounded Cavitating Flows

Morooka et al. [86] performed CT visualization of a vertical bundle of 4 × 4 aluminum tubes using a modified medical X-ray scanner, the relevant application of interest being boiling water reactors. A steam-water mixture was flowing inside

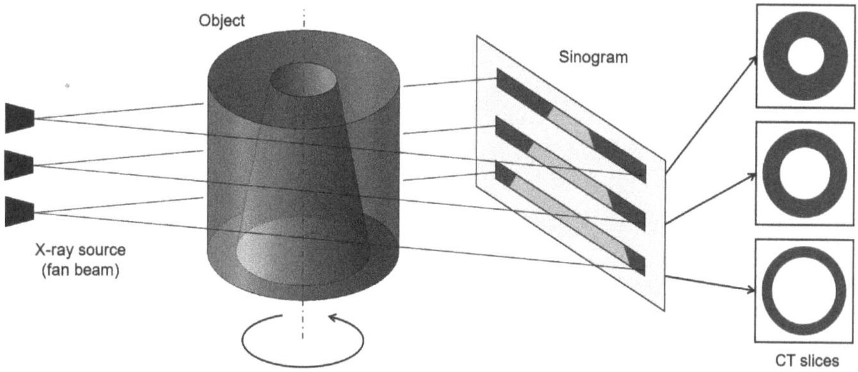

Figure 7.11 Schematic representation of a fan-beam X-ray CT scanner. Multiple radiographs from different projections are obtained to elucidate the internal structure of the sample at different cross sections. Stacking of the multiple CT slices results in the representation of the object's 3D volume.

the 12.3-mm-diameter tubes, which were made of aluminum to minimize X-ray attenuation. The employed X-ray tube had a voltage of 120 kV at a current of 25 mA, and the data set was obtained by rotating the source-detector pair around the object in 6-degree increments. Effects of vapor quality, mass flux, and pressure on the average void fraction were examined. It was verified that the average void fraction increased by increasing quality at constant mass flux and pressure, decreasing the pressure, or increasing the mass flux.

Harvel et al. [87] examined the morphology of a vertical annular flow using both high-speed X-ray CT and neutron radiography (RTNR). The temporal resolution of the RTNR system was limited to a sampling rate of 30 fps. The examined tubular test piece was fabricated by acrylic resin and aluminum for the CT and RTNR experiments, respectively. The high-speed CT setup produced 70 keV X-rays in a 24-degree fan beam, and to increase the temporal resolution, the beam attenuation was measured concurrently at several locations, using 122 cadmium tungstate (CdWO4) scintillators and 18 X-ray sources arranged around the test channel. The spatial resolution was found to be 2.0 mm, and the working fluids were light-heavy water and atmospheric air. Void fraction measurements were obtained at different locations along the tube height, and two regimes of bubbly and slug flow were illustrated. The X-ray CT system showed superior accuracy in resolving the void fraction distribution in cross-flow planes, whereas the RTNR setup was advantageous in resolving the distribution in the axial planes. Moreover, circumferential bubble motion was easily observed through the X-CT setup, while axial bubble motion could not be captured.

Hassan et al. [88] carried out an experiment in which instantaneous behavior of cavitation in a turbopump inducer rotating at 4000 rpm was studied via an X-ray tomographic system. The experimental setup consisted of a fixed generator-detector system, while the inducer was rotating. By synchronizing the irradiation of the beam and the speed of the inducer, the instantaneous void fraction at different locations could be captured despite the constraints imposed by the rotation speed. An X-ray tube operating at 80 kV/30 mA was employed to perform integral measurements of the vapor extent at three axial positions on the inducer with a slice thickness of 20 mm. Ten NaI scintillators were used for a total of 75 angles of incidence to acquire the projection slices, which were then used for the reconstruction process. The acquisition frequency was set to 5000 Hz for each detector, resulting in acquisition of 75 projections during a complete rotation of the inducer. Subsequently, all three axial sections were stacked to obtain a reconstruction of vapor-fraction distribution in pockets attached to the blades of the inducer. Three different cavitation conditions were identified by modifying the inlet pressure while maintaining the inducer rotation speed constant. The CT results were in good agreement with photographic images taken for the same cavitating conditions.

Bauer et al. [89] employed a medical CT machine to measure the void fraction distribution in a nozzle flow in a time-averaged manner. An acrylic test model was fabricated with a geometrical layout that was designed to generate a quasisteady cavitation topology throughout the nozzle cross section. Experiments were conducted with water as the working fluid. The CT system comprising 16 arrays of 1024 detectors allowed the acquisition of 16 slices of 0.75 mm thickness at a 1-Hz rotation frequency. Different types of cavitation, such as cloudlike, incipient, and film cavitation, were examined at high flow rates (~4 L/s), and the structural morphology inside the cavity core, not visible through optical imaging, could be elucidated. It is worth mentioning that as each reconstructed image was taken within one rotation cycle of the X-ray tube, fast phenomena such as bubble breakup could not be captured; however, it was possible to locate the breakup.

Mitroglou et al. [85] quantified the cavitating flow arising inside an enlarged single-orifice replica of a diesel injector using X-ray CT. The custom-made model featured a nonsymmetrical flow arrangement mimicking the prevalent flow features manifested in multihole injector nozzles. More specifically, the geometry comprised an off-axis nozzle with a diameter of 3 mm and a length of 10 mm, whereas an upstream chamber accommodated a fixed nonsymmetrical needle, which promotes a one-sided cloud cavity to emerge in the nozzle. Polyether ether ketone was selected as the model material to induce moderate X-ray attenuation. A prototype cone-beam X-ray source developed at the University of Bergamo operating at 60 kV/40 μA was employed to collect radiographs. Prehardening of the polychromatic beam was inherently achieved, as the model material thickness was deemed to be sufficient for the full absorption of low-energy rays [55]. Tomographic projections were acquired by keeping the detector-source fixed and rotating the test piece every 0.6 degrees over a full revolution, resulting in 600 projections with a voxel size of 15 μm. The FDK algorithm [90] was utilized to perform the tomographic reconstruction of the obtained slices. CT results accentuated the strong points of X-ray over optical imaging as well as of CT compared to line-of-sight methods. As shown in Fig. 7.12A, although cavitation occupies the entire nozzle periphery, its core is hollow, demonstrating, in essence, a "halo effect." This effect cannot be elucidated by optical imaging, which instead exhibits the entire region as opaque. The results also revealed that even a vapor volume fraction of 0.15 was capable of completely obstructing light to penetrate the cavity, as illustrated in Fig. 7.12B.

Bauer et al. [91] investigated the turbulent cavitating flow setting in the orifice that had also been employed in a previous study [89], this time utilizing an ultrafast X-ray CT scanner in which a free electron beam was formed and deflected by means of electron optics to generate a moving X-ray source. A fast multichannel static detector ring was used to measure the beam attenuation. With beam convolution, driven by an alternating magnetic field and a fast synchronized data acquisition,

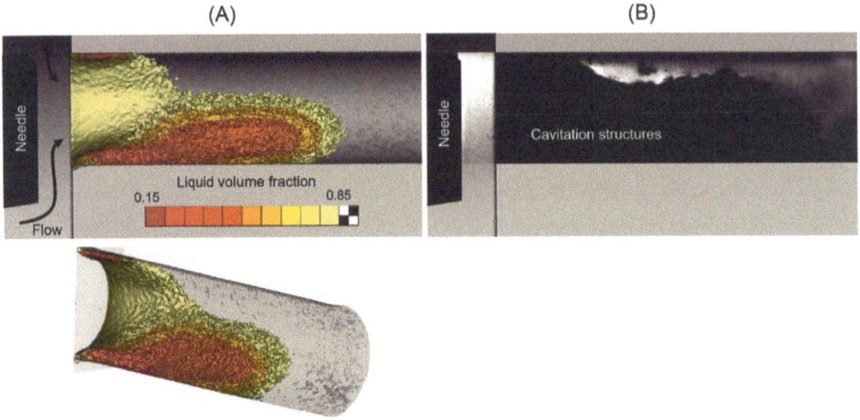

Figure 7.12 (A) Side (top image) and isometric section view (bottom image) at the nozzle plane of symmetry of liquid volume fraction isosurfaces obtained through micro-CT (needle lift of 1 mm, cavitation number of 2.1, Re = 55,500). At the entrance of the nozzle, cavitation sets in as a ring, while pure liquid occupies the nozzle core. Additionally, the asymmetry in the geometry of the test model, directs the flow to enter the nozzle predominantly from the bottom side, resulting in more extensive cavitation compared to the top side. (B) An instantaneous image of the flow acquired through optical imaging, making clear that the cavity region is completely opaque. *Reproduced from N. Mitroglou, M. Lorenzi, M. Santini, M. Gavaises, Application of X-ray micro-computed tomography on high-speed cavitating diesel fuel flows, Experiments in Fluids 57 (11) (2016) 175, licensed under a Creative Commons Attribution 4.0 international (CC BY 4.0), https://creativecommons.org/licenses/by/4.0/.*

frame rates of up to 8000 fps could be achieved. However, since the spatial resolution was inversely dependent on temporal resolution, a lower frame rate of 1000 fps was chosen to increase the spatial resolution to 1 mm/voxel. Therefore reconstructed volumes were derived from projections at an exposure time of 1 ms using a filtered back-projection algorithm. The employed scanner comprised two imaging planes with two independent detector rings. Therefore sequences of slice image pairs were produced at different cross sections of the orifice with an axial distance of 13 mm at a time interval of 0.5 Ms. From the first slice of the pair close to the inlet, a ring cavity attached to the wall was detected, and the development of cavitation could be conceptually visualized, as the void fraction in the ring cavity was found to increase in the second slice. By moving the slice pair farther downstream, an elongated bubble filled completely with vapor could be discerned. Furthermore, a slice in the collapse region of the channel revealed a transient behavior. It should be noted that although the absolute value of the exposure time is low, it is still orders of magnitude higher than the characteristic flow time scale (of the order of microseconds); hence the CT-voxel values represented the time-averaged vapor fraction. Nonetheless, in-nozzle cavitation dynamics were captured to some extent.

Jahangir et al. [28] used an X-ray CT setup to measure the void fraction distribution of the water-vapor flow in an axisymmetric converging-diverging nozzle with a throat diameter of 1.66 cm (Fig. 7.13A), as a follow-up experimental study to that of Jahangir et al. [92], in which the partial cavitation dynamics were investigated through optical imaging. Both experiments were then followed by a computational fluid dynamics (CFD) analysis [93], in which further details on the flow dynamics were extracted. In the CT experiment, a standard industrial type X-ray source (Yxlon-Y.TU 160-D06) was employed at 120 kV and 5 mA. The X-ray sinograms were acquired at 61 Hz for about 1 min. The visible light visualization conducted in the acrylic orifice revealed two distinct forms of cavity shedding, which were governed by the reentrant jet and bubbly shock mechanisms, respectively [92]. 3D reconstruction of the CT images was used to quantify the time-averaged void fraction and to validate the CFD results as illustrated in

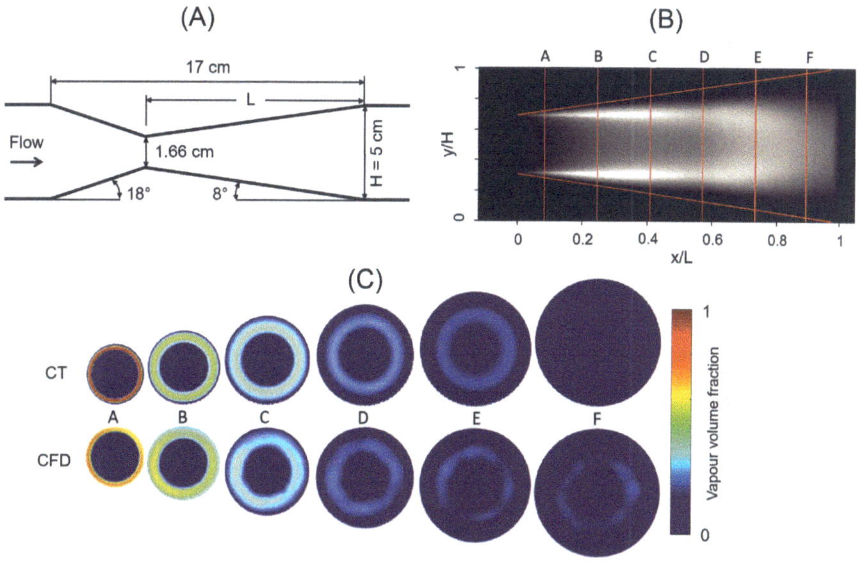

Figure 7.13 (A) Geometry and relevant dimensions of the converging-diverging section. (B) Locations of cross-sectional sampling planes along the nozzle-diverging part. The oblique red lines represent the nozzle walls, and the flow is from left to right. (C) Comparison of the average vapor fraction at different cross sections between CT (top) and CFD results (bottom) for the case in which shedding is governed by the bubbly shock mechanism ($Re = 2 \times 10^5$, $CN = 3.6$). *(A) and (B) are reproduced from S. Jahangir, E.C. Wagner, R.F. Mudde, C. Poelma, Void fraction measurements in partial cavitation regimes by X-ray computed tomography, International Journal of Multiphase Flow 120 (2019) 103085, Copyright © 2019 Elsevier Ltd. (C) is reprinted from M. Brunhart et al., Investigation of cavitation and vapor shedding mechanisms in a venturi nozzle, Physics of Fluids 32 (8) (2020) 083306, with the permission of AIP publishing.*

Fig. 7.13C. It should be pointed out that the CT results could not exhibit any transient features of the flow due to their time-averaged nature. It was found that the void fraction in the bubbly shock–driven shedding downstream the nozzle throat was more than twice that of the reentrant jet-driven cavitation.

Tekawade et al. [94] carried out a tomographic representation of the cavitating flow inside a single-orifice steel diesel injector using synchrotron X-rays delivered to the 7-BM beamline of the APS [95]. The incident beam had a total flux of about 10^{15} photons/s/mm^2 integrated over photon energies of 1–100 keV. As the nozzle had an internal diameter of 200 μm enclosed within 1–2 mm of steel, more than 98.5% of the flux was absorbed within the injector body itself, allowing only 4×10^{12} photons/s/mm^2 to be absorbed by the path of liquid fuel. n-Dodecane was injected at a pressure of 150 MPa, that is, at a realistic injection pressure for the device, while a high-speed camera in sync with the injector trigger captured X-ray projections at different angles during the injection event with an angular resolution of 1.875 degrees over a 180-degree rotation. Three hundred thirty-eight injection events per viewing angle were recorded to capture the distinct flow morphology at an exposure time of 100 μs, which resulted in time- and ensemble-averaged results. 3D snapshots at various time steps were acquired with a voxel size of 2.1 μm and illustrated the prevailing features of cavitation development for increasing values of the needle lift. Moreover, it was confirmed that microscale surface features of the injector internal geometry stemming from manufacturing imperfections had a clear influence on the overall shape of the attached cavity that formed in the injector hole.

7.3.4 X-ray Phase Contrast Imaging

X-ray phase contrast images are produced from interactions of X-rays with matter in two manners: beam extinction due to absorption and refraction caused by a shift of X-ray phase due to a density gradient [96]. In fact, the real part of X-ray refractive index [refer to Eq. (7.1)] corresponds to the phase shift of the X-ray wave, while the imaginary part corresponds to amplitude decrease [45]. Phase shift in the wave occurs as a result of X-ray (Compton and Rayleigh) scattering through the sample mass [97]. The X-ray phase contrast imaging (XPCI) technique has also been referred to as an X-ray alternative to the optical schlieren method [98], as it is highly sensitive to phase boundaries but insensitive to absolute density. As a result, it is suitable for characterizing the geometry of microorifices, as well as the dynamics of cavitating flows [99,100]. To be able to capture refracted rays at the interfaces, the technique requires a partially coherent X-ray beam and an increased distance between the detector and the sample (typically somewhat less than 1 m [51]) compared to the absorption mode as well.

Subsequently, to convert the X-ray beam to visible light, a scintillator screen is used, and images are captured by a high-speed camera [101]. Even though XPCI is qualitative in nature and cannot provide an absolute measurement of vapor content, it has been established as a valuable technique to observe highly transient cavitation regimes in micrometer- and millimeter-scale geometries, owing to its high temporal and spatial resolution.

7.3.4.1 XPCI Examples in Wall-Bounded Cavitating Flows

Duke et al. [102] employed both X-ray raster scan radiography and XPCI to fully elucidate in a qualitative and quantitative manner the topology and dynamics of cavitation arising in an injector replica. An abruptly constricted nozzle with a constant nontapered throat diameter of 500 μm and a length of 3.05 mm made of polycarbonate was used. As in their previous studies, a commercial gasoline surrogate was circulated through the flow loop. For the absorption technique a monochromatic X-ray beam of 8-keV energy was focused to a $5 \times 6\ \mu m^2$ spot, whereas the XPCI experiment employed a polychromatic beam produced by a planar undulator. The undulator gap (top to bottom row magnets) was set between 16 and 19 mm, with a peak irradiance of 10^{14} ph/s/mm^2/0.1%BW at 17 mm. To convert X-rays to visible light, a LuAG:Ce scintillator [103] of 100-μm thickness was used, and images were recorded at a 10-kHz frequency with a spatial resolution of 1.9 μm/pixel. The XPCI radiographs revealed small-scale transient features with a typical lifetime of 5–10 Ms due to micron-scale defects in the nozzle wall. Moreover, it was revealed that pockets of liquid trapped between the vapor film and the nozzle wall traveled upstream at relatively slow velocities (1–2 m/s) in comparison with the average axial velocity of 57 m/s. It was concluded that XPCI was able to elucidate interfacial fluctuations that could not be captured by the raster scan absorption technique.

In a follow-up study by Duke et al. [104], hydraulic and X-ray setups similar to that of Duke et al. [102] were employed to examine the effect of fuel gas content on cavitation evolution. A polychromatic collimated beam at the APS 32-ID beamline [105] provided the necessary high flux for XPCI, which spanned the full cross section of the nozzle (0.5 mm), ensuring a wide field of view. The optical setup resulted in a – μm/pixel spatial resolution at a 40-μs temporal resolution. Small oscillations in the vapor-liquid interface close to the nozzle wall were detected for the degassed fuel, which were attributed to an oscillating reentrant jet flow at the region of cavitation collapse. At a second step, the fuel was exposed to ambient air, which resulted in the emergence of large gas bubbles with diameters of 10–150 μm nucleating at kilohertz time scales from a small machining defect in the nozzle wall.

Karathanassis et al. [97] performed high-speed XPCI on the development of vortical (string) cavitation arising within a throttle orifice. The geometry of the orifice and the presence of a metallic needle further constricting the flow path produced an asymmetrical flow entrance to the 1.5-mm-diameter orifice, thus promoting flow phenomena similar to those prevailing in diesel injectors. The working medium was therefore commercial diesel fuel, and the flow conditions were characterized by Reynolds numbers of 18,000 to 35,500. Measurements were conducted at the APS, using a high-flux white beam of 6 keV and resulted in spatial and temporal resolutions of 5 μm/pixel and 15 μs, respectively. It was demonstrated that underlying flow features, such as vortex intensity and level of turbulence, could be derived, owing to the capturing of highly transient cavitation features in the obtained radiographs. For instance, the intensity of the underlying vortex giving rise to cavitation could be predicted by the theoretical model reported by Pennings et al. [106]. Moreover, measured velocity fluctuations confirmed a suppression of turbulence in the presence of well-established cavitation, as was also found by Koukouvinis et al. [11].

In a follow-up experiment by Karathanassis et al. [107], XPCI was selected to comparatively assess a base diesel fuel and a non-Newtonian blend containing a viscoelasticity-inducing agent in terms of cavitation topology and dynamics. Imaging was conducted at the same carbon-fiber test piece employed in [97] for two fixed values of needle lift (see Fig. 7.14A). A 12-keV white beam from the synchrotron storage ring was utilized to obtain line-of-sight, time-resolved radiographs from two views at an image acquisition frequency of 67,890 Hz with a spatial resolution of 5 μm/pixel. String cavitation was found to be the prevailing regime for the low needle lift (=0.5 mm) examined, while both cloud and vortical cavities could be detected for lift equal to 1.0 mm, as illustrated in Fig. 7.14B and C. It was concluded that viscoelastic additives enhanced vortical yet suppressed cloud cavitation (Fig. 7.14D) as well as the level of turbulence in the orifice.

Moon et al. [108] characterized vortex cavitation in a high-pressure (160 MPa) diesel injector. The three-hole injector under investigation had a hole diameter and length of 112 μm and 825 μm, respectively. Since the injector material was steel, all X-ray photon energies below 40 keV were absorbed by the solid body, and higher energies had to be used for which diesel became transparent to the X-rays. Therefore 1-μm-sized tungsten particles were added as tracers in the liquid matrix (5% v/v) to enhance the absorption level of the fuel so that radiographs with enough contrast can be recorded, as shown in Fig. 7.15. The XPCI setups at beamlines 7-ID and 32-ID of the APS were used, offering polychromatic X-ray beams with sufficient photon flux to achieve high temporal (~68 kHz) and spatial (1.9 μm) resolutions. It was demonstrated that the X-ray tracer method that was used could reveal the transient behavior of vortex flow, string cavitation, and film cavitation inside the steel micronozzle under realistic operating conditions.

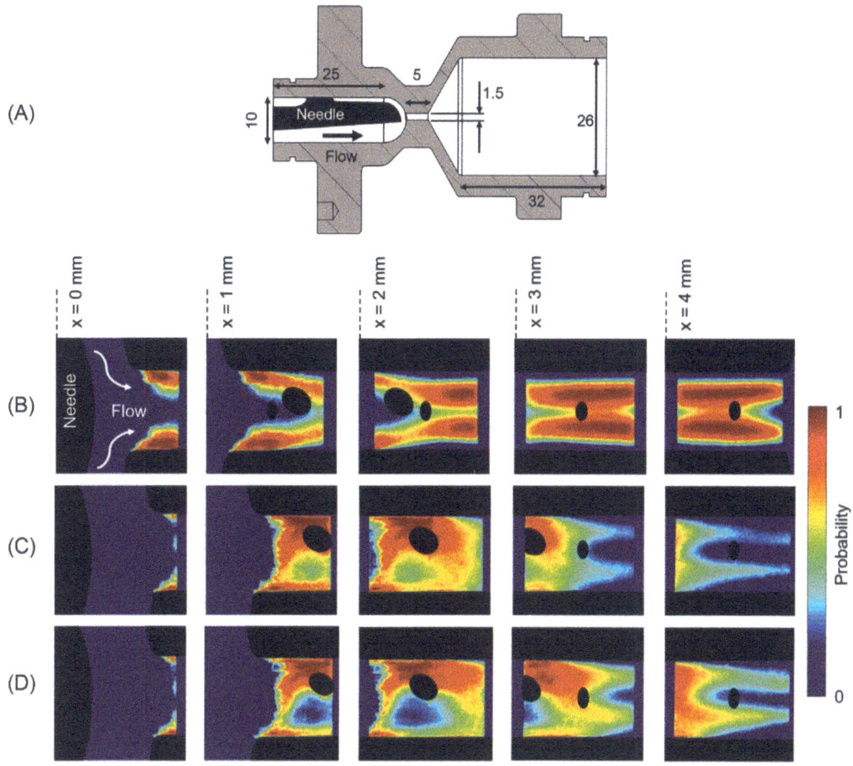

Figure 7.14 (A) Internal geometrical layout of the orifice employed in [97,107]. A static metallic needle is inserted in the upstream chamber to mimic the flow path of a fuel injector. (B–D) Mean probability of cavitation presence calculated by 16,000 radiographs. The orifice was irradiated at successive overlapping locations because of the collimated nature of the beam (interrogation window of 2.56 mm). (B) Base diesel, needle lift = 0.5 mm (Re = 35,500, CN = 7.7): Elongated vaporous structures occupy the entire orifice cross section. The cavities stem from a pair of longitudinal vortices. (C) Base diesel, needle lift = 1.0 mm (Re = 35,500, CN = 6.0): An asymmetrical cloud cavity establishes in the entrance region, while vortical cavitation is also evident in the panels past the orifice midlength ($x \geq 3$ mm). (D) Additized diesel, needle lift = 1.0 mm (Re = 35,500, CN = 6.0). Reduced probability values are evident compared to the base fuel in the regions of cloud cavitation ($x \leq 3$ mm), while the opposite is true in the regions where vortical cavitation prevails ($x > 3$ mm). Black ovals represent artifacts attributed to imperfections of the scintillator. *Reprinted from I.K. Karathanassis, K. Trickett, P. Koukouvinis, J. Wang, R. Barbour, M. Gavaises, Illustrating the effect of viscoelastic additives on cavitation and turbulence with X-ray imaging, Scientific Reports 8 (1) (2018) 14968, licensed under a Creative Commons Attribution 4.0 international (CC BY 4.0), https://creativecommons.org/licenses/by/4.0/.*

(A)

(B)

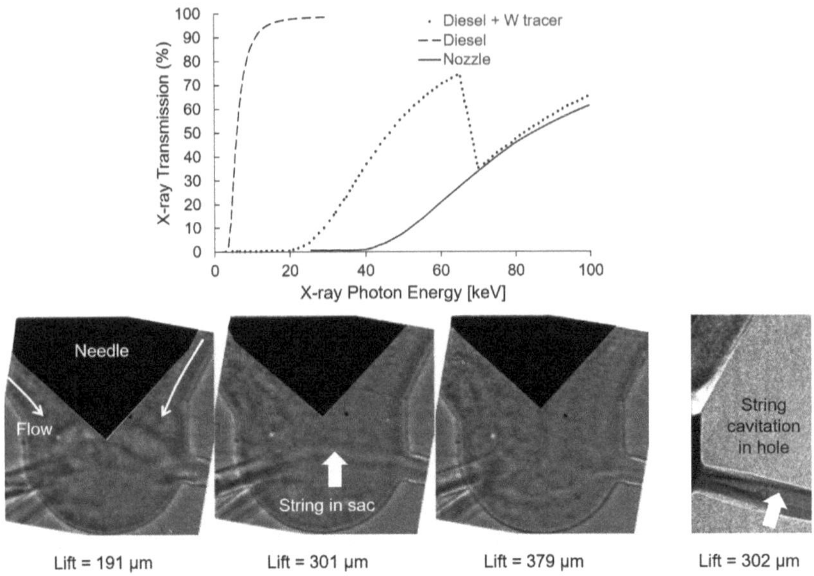

Figure 7.15 (A) X-ray transmission in the injector sac volume filled with air, diesel, and a diesel-tracer mixture. (B) Raw radiographs depicting vortical (string) cavitation in both the injector sac and hole regions. *Reproduced from S. Moon, W. Huang, J. Wang, First observation and characterization of vortex flow in steel micronozzles for high-pressure diesel injection, Experimental Thermal and Fluid Science 105 (2019) 342–348, Copyright © 2019 Elsevier Inc.*

Moreover, it was verified that the effect of needle lift on vortex strength and string cavitation topology resembled that of large-scale nozzle replicas.

Prasetya et al. [109] performed XPCI of the cavitating flow in acrylic rectangular step nozzles, that is, resembling forward-facing steps, with widths of 0.25 to 2.0 mm, all having a length-to-width ratio of 4. Filtered tap water was used as the working fluid at inlet pressure of 2.0 bar and atmospheric outlet. The X-ray setup employed a 12-keV polychromatic X-ray beam, produced by the helical undulator of the Spring-8 synchrotron facility located in Sayo district, Japan. A Lu-Si-O scintillator crystal was used to convert the X-ray beam into visible light. Still images, that is, not in a time sequence, of different cavitation regimes, namely, incipient cavitation, supercavitation, and hydraulic flip [2], were captured through both XPCI and visible light. Based on the dimensions of the nozzles, the range of Reynolds numbers characterizing the flow conditions was 4700–43,700. To increase the spatial resolution of the radiographs, the field of view was limited to $0.4 \times 0.5\ mm^2$, resulting in a sequential visualization along the orifice length, similar to [97,107], at a spatial resolution of 1 μm/pixel. However, the authors mentioned that bubbles less than 4.0 μm in diameter could not be observed clearly. The comparison of visible light and XPCI data confirmed the capability of the latter method to resolve optically dense two-phase mixtures, as opaque regions in the optical visualization were revealed to comprise a multitude of arbitrarily deformed cavitation bubbles. It was also shown that initially at low velocities, cavitation set in as a cloud of nonspherical bubbles, while by increasing the velocity from 11.5 to 17.7 m/s, the topology transformed into a continuous cavitation film. Furthermore, it was found that some degree of scalability existed between nozzle size and cavitation bubble diameter.

7.3.5 X-RAY FLUORESCENCE SPECTROSCOPY

An X-ray technique based on fluorescence photons has been developed similar to visible/UV fluorescence with the major difference that the emitted photons also have energies in the X-ray regime, hence preserving the advantages of X-ray diagnostics compared to optical imaging. When an X-ray photon expels an electron from an inner atomic shell, the created hole leaves the atom in an excited (ionized) state. Subsequently, an electron from the outer shells fills the vacancy. This process is accompanied by releasing energy equal to the difference between the donor and vacant shell energies, which is sufficient to produce an X-ray photon known as X-ray fluorescence. The method has been demonstrated to be applicable to elucidate fluid-mixing processes, and it has been recently employed in cavitation research as well. Typically, the flow is seeded with dopants, and to differentiate the signal produced by fluorescence photons from that of elastically scattered photons, energy-dispersive photon-counting detectors are employed [51].

7.3.5.1 XFS Examples in Wall-Bounded Cavitating Flows

As was discussed in the previous sections focusing on absorption imaging (Section 7.2.1) and XPCI (Section 7.3), Duke et al. [77,79,102] explored different X-ray diagnostic techniques to characterize the cavitating flow of a gasoline surrogate in a 0.5-mm-diameter nozzle. A set of experiments [104] included X-ray fluorescence spectroscopy as well. In the fluorescence test cases, the fluid was doped with tetrabromomethane (CBr4) at a concentration of 400 parts per million. These tracer particles had a high fluorescence yield, which allowed approximately 50% of the fluorescence photons to reach to the detector without being reabsorbed through the surrounding fluid [110]. Fluorescence measurements were conducted with a mean beam energy of 14.7 keV to surpass the K-shell electron-binding energy of bromine. A photon-counting silicon drift detector was placed perpendicular to the beam. By assuming that CBr4 was homogeneously dispersed in the liquid fuel, the detected fluorescence emission could be considered proportional to the beam fraction that passed through the fuel liquid phase. It should be noted that the acquired fluorescence signal was in arbitrary units in proportion to the photon count rate. X-ray absorption data were employed to calibrate the fluorescence results in an attempt to render them quantitative, hence obtaining line-of-sight measurements of the fluid projected density.

Duke et al. [111] applied X-ray fluorescence to measure liquid density and noncondensable gas concentration at the same time in a submerged test nozzle similar to that used by Duke et al. [104] under cavitating conditions. To have high fluorescence energies, dissolved air was removed from the fuel and replaced by Krypton (Kr). A monochromatic X-ray beam excited fluorescence emissions from Br and Kr simultaneously, which were proportional to the total liquid and noncondensable gas mass fractions, respectively. The sample was raster scanned with resolutions of 150 μm and 8 μm in the axial and transverse directions, respectively, to map the fluorescence signal throughout the nozzle. The high fluorescence yields and emission energies of Br and Kr allowed infinitesimal mass fractions of gas down to 10^{-5} to be measured with an uncertainty of 8%. The quantitative data revealed that a significant percentage of gas extent in the orifice outlet region was owed to nucleated gas rather than to cavitation-induced vapor.

In a manner similar to that of the previous two XF studies [104,111], Duke et al. [112] employed a high-energy, monochromatic X-ray beam of 42.5 keV to excite fluorescence from a cerium-doped gasoline surrogate flowing in aluminum nozzles with diameters of 0.36–0.90 mm and wall thicknesses of 2–3 mm. The fluid was doped with 4% by mass cerium nanoparticles, and fluid property measurements revealed that the fluid density and viscosity increased by 6% and 9%, respectively, while vapor pressure was 40% higher compared to the base matrix. Nanoparticles were found to act as additional nucleation sites for the tested fluid, thus modifying the overall cavitation behavior.

7.4 Conclusions

This chapter summarized the basic principles of different X-ray measurement techniques, suitable for the characterization and quantification of wall-bounded cavitating flows, and pointed out their strong points over conventional visible light visualization. The fundamental physics governing X-ray interaction with matter, that is, scattering, absorption, transmission and refraction, as well as the equations describing wave propagation were reported. From a technological state-of-the-art standpoint, the infrastructure and operating principles of laboratory-scale X-ray sources and synchrotron facilities were outlined. Four major imaging techniques were identified: X-ray absorption, X-ray CT, phase contrast imaging, and fluorescence spectroscopy, and the available literature with reference to wall-bounded cavitating flows was reviewed to illustrate the technical challenges that must be addressed to achieve accurate qualitative and/or qualitative results in possibly a time-resolved manner. A range of parameters, including source type, X-ray energy and bandwidth, beam shape, pulsation mode, type of optics and choppers, distance from source to sample, and type of detectors converting X-ray energy to measurable signal designate to a great extent the performance of X-ray diagnostics for a specific irradiated sample.

References

[1] H.K. Suh, C.S. Lee, Effect of cavitation in nozzle orifice on the diesel fuel atomization characteristics, International Journal of Heat and Fluid Flow 29 (4) (2008) 1001–1009.

[2] A. Sou, S. Hosokawa, A. Tomiyama, Effects of cavitation in a nozzle on liquid jet atomization, International Journal of Heat and Mass Transfer 50 (17) (2007) 3575–3582.

[3] J.M. Desantes, R. Payri, F.J. Salvador, J. De la Morena, Influence of cavitation phenomenon on primary break-up and spray behavior at stationary conditions, Fuel 89 (10) (2010) 3033–3041.

[4] C. Badock, R. Wirth, A. Fath, A. Leipertz, Investigation of cavitation in real size diesel injection nozzles, International Journal of Heat and Fluid Flow 20 (5) (1999) 538–544.

[5] M.S. Krivokorytov, et al., Shaping and controlled fragmentation of liquid metal droplets through cavitation, Scientific Reports 8 (1) (2018) 597.

[6] C. Sauter, M.A. Emin, H.P. Schuchmann, S. Tavman, Influence of hydrostatic pressure and sound amplitude on the ultrasound induced dispersion and de-agglomeration of nanoparticles, Ultrasonics Sonochemistry 15 (4) (2008) 517–523.

[7] N. Shimpi, A. Mali, D.P. Hansora, S. Mishra, Synthesis and surface modification of calcium carbonate nanoparticles using ultrasound cavitation technique, Nanoscience and Nanoengineering 3 (1) (2015) 8–12.

[8] A. Karimi, J.L. Martin, Cavitation erosion of materials, International Metals Review 31 (1) (1986) 1–26.

[9] O. Asi, Failure of a diesel engine injector nozzle by cavitation damage, Engineering Failure Analysis 13 (7) (2006) 1126–1133.

[10] M. Gavaises, D. Papoulias, A. Andriotis, E. Giannadakis, A. Theodorakakos, Link between cavitation development and erosion damage in diesel injector nozzles, SAE Technical Paper (2007), No. 2007-01-0246.

[11] P. Koukouvinis, N. Mitroglou, M. Gavaises, M. Lorenzi, M. Santini, Quantitative predictions of cavitation presence and erosion-prone locations in a high-pressure cavitation test rig, Journal of Fluid Mechanics 819 (2017) 21–57.

[12] C. Haosheng, L. Jiang, C. Darong, W. Jiadao, Damages on steel surface at the incubation stage of the vibration cavitation erosion in water, Wear 265 (5–6) (2008) 692–698.

[13] A. Adamkowski, A. Henke, M. Lewandowski, Resonance of torsional vibrations of centrifugal pump shafts due to cavitation erosion of pump impellers, Engineering Failure Analysis 70 (2016) 56–72.

[14] P. Ausoni, M. Farhat, X. Escaler, E. Egusquiza, F. Avellan, Cavitation influence on von Kármán vortex shedding and induced hydrofoil vibrations, Journal of Fluids Engineering 129 (8) (2007) 966.

[15] P. Testud, P. Moussou, A. Hirschberg, Y. Aurégan, Noise generated by cavitating single-hole and multi-hole orifices in a water pipe, Journal of Fluids and Structures 23 (2) (2007) 163–189.

[16] R.H. Mellen, Ultrasonic spectrum of cavitation noise in water, Journal of the Acoustical Society of America 26 (3) (1954) 356–360.

[17] X. Peng, B. Wang, H. Li, L. Xu, M. Song, Generation of abnormal acoustic noise: Singing of a cavitating tip vortex, Physical Review Fluids 2 (5) (2017) 053602.

[18] H. Ganesh, S.A. Mäkiharju, S.L. Ceccio, Bubbly shock propagation as a mechanism for sheet-to-cloud transition of partial cavities, Journal of Fluid Mechanics 802 (2016) 37–78.

[19] H. Ganesh, Bubbly shock propagation as a cause of sheet to cloud transition of partial cavitation and stationary cavitation bubbles forming on a delta wing vortex, Doctoral dissertation, 2015.

[20] S.A. Mäkiharju, H. Ganesh, S.L. Ceccio, The dynamics of partial cavity formation, shedding and the influence of dissolved and injected non-condensable gas, Journal of Fluid Mechanics 829 (2017) 420–458.

[21] S. Barwey, H. Ganesh, M. Hassanaly, V. Raman, S. Ceccio, Data-based analysis of multimodal partial cavity shedding dynamics, Experiments in Fluids 61 (4) (2020) 1–21.

[22] S.A. Mäkiharju, The dynamics of ventilated partial cavities over a wide range of Reynolds numbers and quantitative 2D X-ray densitometry for multiphase flow, Doctoral dissertation, 2012.

[23] A.Y. Kravtsova, D.M. Markovich, K.S. Pervunin, M.V. Timoshevskiy, K. Hanjalić, High-speed visualization and PIV measurements of cavitating flows around a semi-circular leading-edge flat plate and NACA0015 hydrofoil, International Journal of Multiphase Flow 60 (2014) 119–134.

[24] J. Wu, H. Ganesh, S. Ceccio, Multimodal partial cavity shedding on a two-dimensional hydrofoil and its relation to the presence of bubbly shocks, Experiments in Fluids 60 (4) (2019) 1–17.

[25] E.J. Foeth, C.W.H. van Doorne, T. van Terwisga, B. Wieneke, Time resolved PIV and flow visualization of 3D sheet cavitation, Experiments in Fluids 40 (4) (2006) 503–513.

[26] B. Stutz, S. Legoupil, X-ray measurements within unsteady cavitation, Experiments in Fluids 35 (2) (2003) 130–138.

[27] O. Boulon, M. Callenaere, J.-P. Franc, J.-M. Michel, An experimental insight into the effect of confinement on tip vortex cavitation of an elliptical hydrofoil, Journal of Fluid Mechanics 390 (1999) 1–23.

[28] S. Jahangir, E.C. Wagner, R.F. Mudde, C. Poelma, Void fraction measurements in partial cavitation regimes by X-ray computed tomography, International Journal of Multiphase Flow 120 (2019) 103085.

[29] D.D. van der Voort, N.J. Dam, R.P.J. Kunnen, G.J.F. van Heijst, H.J.H. Clercx, Effect of microbubble-induced cavitation on the dispersion of sprays, Physical Review Fluids 2 (3) (2017) 033601.

[30] Y. Saito, K. Sato, Cavitation bubble collapse and impact in the wake of a circular cylinder, Proceedings of the 5th International Symposium on Cavitation 4 (2003) 1–6.

[31] P. Kumar, D. Chatterjee, S. Bakshi, Experimental investigation of cavitating structures in the near wake of a cylinder, International Journal of Multiphase Flow 89 (2017) 207–217.

[32] Y. Wang, C. Xu, X. Wu, C. Huang, X. Wu, Ventilated cloud cavitating flow around a blunt body close to the free surface, Physical Review Fluids 2 (8) (2017) 084303.

[33] D. Lohse, Bubble puzzles: from fundamentals to applications, Physical Review Fluids 3 (11) (2018) 1–42.

[34] W.K. Chan, Correlation between cavitation type and cavitation erosion in centrifugal pumps, International Journal of Heat and Fluid Flow 11 (3) (1990) 269–271.

[35] M. Cudina, Detection of cavitation phenomenon in a centrifugal pump using audible sound, Mechanical Systems and Signal Processing 17 (6) (2003) 1335–1347.

[36] X. Escalar, E. Egusquiza, M. Farhat, F. Avellan, M. Coussirat, Detection of cavitation in hydraulic turbines, Mechanical Systems and Signal Processing 20 (4) (2006) 983–1007.

[37] P.G. Aleiferis, J. Serras-Pereira, A. Augoye, T.J. Davies, R.F. Cracknell, D. Richardson, Effect of fuel temperature on in-nozzle cavitation and spray formation of liquid hydrocarbons and alcohols from a real-size optical injector for direct-injection spark-ignition engines, International Journal of Heat and Mass Transfer 53 (21) (2010) 4588–4606.

[38] C. Arcoumanis, H. Flora, M. Gavaises, M. Badami, Cavitation in real-size multi-hole diesel injector nozzles, SAE Transactions (2000) 1485–1500.

[39] V. Kini, C. Bachmann, A. Fontaine, S. Deutsch, J.M. Tarbell, Flow visualization in mechanical heart valves: occluder rebound and cavitation potential, Annals of Biomedical Engineering 28 (4) (2000) 431–441.

[40] J. Als-Nielsen, D. McMorrow, Elements of Modern X-Ray Physics, second ed., Wiley, 2011.

[41] T.J. Heindel, A review of X-ray flow visualization with applications to multiphase flows, Journal of Fluids Engineering 133 (7) (2011) 074001.

[42] S.A. Zhou, A. Brahme, Development of phase-contrast X-ray imaging techniques and potential medical applications, Physica Medica 24 (3) (2008) 129–148.

[43] B.M. Weon, J.H. Je, Y. Hwu, G. Margaritondo, Phase contrast X-ray imaging, International Journal of Nanotechnology 3 (2–3) (2006) 280–297.

[44] B.L. Henke, E.M. Gullikson, J.C. Davis, X-Ray interactions: photoabsorption, scattering, transmission and reflection E = 50–30,000 eV, Z = 1–92, Atomic Data and Nuclear Data Tables 54 (2) (1993) 181–342.

[45] P. Russo, Physical basis of X-ray imaging, Comprehensive Biomedical Physics, Elsevier, 2014, pp. 1–48.

[46] R.W. James, Optical Principles of the Diffraction of X-rays, Bell Sons, 1948.

[47] T. Grassler, K.-E. Wirth, X-ray computer tomography: potential and limitation for the measurement of local solids distribution in circulating fluidized beds, Chemical Engineering Journal 77 (1–2) (2000) 65–72.

[48] J.H. Hubbell, S.M. Seltzer, Tables of X-ray Mass Attenuation Coefficients and Mass Energy-Absorption Coefficients 1 keV to 20 meV for Elements z = 1 to 92 and 48 Additional Substances of Dosimetric Interest, Ionizing Radiation Division, National Institute Standards Technology, Gaithersburg, MD (United States), 1995.

[49] D.F. Swinehart, The Beer-Lambert law, Journal of Chemical Education 39 (7) (1962) 333–335.

[50] H. Morneburg, Bildgebende Systeme für die medizinische Diagnostik: Röntgendiagnostik und Angiographie, Computertomographie, Nuklearmedizin, Magnetresonanztomographie, Sonographie, integrierte Informationssysteme, Publicis MCD-Verlag, 1995.

[51] A. Kastengren, C.F. Powell, Synchrotron X-ray techniques for fluid dynamics, Experiments in Fluids 55 (3) (2014) 1686.

[52] R.A. Ketcham, W.D. Carlson, Acquisition, optimization and interpretation of X-ray computed tomographic imagery: applications to the geosciences, Computers & Geosciences 27 (4) (2001) 381–400.

[53] E. Storm, D.W. Lier, H.I. Israel, Photon sources for instrument calibration, Health Physics 2 (1974) 179–189.

[54] G.F. Knoll, Radiation Detection and Measurement, third ed., Wiley, 2000.

[55] M. Lorenzi, N. Mitroglou, M. Santini, M. Gavaises, Novel experimental technique for 3D investigation of high-speed cavitating diesel fuel flows by X-ray micro computed tomography, Review of Scientific Instruments 88 (3) (2017) 033706.

[56] M. Piller, D. Casagrande, G. Schena, M. Santini, Pore-scale simulation of laminar flow through porous media, Journal of Physics: Conference Series 501 (1) (2014) 012010.

[57] M. Santini, Fundamentals and recent advances in X-ray micro computed tomography (microCT) applied on thermal-fluid dynamics and multiphase flows, Journal of Physics: Conference Series 655 (1) (2015) 012004.

[58] D.A. Walko, D.A. Arms, A. Miceli, A.L. Kastengren, Empirical dead-time corrections for energy-resolving detectors at synchrotron sources, Nuclear Instruments and Methods in Physics Research Section A Accelerators Spectrometers Detectors and Associated Equipment 649 (1) (2011) 81–83.

[59] A. Snigirev, V. Kohn, I. Snigireva, B. Lengeler, A compound refractive lens for focusing high-energy X-rays, Nature 384 (6604) (1996) 49–51.

[60] K. Evans-Lutterodt, et al., Single-element elliptical hard X-ray micro-optics, Optics Express 11 (8) (2003) 919.

[61] P.J. Eng, M. Newville, M.L. Rivers, S.R. Sutton, Dynamically figured Kirkpatrick Baez X-ray microfocusing optics, International Society for Optics and Photonics, X-Ray Microfocusing: Applications and Techniques 3449 (1998) 145–156.

[62] J.H. Beaumont, M. Hart, Multiple Bragg reflection monochromators for synchrotron x radiation, Journal of Physics E: Scientific Instruments 7 (10) (1974) 823–829.

[63] U. Bonse, G. Materlik, W. Schröder, Perfect-crystal monochromators for synchrotron X-radiation, Journal of Applied Crystallography 9 (3) (1976) 223–230.

[64] M. Cammarata, et al., Chopper system for time resolved experiments with synchrotron radiation, Review of Scientific Instruments 80 (1) (2009) 015101.

[65] M. Gembicky, D. Oss, R. Fuchs, P. Coppens, A fast mechanical shutter for submicrosecond time-resolved synchrotron experiments, Journal of Synchrotron Radiation 12 (5) (2005) 665–669.

[66] A. McPherson, W.K. Lee, D.M. Mills, A synchronized rotating crystal X-ray beam chopper, Review of Scientific Instruments 73 (8) (2002) 2852.

[67] H. Osawa, T. Kudo, S. Kimura, Development of high-repetition-rate X-ray chopper system for time-resolved measurements with synchrotron radiation, Japanese Journal of Applied Physics 56 (4) (2017) 048001.

[68] A.A. Kendoush, A comparative study of the various nuclear radiations used for void fraction measurements, Nuclear Engineering and Design 137 (2) (1992) 249–257.

[69] R.W. Pike, B. Wilkins Jr., H.C. Ward, Measurement of the void fraction in two-phase flow by X-ray attenuation, AIChE Journal 11 (5) (1965) 794–800.

[70] O.C. Jones, N. Zuber, The interrelation between void fraction fluctuations and flow patterns in two-phase flow, International Journal of Multiphase Flow 2 (3) (1975) 273–306.

[71] M.A. Vince, R.T. Lahey, On the development of an objective flow regime indicator, International Journal of Multiphase Flow 8 (2) (1982) 93–124.

[72] B. Stutz, J.L. Reboud, Experiments on unsteady cavitation, Experiments in Fluids 22 (3) (1997) 191–198.

[73] O. Coutier-Delgosha, J.-F. Devillers, T. Pichon, A. Vabre, R. Woo, S. Legoupil, Internal structure and dynamics of sheet cavitation, Physics of Fluids 18 (1) (2006) 017103.

[74] B. Stutz, J.-L. Reboud, Measurements within unsteady cavitation, Experiments in Fluids 29 (6) (2000) 545–552.

[75] V. Aeschlimann, S. Barre, S. Legoupil, X-ray attenuation measurements in a cavitating mixing layer for instantaneous two-dimensional void ratio determination, Physics of Fluids 23 (5) (2011) 055101.

[76] S.A. Mäkiharju, C. Gabillet, B.-G. Paik, N.A. Chang, M. Perlin, S.L. Ceccio, Time-resolved two-dimensional X-ray densitometry of a two-phase flow downstream of a ventilated cavity, Experiments in Fluids 54 (7) (2013) 1561.

[77] D.J. Duke, A.L. Kastengren, F.Z. Tilocco, A.B. Swantek, C.F. Powell, X-ray radiography measurements of cavitating nozzle flow, Atomization and Sprays 23 (9) (2013) 841–860.

[78] S. Gopalan, J. Katz, Flow structure and modeling issues in the closure region of attached cavitation, Physics of Fluids 12 (4) (2000) 895–911.

[79] D.J. Duke, et al., Time-resolved X-ray radiography of cavitation in a metal nozzle, in: Proceedings of the ILASS – Europe, 2016.

[80] L. Barbaca, B.W. Pearce, H. Ganesh, S.L. Ceccio, P.A. Brandner, On the unsteady behaviour of cavity flow over a two-dimensional wall-mounted fence, Journal of Fluid Mechanics 874 (2019) 483–525.

[81] R.A. Williams, C.-G. Xie, Tomographic techniques for characterising particulate processes, Particle and Particle Systems Characterization 10 (5) (1993) 252–261.

[82] J. Hsieh, Computed Tomography: Principles, Design, Artifacts, and Recent Advances, vol. 114, SPIE Press, 2003.

[83] H.H. Barrett, W. Swindel, Radiological Imaging: The Theory of Image Formation, Detection, and Processing, Academic Press, 1996.

[84] A.C. Kak, M. Slaney, G. Wang, Principles of computerized tomographic imaging, Medical Physics 29 (1) (2002) 107.

[85] N. Mitroglou, M. Lorenzi, M. Santini, M. Gavaises, Application of X-ray micro-computed tomography on high-speed cavitating diesel fuel flows, Experiments in Fluids 57 (11) (2016) 175.

[86] S. Morooka, T. Ishizuka, M. Iizuka, K. Yoshimura, Experimental study on void fraction in a simulated BWR fuel assembly (evaluation of cross-sectional averaged void fraction), Nuclear Engineering and Design 114 (1) (1989) 91–98.

[87] G.D. Harvel, K. Hori, K. Kawanishi, J.S. Chang, Cross-sectional void fraction distribution measurements in a vertical annulus two-phase flow by high speed X-ray computed tomography and real-time neutron radiography techniques, Flow Measurement and Instrumentation 10 (4) (1999) 259–266.

[88] W. Hassan, S. Legoupil, D. Chambellan, S. Barre, Dynamic localization of vapor fraction in turbo pump inducers by X-ray tomography, IEEE Transactions on Nuclear Science 55 (1) (2008) 656–661.

[89] D. Bauer, H. Chaves, C. Arcoumanis, Measurements of void fraction distribution in cavitating pipe flow using X-ray CT, Measurement Science and Technology 23 (5) (2012) 055302.

[90] L.A. Feldkamp, L.C. Davis, J.W. Kress, Practical cone-beam algorithm, Journal of the Optical Society of America A 1 (6) (1984) 612.

[91] D. Bauer, F. Barthel, U. Hampel, High-speed X-ray CT imaging of a strongly cavitating nozzle flow, Journal of Physics Communications 2 (7) (2018) 075009.

[92] S. Jahangir, W. Hogendoorn, C. Poelma, Dynamics of partial cavitation in an axisymmetric converging-diverging nozzle, International Journal of Multiphase Flow 106 (2018) 34–45.

[93] M. Brunhart, et al., Investigation of cavitation and vapor shedding mechanisms in a venturi nozzle, Physics of Fluids 32 (8) (2020) 083306.

[94] A. Tekawade, B.A. Sforzo, K.E. Matusik, K. Fezzaa, A.L. Kastengren, C.F. Powell, Time-resolved 3D imaging of two-phase fluid flow inside a steel fuel injector using synchrotron X-ray tomography, Scientific Reports 10 (1) (2020) 1–9.

[95] A. Kastengren, C.F. Powell, D. Arms, E.M. Dufresne, H. Gibson, J. Wang, The 7BM beamline at the APS: a facility for time-resolved fluid dynamics measurements, Journal of Synchrotron Radiation 19 (4) (2012) 654–657.

[96] K.A. Nugent, T.E. Gureyev, D.F. Cookson, D. Paganin, Z. Barnea, Quantitative phase imaging using hard X rays, Physical Review Letters 77 (14) (1996) 2961–2964.

[97] I.K. Karathanassis, et al., High-speed visualization of vortical cavitation using synchrotron radiation, Journal of Fluid Mechanics 838 (2018) 148–164.

[98] G.S. Settles, Schlieren and Shadowgraph Techniques, Springer, Berlin Heidelberg, 2006.

[99] W.-K. Lee, K. Fezzaa, J. Wang, Metrology of steel micronozzles using X-ray propagation-based phase-enhanced microimaging, Applied Physics Letters 87 (8) (2005) 084105.

[100] D. Duke, et al., X-ray imaging of cavitation in diesel injectors, SAE International Journal of Engines 7 (2) (2014). 2014−01−1404.

[101] T. Martin, A. Koch, Recent developments in X-ray imaging with micrometer spatial resolution, Journal of Synchrotron Radiation 13 (2) (2006) 180−194.

[102] D.J. Duke, et al., Comparing simulations and X-ray measurements of a cavitating nozzle, in: Proceedings of ILASS-Americas, 2014.

[103] J.A. Mares, et al., Scintillation response of Ce-doped or intrinsic scintillating crystals in the range up to 1 MeV, Radiation Measurements 38 (4−6) (2004) 353−357.

[104] D. Duke, A. Swantek, A. Kastengren, K. Fezzaa, C. Powell, Recent developments in X-ray diagnostics for cavitation, SAE International Journal of Fuels and Lubricants 8 (1) (2015). 2015−01−0918.

[105] Q. Shen, et al., Dedicated full-field X-ray imaging beamline at advanced photon source, Nuclear Instruments and Methods in Physics Research Section A Accelerators Spectrometers Detectors and Associated Equipment 582 (1) (2007) 77−79.

[106] P.C. Pennings, J. Bosschers, J. Westerweel, T.J.C. van Terwisga, Dynamics of isolated vortex cavitation, Journal of Fluid Mechanics 778 (2015) 288−313.

[107] I.K. Karathanassis, K. Trickett, P. Koukouvinis, J. Wang, R. Barbour, M. Gavaises, Illustrating the effect of viscoelastic additives on cavitation and turbulence with X-ray imaging, Scientific Reports 8 (1) (2018) 14968.

[108] S. Moon, W. Huang, J. Wang, First observation and characterization of vortex flow in steel micronozzles for high-pressure diesel injection, Experimental Thermal and Fluid Science 105 (2019) 342−348.

[109] R. Prasetya, A. Sou, S. Moon, R.H. Pratama, Y. Wada, H. Yokohata, X-ray phase contrast imaging of cavitation and discharged liquid jet in nozzles with various sizes, Atomization and Sprays 29 (1) (2019) 59−78.

[110] M.S. Del Río, R.J. Dejus, XOP 2.1− a new version of the X-ray optics software toolkit, AIP Conference Proceedings 705 (1) (2004) 784−787.

[111] D.J. Duke, A.L. Kastengren, A.B. Swantek, K.E. Matusik, C.F. Powell, X-ray fluorescence measurements of dissolved gas and cavitation, Experiments in Fluids 57 (10) (2016) 162.

[112] D.J. Duke, A.L. Kastengren, K.E. Matusik, C.F. Powell, Hard X-ray fluorescence spectroscopy of high pressure cavitating fluids in aluminum nozzles, International Journal of Multiphase Flow 108 (2018) 69−79.

Chapter 8

Interaction of Cavitation with Sprays in High-Pressure Diesel Injection Systems

Joonsik Hwang[1,2]

[1]Department of Mechanical Engineering, Mississippi State University, Starkville, MS, United States, [2]Center for Advanced Vehicular Systems (CAVS), Mississippi State University, Starkville, MS, United States

8.1 Introduction

Spray atomization and mixing processes dominate combustion and emission characteristics in modern diesel engines that employ direct fuel injection into the combustion chamber [1–3]. High-pressure injection systems with sophisticated injection strategies are applied to reduce emissions while maintaining high fuel efficiency [4,5]. Meanwhile, as injection pressure increases, there tends to be a large pressure gradient in the nozzle. The pressure drop may cause local boiling-off of liquid fuel and create vapor pockets called cavitation in the sac or nozzle orifice. This is usually initiated at a low-pressure region in the nozzle, such as a sharp turn at the upper corner of the nozzle orifice, especially during needle opening and closing periods. With cavitating flow, the effective flow area decreases, owing to the cavitation bubbles, so the flow rate decreases. Cavitating flow employs high turbulence with a complex liquid-vapor distribution, so it can improve air-fuel mixing but also cause fluctuations in fuel injection quantity.

Cavitation and Bubble Dynamics. DOI: https://doi.org/10.1016/B978-0-12-823388-7.00007-2

External flow characteristics can be altered by different in-nozzle flow structure with cavitation bubbles and unpredicted mass flow rate. Thus an in-depth understanding of the cavitation process is essential to achieve better nozzle configuration and injection strategies. Yet the behind the cavitation process in high-pressure diesel injector flow has not been fully understood. Fuel flow in a nozzle orifice moves at a rate of several hundred meters per second with highly turbulent characteristics; therefore capturing instant motion of flow is challenging. Typical diesel injector orifices are extremely small, about 1 mm long and 100 μm in diameter; thus the light setup for visualization is also limited, owing to the small nozzle size and rounded shape. Nevertheless, recent studies with advanced optical diagnostics have showed noticeable progress in understanding internal nozzle flow. This chapter aims to review the current state of experimental studies on cavitation and injection processes in high-pressure diesel. The first part of the chapter presents cavitation processes and the effects of nozzle geometry. Then bubble dynamics during transient needle motion and interaction between cavitation bubbles will be discussed. The last part will focus on the interaction between internal nozzle flow and spray combustion.

8.2 Cavitation in Internal Nozzle Flow

Several studies have investigated cavitation behavior using visible light diagnostics with transparent nozzles. Notable work was done by Gavaises et al., who tried to correlate cavitation structure and flow turbulence through a detailed study of cavitating flow in a 20 × enlarged multihole transparent nozzle [6]. The experiment was carried out at a fixed needle-lift condition with different injection pressure and ambient condition to set various Reynolds number (Re) and cavitation number conditions. A mixture of 32% tetraline and 68% turpentine oil was used as a fuel to match the refractive index with an acrylic nozzle. The researchers noted that the cavitation inception did not occur simultaneously in all orifices but rather occurred randomly, hole by hole in a multihole injector. They attributed this to hole-to-hole variations of the recirculation pattern in the sac volume. As cavitation numbers increased, the bubble pattern in the nozzle developed from inception to film-type cavitation. The transition of cavitation structure was also observed by Arcoumanis et al., who noted that individual bubbles formed large vapor pockets and finally film cavitation as the flow rate increased [7]. Meanwhile, string cavitation was also found between two adjacent holes, as shown in Fig. 8.1. The mechanism of string cavitation was explained as induction of a low-pressure zone in the core region of the sac volume by vortex flow, which created nuclei, and the existing vapor bubbles in the nozzle orifice was ingested into the core and finally connected as string cavitation.

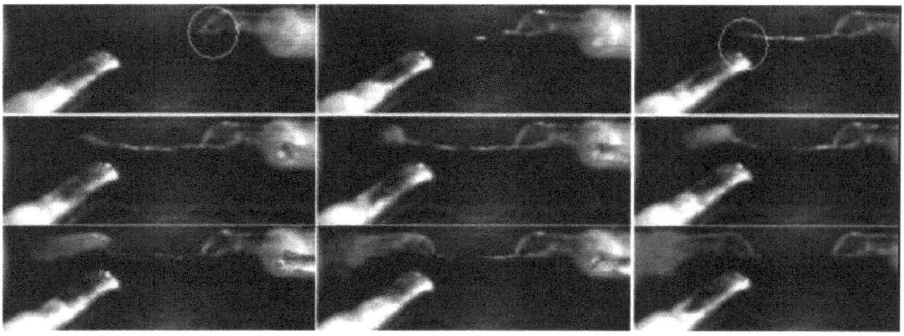

Figure 8.1 String cavitation in a multihole nozzle at a needle lift of 6 mm and a cavitation number of 4.34. Images show sequential time steps from left to right and from top to bottom with an interval of 0.2 ms. *Credit: M. Gavaises, C. Arcoumanis, H. Roth, Y.S. Choi, A. Theodorakakos, Nozzle flow and spray characteristics from VCO diesel injector nozzles, Thermo- and Fluid Dynamic Processes in Diesel Engines 2 (2004) 31–48. https://doi.org/10.1007/978-3-662-10502-3_3.*

It is known that nozzle geometry can affect the cavitation process. The shape of the nozzle orifice is usually described with a k-factor that is defined as follows:

$$k = \frac{D_{\mathrm{inlet}}[\mu\mathrm{m}] - D_{\mathrm{outlet}}[\mu\mathrm{m}]}{10} \tag{8.1}$$

According to the correlation above, convergent nozzles have positive k-factors, and divergent nozzles have negative k-factors. The effects of the k-factor on cavitation have been extensively investigated. Blessing et al. applied three different real-size transparent nozzles having k-factors of -2.5, 0, and 2.5 for cavitation imaging and found that cavitation formation was suppressed with convergent nozzles [8]. With less cavitation in the nozzle, the convergent nozzle showed a smaller spray cone angle than that of the straight nozzle. While the divergent nozzle had an opposite trend having a larger spray cone angle due to flow detachment and cavitation in the nozzle. The difference in internal nozzle flow has a direct impact on external flow; it is expected to have a longer liquid penetration length with a conical nozzle. Payri et al. demonstrated how nozzle design influences external sprays in diesel injection systems [9]. They applied two different nozzles with k-factors of 0 and 1.8 in nonevaporating ambient conditions. The major finding was that even though the mass flow rate decreased with conical nozzle, owing to the smaller orifice exit diameter, it showed smaller plume dispersion but longer liquid penetration length than those of straight nozzle. These results were consistent with previous studies.

Previously introduced studies used replicas of diesel nozzle made with transparent material and visible light sources for cavitation imaging. In contrast to visible light, which limits investigation in qualitative manners, X-rays enable quantifications of cavitating flow, since it has high penetrating power with negligible scattering issues. A recent study by Aniket et al. successfully addressed two-phase flow features in a time-resolved three-dimensional (3D) domain with X-ray tomography [10]. They used the Engine Combustion Network (ECN) Spray C injector. The injector was installed in a rotational stage and operated with n-dodecane at an injection pressure of 1500 bar. A high-speed camera was utilized to capture a two-dimensional projected image and acquired an ensemble-averaged image through 200 injection events per viewing angle. High-resolution X-ray tomography of the ECN Spray C injector conducted by [28], revealed that the orifice was drilled off-center concerning the sac volume, so it has a 20-μm-wide ridge on the nozzle inlet (Matusik et al., 2017). Detailed features of the injector surface are shown in Fig. 8.2.

Since the nozzle orifice is asymmetric and has a sharp inlet at a side, a cavitation layer has been observed in this edge, and it was assumed to have a significant portion of cavitation bubbles at the orifice exit [12]. This hypothesis was confirmed by 3D computed tomography (CT) of internal nozzle flow, as shown in Fig. 8.3. It indicates that the cavitation layer was initiated from the upper side, where it has a sharp edge in the orifice inlet, and it developed through the orifice. Detailed features such as cavitation layer thickness and wrinkles were able to be captured by the CT. One interesting feature is the presence of a liquid jet in the middle of the gas layer, as shown in the slice image at $x = -0.638$ μm. The liquid jet appeared between 0.8 ms and 0.9 ms after the injection command and disappeared at approximately 1.4 ms. On the basis of the weaker contrast and blurrier pixels at this spot, it is believed to be highly transient but repeatable, since it appears in statistically averaged images from several thousand injections. Regardless of the liquid jet in the upper orifice region, the effective flow area was reduced to 60%–70% compared to the total cross-sectional area of the orifice

Figure 8.2 Surface features of the engine combustion network (ECN) Spray C injector from high-resolution X-ray computed tomography (CT). Planar image from synchrotron X-ray CT scan (left), entire isosurface of the injector (middle), and orifice nozzle inlet image (right). *Credit: A. Tekawade, B.A. Sforzo, K.E. Matusik, K. Fezzaa, A.L. Kastengren, C.F. Powell, Time-resolved 3D imaging of two-phase fluid flow inside a steel fuel injector using synchrotron X-ray tomography, Scientific Reports 10 (2020) 1–9. https://doi.org/10.1038/s41598-020-65701-x.*

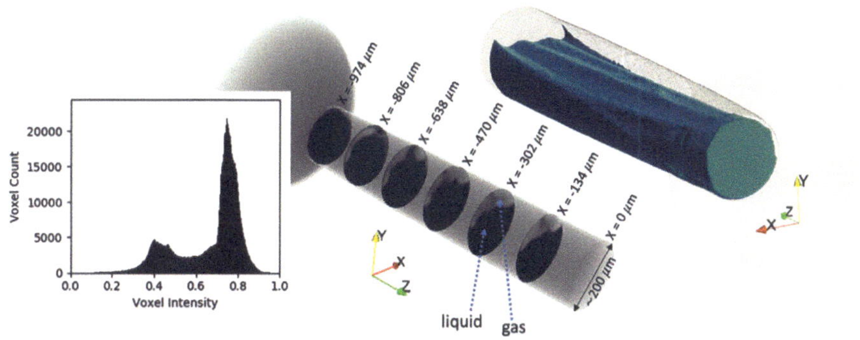

Figure 8.3 Six different slices of information from fully reconstructed computed tomography (CT) volume (left) and 3D volume rendering of the liquid phase in the nozzle. Note the axis directions in the figure. The nozzle orifice was set as $x = 0$ μm. *Credit: Tekawade, A., Sforzo, B.A., Matusik, K.E., Fezzaa, K., Kastengren, A.L., Powell, C.F., Time-resolved 3D imaging of two-phase fluid flow inside a steel fuel injector using synchrotron X-ray tomography, Scientific Reports 10 (2020) 1–9. https://doi.org/10.1038/s41598-020-65701-x.*

outlet. This result is consistent with a previous study that showed a lower mass flow rate with ECN spray C injector compared to Spray D injector that has even smaller nozzle diameter [13]. This implies that the flow rate was restricted by cavitating flow in the nozzle orifice.

Not only injector geometry but also fuel type has a great influence on internal nozzle flow, so there has been attention to investigate fuel effects on cavitation process in the diesel injection system. Park et al. applied soybean-based biodiesel in two different transparent nozzles (different *L/D* ratios of the nozzles) to compare internal nozzle flow with baseline diesel fuel [16]. Droplet diameter was measured by using a phase Doppler particle analyzer to find a correlation between cavitation and atomization characteristics. They found that the lower *L/D* ratio case showed stronger cavitation in the nozzle, which induced a larger spray cone angle and smaller drop size compared to that of the higher *L/D* ratio case. In the aspect of fuel, biodiesel suppressed cavitation, and a larger droplet size was measured than with diesel fuel. Suh et al. also presented a lower level of cavitation with biodiesel compared to diesel [27]. They attributed this to different fuel properties, such as higher viscosity and surface tension, with biodiesel than those of diesel. Apparently, the general behavior of an external flow can be expected with nondimensional numbers such as Re and Weber number (We), defined as in the following equations:

$$\mathrm{Re} = (\rho_f V d_o)/\mu = (V d_o)/\nu \quad (8.2)$$

$$\mathrm{We} = (\rho_f V^2 d_o)/\sigma \quad (8.3)$$

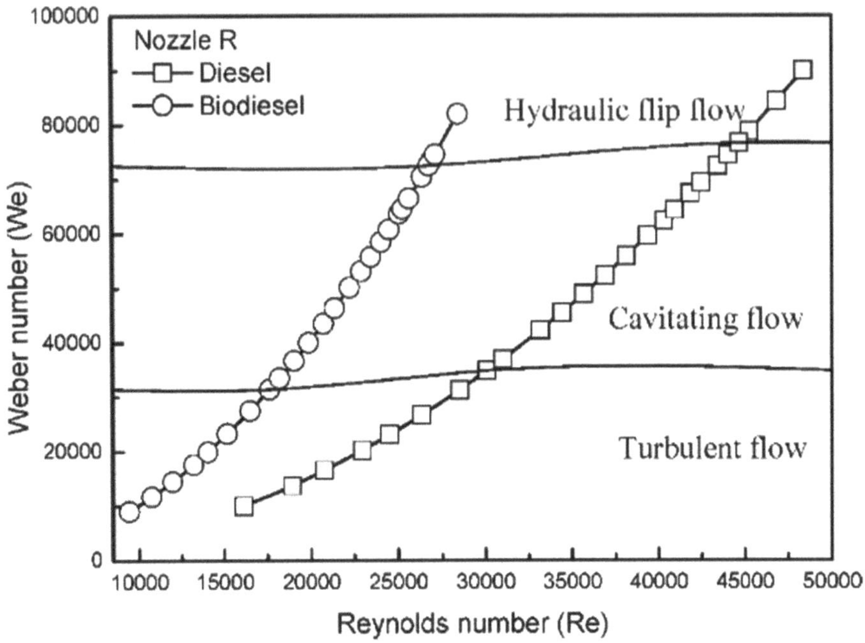

Figure 8.4 Comparison of flow behavior between biodiesel and diesel according to We and Re numbers. *Credit: S.H. Park, H.K. Suh, C.S. Lee, Effect of cavitating flow on the flow and fuel atomization characteristics of biodiesel and diesel fuels, Energy and Fuels 22 (2008) 605–613. https://doi.org/10.1021/ef7003305.*

where ρ_f is the fuel liquid density, V is the mean injection flow velocity, d_o is the nozzle hole diameter, μ is the dynamic viscosity, ν is the kinematic viscosity, and σ is the surface tension.

Biodiesel fuels are known to have larger fuel density, viscosity, and surface tension. Viscosity and surface tension are almost 2 and 1.2 times larger, respectively, than those of diesel. Thus the Re and We numbers of biodiesel are approximately 50% and 80% of those of diesel, respectively [16]. On the basis of a diagram suggested by Park et al., as shown in Fig. 8.4, biodiesel would not have cavitating flow, while diesel would at high Re numbers.

8.3 Bubble Dynamics in Transient Needle Motion

There has been less focus on bubble dynamics during transient injection events such as needle opening and closing periods. However, this aspect is gaining in importance in modern diesel engines that use short multiple injection strategies. In this situation, quasisteady flow does not represent flow characteristics, since the injector needle may not reach a steady state but may keep moving back and forth in a short period of time. Thanks to progress in imaging devices and manufacturing techniques, the capturing of the cavitation process with transient needle motion have been explored. Manin et al. visualized the gas exchange process in an ECN Spray D injector with a transparent acrylic nozzle installed at the machined injector tip [17]. The acrylic nozzle was investigated carefully to validate its geometry compared to original metal injector. A stereoscopic imaging system using two line-of-sight back-illumination systems was utilized in an orthogonal configuration. The illumination system relies on a custom-built nanosecond LED that is able to create a 10-ns pulse light at megahertz repetition [18]. This system enabled the capturing of the detailed gas exchange process and bubble dynamics in the injector. n-Dodecane fuel was injected with injection pressures of 500 and 1000 bar at ambient pressures from 1 to 20 bar. The major finding of this study was that the sac volume contained a large number of gas bubbles after the end of injection under low ambient pressure conditions. The difference of sac condition after the end of injection under 20 and 1 bar ambient conditions is shown in Fig. 8.5.

This image clearly shows the difference in the gas exchange process according to ambient pressure conditions. The low ambient pressure condition showed bulk cavitation (fuel vapor) of liquid fuel at 74 μs after the needle close. This bulk cavitation is known to occur with an intense pressure drop below vapor pressure in the fluid. However, at 124 μs after the needle close, this bulk cavitation disappeared as sac pressure recovered above the vapor pressure. The volume change inside the nozzle pulled ambient gas into the sac and orifice, as shown in 224 μs after needle close. However, this gas exchange process could not be seen under high ambient pressure conditions, owing to the absence of the bulk cavitation process. This result is consistent with previous studies that have shown gas bubbles after the end of injection [19]. Still, there is an open question as to whether the bubbles are fuel vapor or ambient gas. However, research by Swantek et al. clearly shows that the bubbles came from ambient gas through the gas exchange process rather than fuel vapor [20]. Recent computational fluid dynamics (CFD) simulation results of Battistoni also proved ambient gas ingestion after the end of the injection [21]. On the other hand, Manin et al. described the effects of residual gas on early spray development. The start of injection was visualized in an identical optical setup used for end-of-injection imaging. Fig. 8.6 shows the difference in spray morphology at the start of injection with full gas and liquid for initial sac conditions. With full gas as an initial condition, a mixture of gas and liquid was injected. It showed erratic spray morphology and a wider spray plume angle compared to the fully liquid condition in which very smooth geometries, a clear intact core, and a mushroom-shaped spray head were seen. The full gas condition also showed slower penetration because of the time required to fill the sac and orifice with liquid. This result again emphasizes how important it is to understand transient gas exchange to predict spray and mixture formation processes.

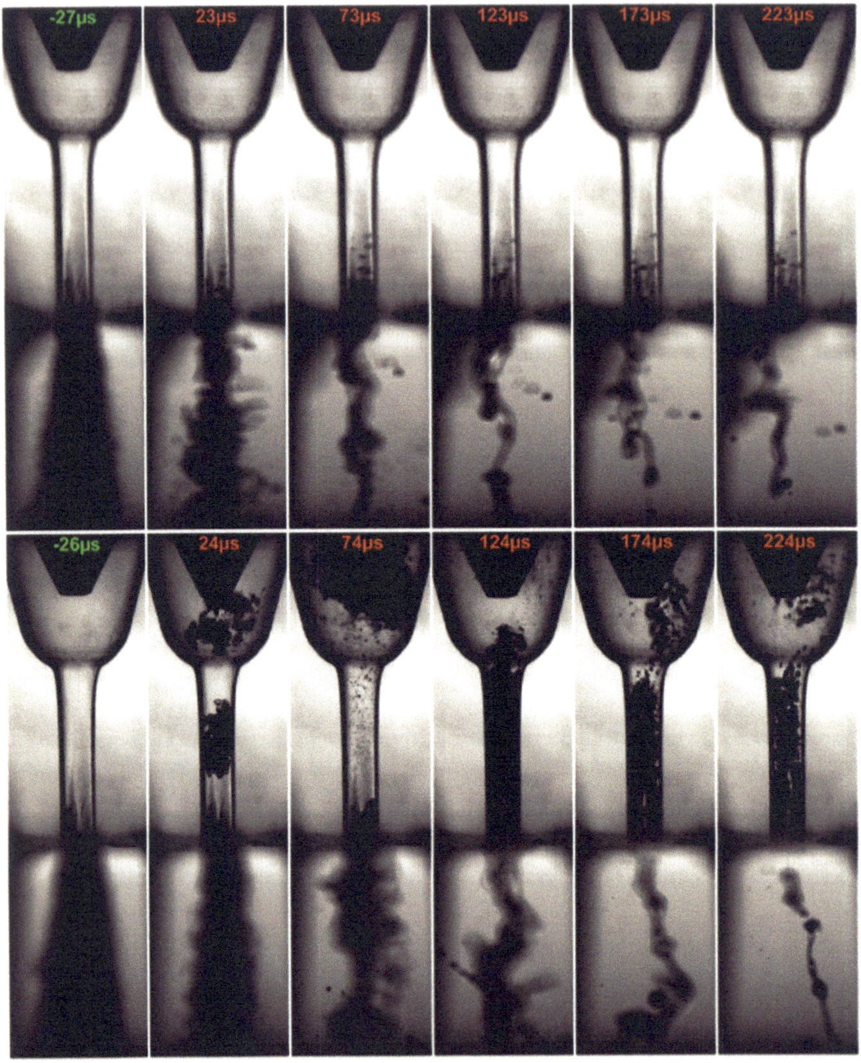

Figure 8.5 End of injection gas exchange processes with injection pressure of 500 bar and ambient pressures of 20 bar (top), and 1 bar (bottom) according to time after needle closing. Image focused on internal flow so external spray is blurred. *Credit: J. Manin, L.M. Pickett, K. Yasutomi, Stereoscopic high-speed microscopy to understand transient internal flow processes in high-pressure nozzles, Experimental Thermal and Fluid Science 114 (2020) 110027. https://doi.org/10.1016/j.expthermflusci.2019.110027.*

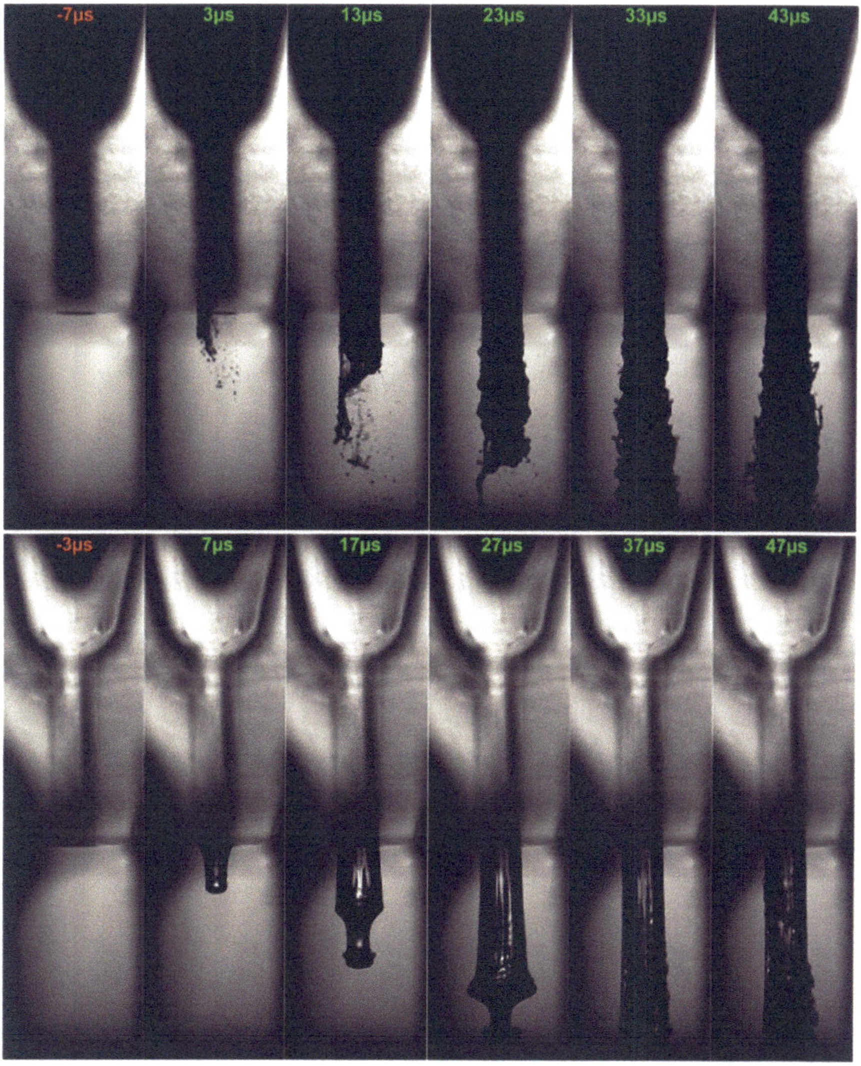

Figure 8.6 Start of injection with injection pressure of 250 bar and ambient pressure of 1 bar (full gas in sac, top), injection pressure of 500 bar and ambient pressure of 1 bar (full liquid fuel in sac, top). *Credit: J. Manin, L.M. Pickett, K. Yasutomi, Stereoscopic high-speed microscopy to understand transient internal flow processes in high-pressure nozzles, Experimental Thermal and Fluid Science 114 (2020), 110027. https://doi.org/10.1016/j.expthermflusci.2019.110027.*

Along with the efforts to understand the transient gas exchange process, transient bubble dynamics were investigated under simulated engine cycle conditions. Abers et al. utilized an imaging setup identical to that of the previous study and applied a programmable solenoid valve system to emulate engine compression and expansion cycles [22]. The ambient pressure was kept at 5, 10, and 20 bar and decreased to atmospheric conditions for the expansion cycle. Similarly, the compression cycle was performed by pressurizing the spray chamber from 1 to 5, 10, and 20 bar. They tried to quantify bubble volume on the basis of the projected area in line-of-sight measurement. Fig. 8.7 shows the binarized bubble area and

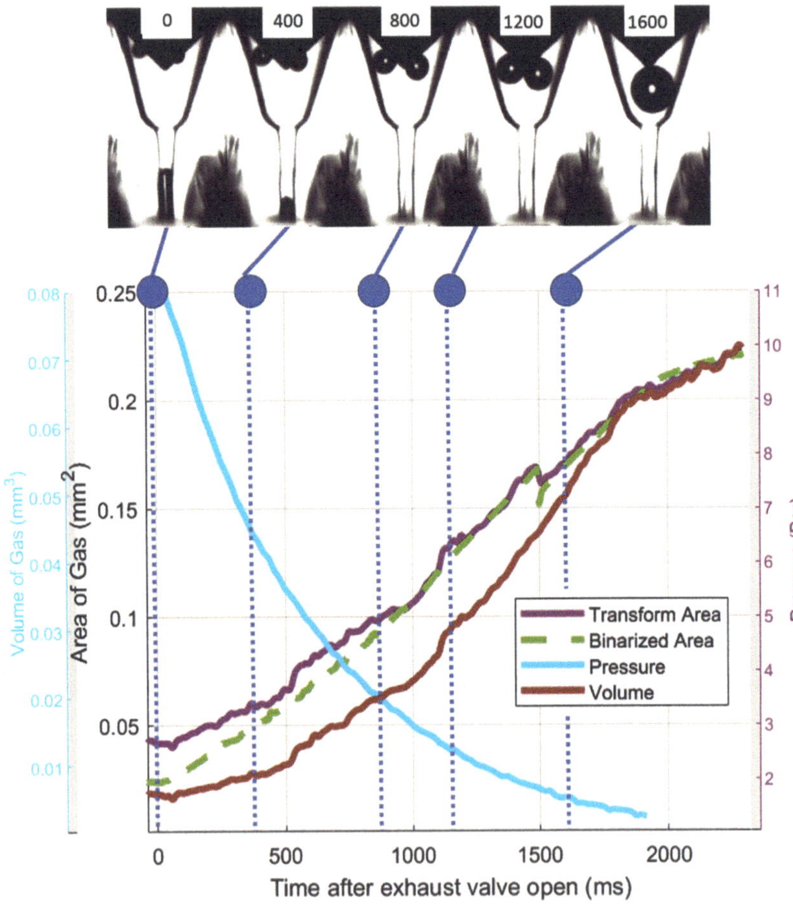

Figure 8.7 Measured projected bubble area, volume, and ambient pressure during the expansion cycle from 10 to 1 bar. *Credit: P.M. Abers, E. Cenker, K. Yasutomi, J. Hwang, L.M. Pickett, Effect of pressure cycling on gas exchange in a transparent fuel injector, SAE Technical Papers (2019) 1–11. https://doi.org/10.4271/2019-01-2280.*

volume under the expansion cycle. The timestamps shown at the top of each raw image indicate the time after the valve opens (in milliseconds). Residual gas after the end of injection is shown near the needle, and expansion can be observed as ambient pressure decreases. This result explains previously observed fuel dribbling during the expansion stroke in engines [23]. Liquid fuel in the nozzle could be pushed out to ambient by an expanding bubble in the sac based on this observation.

The interesting finding in this study is that bubble expansion followed an isothermal process rather than an isentropic (adiabatic) one. It is possible to have heat exchange between the bubble and surrounding fuel liquid by large fuel heat capacity and high surface area of the bubble. The result also showed that the final volume after the expansion cycle was greater than the theoretical volume by isothermal expansion. One explanation is the desorption of dissolved gas in the liquid fuel added to volume increase by expansion. The evidence of dissolved gas showed that the volume of the bubble continued to increase even after pressure reached 1 bar. The volume increase by desorption of dissolved gas also showed agreement in terms of initial ambient pressure. The volume increase by dissolved gas got larger for the expansion cycle that has higher initial ambient pressure. This study showed a guideline for bubble modeling in the engine cycle and emphasized the importance of the dissolved gas assumption for CFD simulation.

8.4 Interaction Between Cavitation and External Nozzle Flow

Literature studies have emphasized the importance of the cavitation process in primary spray breakup and development in external flow. The existence of two-phase flow in the nozzle implies initiation of mixture preparation and spray breakup. This can advance secondary breakup and alter the mixing process in a favorable way. Common observations with cavitating internal nozzle flow are shorter jet breakup length, larger spray plume cone angle, and smaller droplet diameter. Desantes et al. investigated near-nozzle spray under various cavitation number conditions [24]. A single axis-symmetric diesel injector was used under nonevaporating conditions. The results shown in Fig. 8.8 indicate that the spray cone angle increased by 1.5 degrees with higher injection pressure, even though ambient density was kept identical. The researchers explained that this is attributed to an interaction between cavitation bubbles and external spray. Oscillation by cavitation bubbles could cause more turbulence at the nozzle exit and enhance atomization of the external spray.

Meanwhile, Watanabe et al. investigated the link between cavitation structure in nozzle and external spray [25]. Three needle tip designs were applied in a transparent

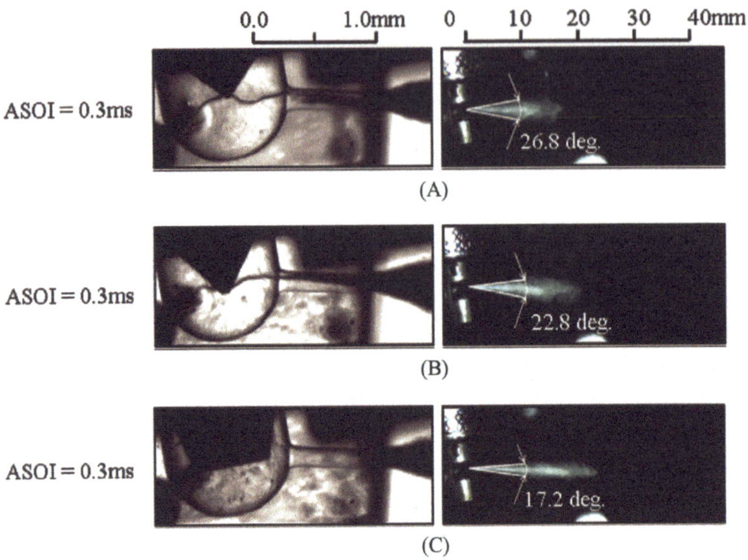

Figure 8.8 Microparticle image velocimetry results (top) and internal/external flow characteristics at 0.3 ms after start of injection (ASOI) with different needle geometry. *Credit: H. Watanabe, M. Nishikori, T. Hayashi, M. Suzuki, N. Kakehashi, M. Ikemoto, Visualization analysis of relationship between vortex flow and cavitation behavior in diesel nozzle, International Journal of Engine Research 16 (2015) 5–12. https://doi.org/10.1177/1468087414562459.*

nozzle to create different vortex flow in the nozzle. The flow in the nozzle was visualized by microparticle image velocimetry techniques and supported by CFD simulation. The major findings of this study are presented in Fig. 8.8.

It is shown that with a 90-degree needle, a large vortex was created in the core area of the sac volume. The vortical flow reached the bottom of the sac and formed a large recirculation zone. A 70-degree needle still showed vertical motion with a low level of scale. Nozzle orifices are affected by this vortex and indicated a possibility of string cavitation for both cases. On the other hand, at 30 degrees, it was not affected by the vortex, and the fuel from upstream flowed directly to the nozzle orifice. The weaker vortex flow was not sufficient to cause string cavitation in this case. Spray experiments with larger needle angles showed string cavitation across the sac volume and finally resulted in a larger spray cone angle. These results confirmed that the spray cone angle was proportionally increased with respect to the thickness of string cavitation in the nozzle orifice. The spray tip penetration showed a gradual decrease as the thickness of string cavitation increased. Hayashi et al. continued to study the effects of string cavitation on combustion [26]. Instead of modifying the needle tip geometry, they tried using two transparent nozzles with different nozzle orifice inlet radii. A rapid compression machine was used to create high temperature and pressure conditions for autoignition of the spray. A piezo-driven three-hole injector fueled by conventional diesel was utilized for the test. The internal nozzle flow and spray combustion results are shown in Fig. 8.9. The researchers found that the internal nozzle flow with a smaller inlet radius (sharp turn) had only film-type cavitation, but the nozzle with the larger inlet radius showed additional string cavitation that connected with the adjacent orifice. Accordingly, string cavitation facilitated plume dispersion, so the external spray had a spray cone angle that was approximately 8 degrees larger than that of smaller

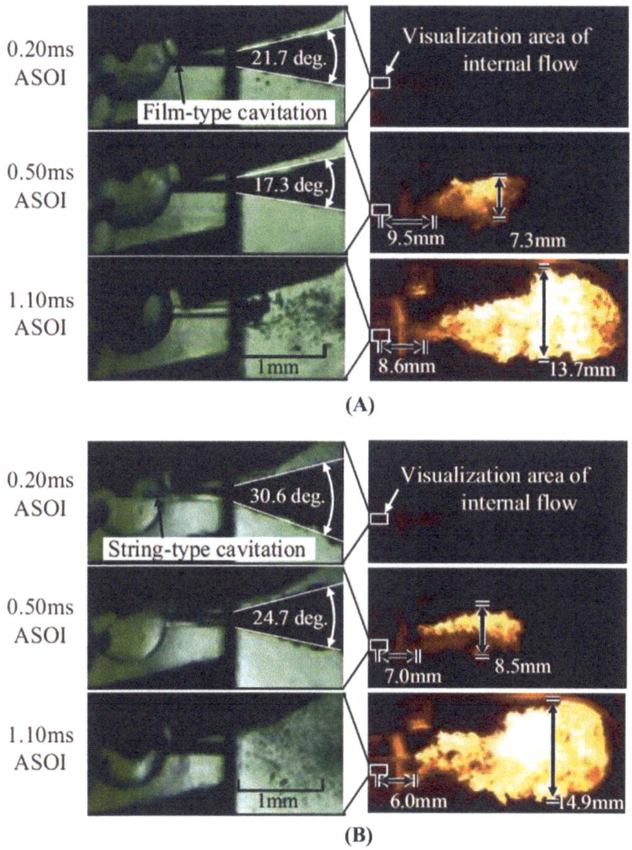

Figure 8.9 Comparison of cavitation and spray combustion process with different inlet roundness of nozzle orifice (three-hole injector, evaporating condition, ambient temperature of 873 K, and injection pressure of 500 bar). (A) Inlet radius of 0.016 mm. (B) Inlet radius of 0.033 mm. The time indicated in the left column shows the time after the start of injection (ASOI). *Credit: T. Hayashi, M. Suzuki, M. Ikemoto, Effects of internal flow in a diesel nozzle on spray combustion, International Journal of Engine Research, 14 (2013) 646–654. https://doi.org/10.1177/1468087413494910.*

inlet radius case. A larger spray cone angle implies facilitated air-fuel mixing, so it turned out to have shorter ignition delay and flame lift-off length with a larger orifice inlet diameter. This result indicates that the internal flow can have a crucial influence on diesel combustion emission characteristics in real engines, as was confirmed in the study by Benajes et al [27].

8.5 Summary

Cavitation in modern diesel injection systems has been widely investigated by using various optical diagnostics. Studies of internal nozzle flow using single-hole or multihole nozzles revealed that the internal nozzle flow features, such as cavitation structure and vortex motion, play a dominant role in determining external spray and combustion characteristics. Common findings are a wider spray angle and a shorter liquid penetration length with the cavitating in-nozzle flow. High turbulence in the nozzle enhances atomization and the air-fuel mixing process, so it shortens ignition delay and flame lift-off length. On the other hand, the cavitation bubble in the nozzle can reduce the flow rate and cause cavitation erosion on the injector surface [28]. Therefore a careful understanding of cavitation physics and prediction of two-phase flow have great importance for the development of injection components.

Even though the literature showed progress in the cavitation process and two-phase flow identification, currently available studies are limited in their visible optical diagnostics, making it difficult to quantify results. Experimental conditions are also confined to the nonevaporating condition and low injection pressure, owing to the failure of the transparent nozzle. To obtain a detailed understanding of the cavitation mechanism and quantified information, penetrating diagnostics such as X-ray or neutron imaging have been considered to be promising techniques. CFD simulation with realistic nozzle geometry will also contribute to a better understanding of the cavitation process in the nozzle.

References

[1] L.M. Pickett, J. Manin, C.L. Genzale, D.L. Siebers, M.P.B. Musculus, C.A. Idicheria, Relationship between diesel fuel spray vapor penetration/dispersion and local fuel mixture fraction, SAE International Journal of Engines 4 (2011) 764–799. Available from: https://doi.org/10.4271/2011-01-0686.

[2] J.E. Dec, A conceptual model of DI diesel combustion based on laser-sheet imaging, SAE Technical Papers (1997). Available from: https://doi.org/10.4271/970873.

[3] J.E. Dec, Advanced compression-ignition engines—understanding the in-cylinder processes, Proceedings of the Combustion Institute 32 (II) (2009) 2727–2742. Available from: https://doi.org/10.1016/j.proci.2008.08.008.

[4] X. Wang, Z. Huang, O.A. Kuti, W. Zhang, K. Nishida, Experimental and analytical study on biodiesel and diesel spray characteristics under ultra-high injection pressure, International Journal of Heat Fluid Flow 31 (2010) 659–666. Available from: https://doi.org/10.1016/j.ijheatfluidflow.2010.03.006.

[5] P. Ye, A.L. Boehman, An investigation of the impact of injection strategy and biodiesel on engine NO_x and particulate matter emissions with a common-rail turbocharged di diesel engine, Fuel 97 (2012) 476–488. Available from: https://doi.org/10.1016/j.fuel.2012.02.021.

[6] M. Gavaises, C. Arcoumanis, H. Roth, Y.S. Choi, A. Theodorakakos, Nozzle flow and spray characteristics from VCO diesel injector nozzles, Thermo- and Fluid Dynamic Processes in Diesel Engines 2 (2004) 31–48. Available from: https://doi.org/10.1007/978-3-662-10502-3_3.

[7] C. Arcoumanis, H. Flora, M. Gavaises, N. Kampanis, R. Horrocks, Investigation of cavitation in a vertical multi-hole injector, SAE Technical Papers (1999). Available from: https://doi.org/10.4271/1999-01-0524.

[8] M. Blessing, G. König, C. Krüger, U. Michels, V. Schwarz, Analysis of flow and cavitation phenomena in diesel injection nozzles and its effects on spray and mixture formation, SAE Technical Papers (2003) 7191. Available from: https://doi.org/10.4271/2003-01-1358.

[9] F. Payri, V. Bermúdez, R. Payri, F.J. Salvador, The influence of cavitation on the internal flow and the spray characteristics in diesel injection nozzles, Fuel 83 (2004) 419–431. Available from: https://doi.org/10.1016/j.fuel.2003.09.010.

[10] A. Tekawade, B.A. Sforzo, K.E. Matusik, K. Fezzaa, A.L. Kastengren, C.F. Powell, Time-resolved 3D imaging of two-phase fluid flow inside a steel fuel injector using synchrotron X-ray tomography, Scientific Reports 10 (2020) 1–9. Available from: https://doi.org/10.1038/s41598-020-65701-x.

[11] Katarzyna E Matusik, Daniel J Duke, Alan L Kastengren, Nicholas Sovis, Andrew B Swantek and Christopher F Powell. High-resolution X-ray tomography of Engine Combustion Network diesel injectors, International Journal of Enginer Research Vol 19(9) 963–976. Available from: https://doi.org/10.1177/1468087417736985

[12] B.A. Sforzo, K.E. Matusik, C.F. Powell, A.L. Kastengren, S. Daly, E. Cenker, et al., Fuel nozzle geometry effects on cavitation and spray behavior at diesel engine conditions, Proceedings 10th of the International Symposium Cavitation (2019) 474–480. Available from: https://doi.org/10.1115/1.861851_ch90.

[13] J. Gimeno, G. Bracho, P. Martí-Aldaraví, J.E. Peraza, Experimental study of the injection conditions influence over n-dodecane and diesel sprays with two ECN single-hole nozzles. Part I: inert atmosphere, Energy Conversion and Management 126 (2016) 1146–1156. Available from: https://doi.org/10.1016/j.enconman.2016.07.077.

[14] S.H. Park, H.K. Suh, C.S. Lee, Effect of cavitating flow on the flow and fuel atomization characteristics of biodiesel and diesel fuels, Energy and Fuels 22 (2008) 605–613. Available from: https://doi.org/10.1021/ef7003305.

[15] H.K. Suh, S.H. Park, C.S. Lee, Experimental Investigation of nozzle cavitating flow characteristics for diesel and biodiesel fuels, International Journal of Automotive Technology 9 (2) (2008) 217–224. Available from: https://doi.org/10.1007/s12239-008-0028-3.

[16] J. Hwang, C. Bae, C. Patel, R.A. Agarwal, T. Gupta, A. Kumar Agarwal, Investigations on air-fuel mixing and flame characteristics of biodiesel fuels for diesel engine application, Applied Energy 206 (2017) 1203–1213. Available from: https://doi.org/10.1016/j.apenergy.2017.10.027.

[17] J. Manin, L.M. Pickett, K. Yasutomi, Stereoscopic high-speed microscopy to understand transient internal flow processes in high-pressure nozzles, Experimental Thermal and Fluid Science 114 (2020) 110027. Available from: https://doi.org/10.1016/j.expthermflusci.2019.110027.

[18] J. Manin, L.M. Pickett, K. Yasutomi, Transient cavitation in transparent diesel injectors, in: ICLASS Triennial International Conference on Liquid Atomization and Spray Systems, 2018, pp. 1–9.

[19] J. Manin, M. Bardi, L.M. Pickett, R.N. Dahms, J.C. Oefelein, Microscopic investigation of the atomization and mixing processes of diesel sprays injected into high pressure and temperature environments, Fuel 134 (2014) 531–543. Available from: https://doi.org/10.1016/j.fuel.2014.05.060.

[20] A.B. Swantek, D.J. Duke, F.Z. Tilocco, N. Sovis, C.F. Powell, End of injection, mass expulsion behaviors in single hole diesel fuel injectors, in: ILASS Americas 26th Annual Conference on Liquid Atomization and Spray Systems, Portland, OR, May, 2014.

[21] M. Battistoni, S. Som, C.F. Powell, Highly resolved Eulerian simulations of fuel spray transients in single and multi-hole injectors: nozzle flow and near-exit dynamics, Fuel 251 (2019) 709–729. Available from: https://doi.org/10.1016/j.fuel.2019.04.076.

[22] P.M. Abers, E. Cenker, K. Yasutomi, J. Hwang, L.M. Pickett, Effect of pressure cycling on gas exchange in a transparent fuel injector, SAE Technical Papers (2019) 1–11. Available from: https://doi.org/10.4271/2019-01-2280.

[23] W.E. Eagle, M.P.B. Musculus, Cinema–stereo imaging of fuel dribble after the end of injection in an optical heavy–duty diesel engine, in: THIESEL 2014 Thermo- and Fluid Dynamic Processes in Direct Injection Engines, 2014, pp. 1–20.

[24] J.M. Desantes, R. Payri, F.J. Salvador, J. De La Morena, Influence of cavitation phenomenon on primary break-up and spray behavior at stationary conditions, Fuel 89 (2010) 3033–3041. Available from: https://doi.org/10.1016/j.fuel.2010.06.004.

[25] H. Watanabe, M. Nishikori, T. Hayashi, M. Suzuki, N. Kakehashi, M. Ikemoto, Visualization analysis of relationship between vortex flow and cavitation behavior in diesel nozzle, International Journal of Engine Research 16 (2015) 5–12. Available from: https://doi.org/10.1177/1468087414562459.

[26] T. Hayashi, M. Suzuki, M. Ikemoto, Effects of internal flow in a diesel nozzle on spray combustion, International Journal of Engine Research 14 (2013) 646–654. Available from: https://doi.org/10.1177/1468087413494910.

[27] J. Benajes, S. Molina, C. González, R. Donde, The role of nozzle convergence in diesel combustion, Fuel 87 (2008) 1849–1858. Available from: https://doi.org/10.1016/j.fuel.2007.12.016.

[28] J.P. Franc, M. Riondet, A. Karimi, G.L. Chahine, Material and velocity effects on cavitation erosion pitting, Wear 274–275 (2012) 248–259. Available from: https://doi.org/10.1016/j.wear.2011.09.006.

Chapter 9

Experimental Diagnostics and Investigations on Cavitation in Engine Injector Nozzles

Michael Wensing, Sebastian Riess, Dmitrii Mamaikin, Chris Conrad and Alexander Neubauer
Institute of Engineering Thermodynamics, Friedrich-Alexander-University Erlangen-Nuremberg, Erlangen, Germany

9.1 Appearance and Diagnostics of Cavitation in Injector Nozzles

Cavitation may appear in various hydraulic components, such as injectors, pumps, and valves. These components are used in various technical fields, such as industrial hydraulic systems or automotive systems (steering, braking, and injection systems). The phenomena of cavitation play an important role in the performance of modern fuel injectors. Researchers in both academia and industry have intensively studied this field of hydrodynamics in recent decades. Such a strong interest is due to the appearance of cavitation inside fuel injector nozzles, which influences the spray characteristics and overall function of the injectors.

Generally, the phenomena of cavitation can be investigated experimentally as well as numerically. Chapter 1 and Chapter 4 described numerical approaches and models for cavitation, so these are not discussed further in this chapter.

Cavitation and Bubble Dynamics. DOI: https://doi.org/10.1016/B978-0-12-823388-7.00002-3

Generally, the experimental approach to cavitation phenomena can be divided into several steps:

- The general behavior of cavitation in a liquid of interest can be studied where cavitation can be produced by specifically designed instruments (such as a sonotrode or a fast flow channel).
- The cavitation and its impact on a component and flow behavior can be studied in simplified model test facilities.
- Damage analysis can be performed to observe the impact of cavitation erosion on the components. For that, endurance test facilities are often employed. Among the analytic methods used are REM imaging, white light interferometry, and profilometry. Here, the effects of cavitation erosion on a component of interest are evaluated. Endurance test facilities enable quantitative assessment of erosion. For instance, the influence of fuel properties on cavitation erosion are investigated. The resistance of material properties to erosion aggressiveness can also be evaluated in this type of test.

9.1.1 Magnitude and Appearance

Cavitation is formed when the local pressure in a liquid drops below the saturation pressure of the liquid at a working temperature. The local pressure reduction can be caused by a hydraulic and acoustic reasons. Hydraulic cavitation occurs when the pressure reduction is caused by the hydrodynamic motion of the bulk liquid, which is normally attributed to the flow path. In turn, acoustic cavitation is formed from the pressure differences caused by pressure wave propagation through a liquid flow [1].

In relation to fuel injectors, cavitation occurs inside injector nozzles, particularly in the nozzle holes, owing to very fast turbulent flows in this area with high-pressure gradients. Furthermore, Salvador et al. performed a numerical study of the internal flow of a diesel nozzle under transient needle lift conditions and observed cavitation formations in the needle seat for the initial injection phase with a needle lift of up to 10 μm [2]. In other words, it appears for a very short time. In turn, cavitation that appears in the nozzle holes can be present almost during the whole injection event. Furthermore, it plays a more relevant role in the formation of the spray. Because of that, this chapter deals with the cavitation types that appear inside the nozzle holes.

Cavitation has various effects. It can induce positive effects as well as negative ones on the performance of various components. For instance, cavitation may lead to the reduction of the Sauter mean diameter (SMD) in sprays. The reason for this is that the cavitation bubbles that are formed inside the nozzle explode as they reach the injector nozzle exit, owing to the increase in the static pressure.

It induces micro-scale turbulences in the flow, which in turn enhances the atomization performance. The reduction of SMD is a desirable effect in direct injection engines, as it contributes to lower penetration depths of the spray.

On the other hand, the collapse of the cavitation bubbles near nozzle walls can cause erosion. Erosion is usually an undesirable effect that often destroys the component material and leads to deterioration of the component's performance. This can lead to reduction of the spray performance and significantly lower the lifetime of the injector. Moreover, cavitation may induce corrosion on the material surface, especially considering the differences in the fuel composition, which in turn depends on the geographical region and the season of the year.

These factors underline the significance of the cavitation phenomenon in the engineering of advanced fuel injection systems. The main goal is to minimize the negative effects of cavitation in injector nozzles. For that, an in-depth understanding of cavitation behavior is required. Various experimental diagnostics are used to investigate the cavitation phenomenon in injector nozzles.

9.1.2 Diesel and Gasoline Injection

Both gasoline and diesel injection processes are influenced by cavitation. Here, some important differences in these injection systems are underlined. First, the injector design of a diesel injector is significantly different from that of a gasoline direct injection (GDI) injector. This is due to the different requirements for the fuel injectors. For instance, the diesel injector has to be able to inject the fuel into a much denser environment than the gasoline injector does, owing to the difference in the compression ratio of these engine systems. In fact, the injection pressure of modern series production diesel common rail systems is one order higher than that for GDI systems. Second, the ignition of the air fuel mixture for the gasoline engine is initiated by a spark plug, whereas the ignition in the diesel engine is initiated by the elevated temperature in the cylinder caused by the high compression. Because of that, the requirements for the spray propagation in the combustion chamber differ for gasoline and diesel engines. For gasoline engines, spray penetration is desired to be minimized to avoid piston wetting or other formations of liquid films inside the combustion chamber. These formations lead to a locally high concentration of the fuel and deteriorate the fuel-air mixture quality in the combustion chamber. As a consequence, it contributes to particulate matter formation [3]. On the other hand, the reduction of the spray penetration for diesel engines does not represent a technical goal in the design and optimization of diesel engine components. However, better atomization performance and smaller droplet sizes are of interest for both gasoline and diesel injection systems.

There are various nozzle designs for diesel injectors, such as minisac nozzles, microsac nozzles, and valve-covered orifice nozzles. Another nozzle type design concept, a pintle nozzle, was introduced by Payri et al. [4].

GDI injectors also vary in nozzle design. For instance, there are swirl-type, slit-type, multihole, and outward-opening injectors. Because of the differences in the design, the flow physics can significantly vary among these types. The multihole injector is the most common choice for modern GDI systems; therefore it has attracted more attention in the spray community in comparison to the other injector types. This chapter also considers this type of the injector nozzle.

Fig. 9.1 depicts a schematic illustration of a diesel injector nozzle and a gasoline injector nozzle. The difference in the nozzle design and the needle form are evident. The holes of the diesel nozzle are quite long in relation to the diameter, typically by more than a factor of 5. The hole length of the GDI nozzle is significantly shorter and is typically not much longer than the diameter, which is typically in the range between 0.1 mm and 0.25 mm.

It is important to highlight the main differences in the thermophysical properties between diesel and gasoline, which contributes to the difference in the flow behavior and spray form in addition to the significant variation in the operating conditions and different injector designs mentioned earlier. First, the viscosity of diesel fuel is significantly higher than that for gasoline. Second, the vapor pressure of diesel is notably lower than that of gasoline. Diesel and gasoline also vary in density. The deviations in these parameters contribute to the variations in turbulence level and other characteristics of the flow, which also affect cavitation phenomena.

Fig. 9.2 shows single liquid jets emitted by a diesel injector and by a GDI injector. Each nozzle has three holes. The images were taken by X-ray phase-contrast imaging, which allows for visualization of dense spray structures, such as those appearing in the near-field region. The technique will be discussed in detail later in

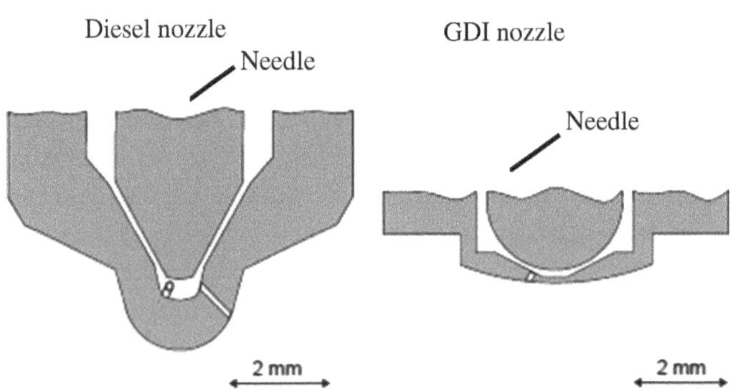

Figure 9.1 Schematic illustration of a diesel injector nozzle and a gasoline injector nozzle.

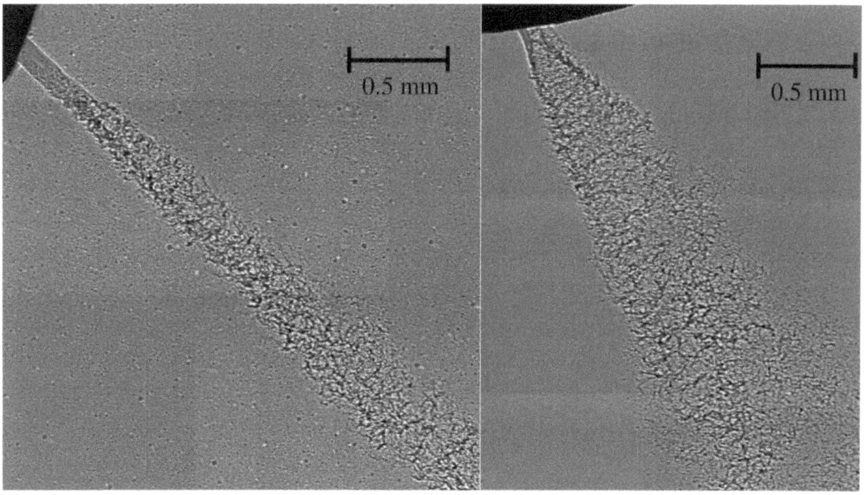

(A) (B)

Figure 9.2 (A) Near-nozzle spray of a diesel injector (100 MPa, 0.1 MPa, cold, CF ISO 4113). (B) Near-nozzle spray of a gasoline injector (20 MPa, 0.1 MPa, cold, *n*-heptane).

the chapter. Calibration fluid (ISO 4113) and *n*-heptane were used as surrogate for diesel and gasoline, respectively. The differences in the liquid jet appearance are obvious from these images. In Fig. 9.2A the diesel jet appears as an intact core as it leaves the nozzle up to a distance of approximately 0.5 mm from the tip. Farther downstream, the jet starts to disintegrate into liquid ligaments as a result of the high turbulence of the liquid jet and instabilities induced by aerodynamic forces. For a gasoline injection the jet leaves the injector nozzle in the form of liquid ligaments, as can be seen in Fig. 9.2B. From this image, it can be concluded that the aerodynamic forces were not the initial factor in the liquid jet breakup. Owing to the relatively short nozzle hole length, it can be assumed that the primary breakup of the gasoline liquid jet is initiated by high turbulence of the internal flow and cavitation in the nozzle hole. It can be expected that cavitation bubbles collapse in the nozzle holes, inducing additional microturbulence in the nozzle hole and leading to an increase in instabilities of the liquid jet at the nozzle outlet. The degree of turbulence of the internal flow for diesel and gasoline injectors is also different. In the case of diesel, a higher pressure difference contributes to the higher degree of turbulence of the flow as a function of the square root (according to the Bernoulli principle). On the other hand, a significantly lower viscosity of diesel in comparison to gasoline reduces the degree of flow turbulence. As a result, the Reynolds number for the diesel case is approximately 13,000, while that for the gasoline case is around 70,000 as calculated for the experimental conditions referring to the presented images. Because the Reynolds number for a gasoline injection is about five times higher than that for the diesel case, the differences in the liquid jet appearance cannot be attributed solely to the cavitation influence.

Gilles-Birth et al. conducted imaging of the internal flow in transparent nozzles using a backlight illumination approach [5]. The authors categorize the cavitation types that appear in an injector nozzle into three groups: homogeneous (bubble) cavitation, heterogeneous (film) cavitation, and string cavitation.

For homogeneous cavitation, temporary microscopic voids in the liquid can include the nuclei for cavitation that further transform into macroscopic bubbles. In the case of heterogeneous cavitation the surface roughness of a nozzle hole acts as a nuclei reservoir for cavitation occurrence. The rapid flow deflection that is induced in the nozzle hole inlet leads to the formation of cavitation films. Depending on the nozzle hole configuration, operating conditions, and fuel properties, the cavitation films can vary in size and occupy different fractions of the nozzle hole. If the cavitation film extends to the exit of the hole, this type of cavitation is referred to as supercavitation. String cavitation is a result of the vortical flow structure inside the nozzle hole. The centrifugal forces lead to reduction in the static pressure in the center of the vortex, and if the vortex is strong enough, the pressure drops below the vapor pressure of the fuel, which leads to the formation of string cavitation.

9.1.3 Experimental Techniques

9.1.3.1 Nonoptical Techniques

There are different methods to investigate cavitation phenomena in injector nozzles. They can be divided into nonoptical and optical methods.

A common nonoptical method to characterize the internal nozzle flow is HDA (hydraulischer Drucksanstiegs-Analyzator), which is a measuring unit for the injection quantity and rate. In this method the fuel is injected into a liquid-filled reservoir [6,7]. The simultaneous measurement of the pressure increase and speed of sound in the liquid allows evaluation of the mass flow rate provided by the injector. The measurement is performed under various elevated back pressure levels. A sketch of a typical HDA setup is depicted in Fig. 9.3.

Because the mass flow rate provided by the injector is linked to the degree of hole filling and cavitation, it is also used to estimate the degree of cavitation in the injector nozzle. This technique has also been used to evaluate the influence of the nozzle geometry on cavitation tendency [8]. Various nozzle features, such as hole cone angle and hole length, were found to influence the cavitation behavior. The advantage of this system is that it is able to characterize the cavitation for realistic nozzle designs. The disadvantage is that the flow conditions are changed in comparison to the application, owing to the principle of this experimental approach based on the injection into liquid fuel.

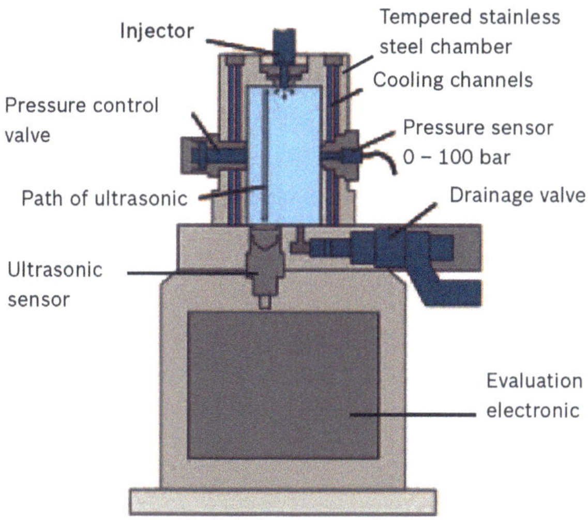

Figure 9.3 Sketch of the HDA setup. *From Moehwald GmbH, Injection rate and quantity measurement, technical brochure. https://moehwald.de.*

9.1.3.2 Optical Nozzles and Techniques

Optical methods for the investigation of the internal flow and associated liquid jets imply optical access into the nozzle flow and liquid jets. The investigation of the far-field spray is commonly performed in optically accessible injection chambers. These chambers include windows and allow various operating conditions. Spray properties have also been investigated in optically accessible engines [9,10]. However, the study of the internal flow and near-field spray requires further demands on the optical methods. First, the internal flow has to be optically accessible for visualization of the flow structures and properties. Second, the near-field spray represents optical density that is too high for conventional optical experimental diagnostics. To study the development of liquid instabilities and spray structures for these particular flow regions, advanced measurement techniques with high temporal and spatial resolution have to be applied.

In this chapter the optical experimental diagnostics for the investigation of the internal nozzle flow and dense spray region are categorized into two approaches. The first approach deals with optical nozzles. These nozzles are made of a transparent material and enable optical access for experimental techniques that involve visible light. The second approach involves the use of X-ray radiation, which allows optical access into metal nozzles and near-field spray areas. In the next section these approaches are discussed in detail.

Various optical nozzles have been designed to investigate cavitation phenomena, such as planar nozzles, large-scale nozzles, and real-size nozzles. They are typically used in combination with transmitted light experimental techniques, such as shadowgraphy. Also, some works applied laser-induced fluorescence (LIF) to visualize the internal flow and to retrieve flow characteristics.

A review of measurements on optical nozzles is given in the following subsections, along with examples for different optical nozzle designs. Experimental studies based on shadowgraphy and LIF are discussed separately. Furthermore, the working principle of these experimental approaches is introduced.

9.1.3.3 Shadowgraphy

Shadowgraph imaging of the internal flow is realized by placing the nozzle flow area in the line between a light source and the camera. The liquid and vapor phases are distinguished on shadowgraph images as a result of the high change in the refractive indices of these phases. Thus it allows visualization of cavitation formations with high contrast.

Some optically accessible nozzles are formed by two plates constraining the flow, thus representing a two-dimensional (2D) flow field. These planar nozzles simplify the optical access into the nozzle as they avoid optical aberrations. Fig. 9.4 shows examples of the images obtained in planar nozzles. This type of nozzle was used in [11–13].

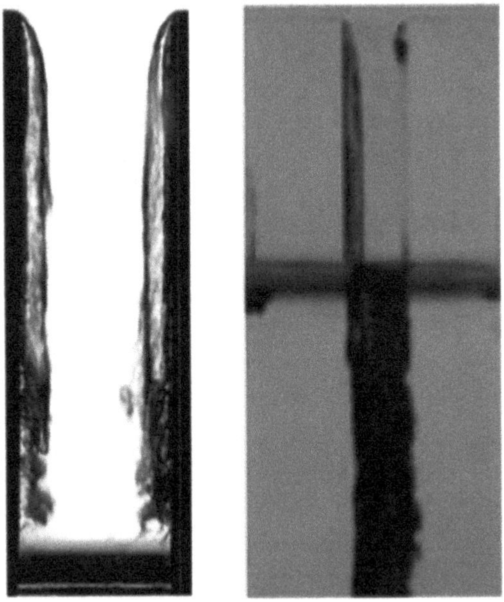

Figure 9.4 Planar nozzle investigated with water. *(Left) From A. Sou, S. Hosokawa, A. Tomiyama. Effects of cavitation in a nozzle on liquid jet atomization. International Journal of Heat and Mass Transfer 50 (2007) 3575–3582. (Right) From A. Sou, S. Minami, R. Prasetya, R. H. Pratama, S. Moon, Y. Wada, H. Yokohata (2015) X-Ray Visualization of Cavitation in Nozzles with Various Sizes. 13th Triennial International Conference on Atomization and Spray Systems.*

Large-scale nozzles are made of a transparent material and are manufactured in upscaling sizes in relation to the original nozzle dimensions. They improve optical access into the nozzle and facilitate the manufacturing process. Furthermore, the mechanical stress on the nozzle is significantly reduced. The experimental conditions on such models are set according to the hydrodynamic similarity concept. The flow characteristics are maintained by considering various dimensionless numbers of the flow, such as Reynolds number and cavitation number. In practice, maintaining all necessary flow conditions is rather challenging; therefore many experimental studies neglect one or more of criteria, depending on the research interest. In contrast to planar nozzles, large-scale nozzles have curved surfaces. Therefore the refractive index of a test fuel and that of the nozzle material have to be considered in order to obtain full access into the flow and to avoid optically inaccessible areas due to total internal reflection. This type of enlarged nozzle has been used in many research activities. The scaling factor can vary depending on the study. For instance, Nouri et al. performed measurements on an enlarged nozzle with a scale of 29, Arcoumanis et al. used a nozzle with a scale of 20, Aleiferis et al. used a nozzle with a scale of 10, and Andriotis et al. used a nozzle with a scale of 2 [14–17]. In contrast to planar nozzles, this nozzle type allows study of real three-dimensional (3D) flow structures that appear in the nozzle. For instance, this is crucial for string cavitation, which is the result of strong vortices that can occur in the nozzle during injection [18]. Furthermore, this nozzle type enables flow conditions upstream from the nozzle hole to be more representative of the real case, owing to the introduction of the nozzle sac and the injector needle.

Fig. 9.5 shows some typical cavitation structures that can be observed inside enlarged nozzles. Andriotis et al. observed film cavitation (called geometric-induced cavitation by the authors) in a nozzle hole, as can be seen in Fig. 9.5A

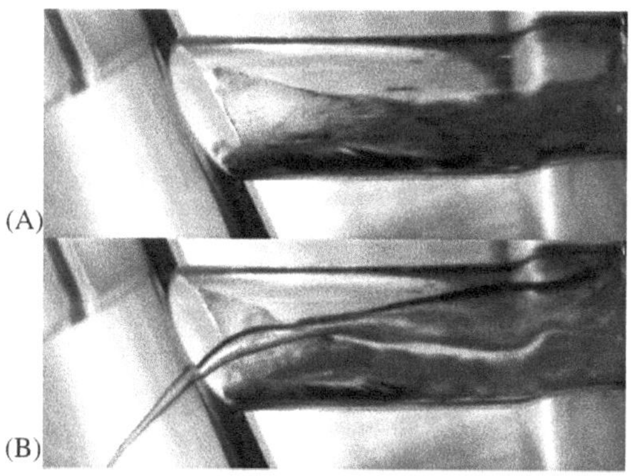

Figure 9.5 Cavitation structures inside large-scale nozzles. (A) Flow inside a nozzle hole without string cavitation. (B) Flow inside a nozzle hole with string cavitation. *From A. Andriotis, M. Gavaises, C. Arcoumanis Vortex flow and cavitation in diesel injector nozzles. Journal of Fluid Mechanics 610 (2008) 195–215.*

[17]. In that case the cavitation formation occupies one side of the nozzle hole, where it starts at the hole inlet and extends to the hole exit. Fig. 9.5B shows the flow inside the same nozzle but with the presence of the string cavitation. As can be seen, string cavitation influences the flow structure inside the nozzle. As the authors state, both types of cavitation become indistinguishable in the image, and cavitation does not occupy only one side of the hole, as in Fig. 9.5A, but rather occupies the cross-sectional area. Therefore film cavitation and string cavitation can occur simultaneously inside the nozzle holes. Nouri et al. [14] demonstrated in the experiment performed in a GDI nozzle that the string cavitation can extend to another nozzle hole through the nozzle sac. It is an interesting phenomenon that again underlines the differences between this type of cavitation and film cavitation. It is interesting that these "bridging strings," as they are called by Reid et al., are influenced by the spacing between the neighbor holes [19]. The effect of string cavitation on the emerging jet will be discussed later in the chapter.

Real-size optical nozzles are made of a transparent material and are manufactured with the original scale and dimensions. Various works have used real-size transparent nozzles [5,20–24]. Arcoumanis et al. performed measurements on a six-hole diesel nozzle with the nozzle design modified to obtain optical access into one of the holes [20]. The measurements were performed under steady-state needle conditions. Badock et al. conducted measurements on a single-hole diesel nozzle under transient conditions [21]. The nozzle tip of a diesel injector was replaced with a Perspex tip modeling a sac and a nozzle hole. In this work, shadowgraph imaging was compared to a laser light sheet technique, which basically relies on light scattering. A comparison of these techniques revealed the same length of film cavitation in the nozzle. The authors state, however, that the laser light sheet has the advantage of being able to view the core of the flow covered by cavitation films, which remains inaccessible for shadowgraphy.

The principle differences in design between a diesel nozzle and a GDI nozzle was schematically shown in Fig. 9.1. The substitution of a part of a GDI nozzle with a transparent material is quite challenging, owing to the location of the needle seat and nozzle holes. A transparent nozzle was used in the work of Gilles-Birth et al. to represent a Valve-covered orifice nozzle design for GDI application [5]. The authors investigated two single-hole nozzles made of acrylic glass. The injection pressure varied from 2 to 10 MPa. The fuel is injected for transient needle conditions. Mamaikin et al. performed visualization of internal flow in a two-hole transparent nozzle made of acrylic glass [23]. The nozzle block was specifically designed to withstand high-pressure loads during injections and was mounted to a modified GDI series-type injector that allowed representation of actual needle dynamics and therefore transient flow conditions during the experiments. A mixture of calibration fluid (ISO 4113) and 1-bromnaphthalene was used as test fuel. This mixture allows great optical insight into the internal flow of the nozzle due to the same refractive index as the acrylic glass; however, some thermophysical properties, such as viscosity and vapor pressure, differ from those of gasoline. Conrad et al. introduced a transparent nozzle design that could be used to investigate the flow under steady-state conditions [24]. The nozzle was made of fused silica and produced by a selective laser-etching method. The nozzle showed reliable performance at high-pressure conditions. To match the refractive index, a mixture of *n*-decane and toluene was employed. The experimental facility allowed investigation of the nozzle flow at various operating conditions, among which were flash-boiling conditions. In addition, LIF was used to visualize the flow and compare it to shadowgraph images. The LIF technique will be discussed further later in this chapter.

Fig. 9.6 shows the internal flow images from some of the abovementioned studies for GDI applications. It can be noticed that the design of transparent nozzles is remarkably different between these works. This is the consequence of a number of difficulties during the manufacturing process of real-size nozzles and challenging experimental conditions. Therefore various concepts of the experimental design have been applied to this topic.

Although the real-size nozzles come closer to describing the real behavior of the internal nozzle flow, some assumptions are made to facilitate the studies based on shadowgraph imaging. One of the assumptions that is commonly made is the reduced number of nozzle holes, as the shadowgraphy represents a line-of-sight imaging technique.

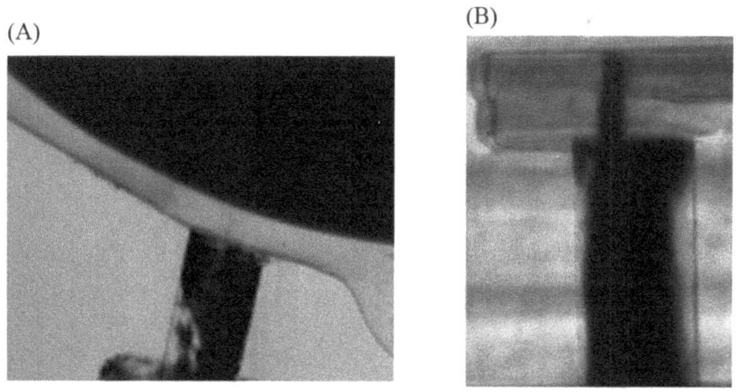

Figure 9.6 The experimental images obtained from the shadowgraph imaging of different experimental studies. *(a) From D. Mamaikin, T. Knorsch, P. Rogler, P. Leick, M. Wensing High speed shadowgraphy of transparent nozzles as an evaluation tool for in-nozzle cavitation behavior of GDI injectors. 28th Conference on Liquid Atomization and Spray Systems, Spain (2017) 1027–1034. (b) From C. Conrad, S. Bornschlegel, A. Durst, D. Jordan, M. Wensing Influence of the nozzle geometry on in-nozzle cavitation investigated in real-size glass nozzles with shadowgraphy and LIF. 14th Triennial International Conference on Liquid Atomization and Spray Systems, USA (2018) 1–6*

9.1.3.4 Laser-Induced Fluorescence

LIF was also used to study the internal nozzle flow. However, this technique has been applied in just a few studies. These studies will be briefly reviewed here to show that the LIF technique has potential for nozzle flow studies. The measurement principle will be provided based on the experimental facility presented by Conrad et al. [24]. In addition, the shadowgraphy imaging technique is introduced. The short discussion on comparison between these techniques will be provided.

The setup for the visualization of the nozzle flow by shadowgraph and LIF imaging is shown in Fig. 9.7. A pulsed LED with a collimator optic is used as back illumination for the shadowgraphy image. This is captured in the first frame of a double-shutter camera (pco.2000) equipped with a long-distance microscope (LDM). In parallel a pulsed ND:YAG laser operates at 532 nm. With a temporal delay of 200 ns in relation to the LED a laser pulse is coupled into the glass nozzle, where it excites the tracer dissolved in the fuel. The resulting fluorescence is captured by the second frame of the camera, while the refracted light at 532 nm is absorbed by the longpass filter mounted between the LDM and the camera sensor.

Fig. 9.8 shows a side-by-side visualization of the internal nozzle flow with shadowgraphy and LIF obtained by Conrad et al. [24]. Here, the LIF tracer is dissolvable in the fuel. The laser excites the tracer, and with more fuel in the line of sight, the signal increases. In addition, the emitted light scatters at the vapor liquid interface of the cavitation regions. The authors state that the images obtained by shadowgraph imaging and LIF can provide better views into the

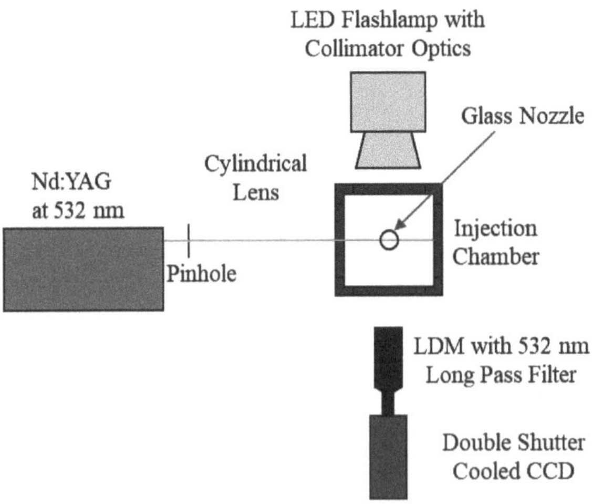

Figure 9.7 Experimental setup for shadowgraph imaging and LIF. CCD = charge-coupled device.

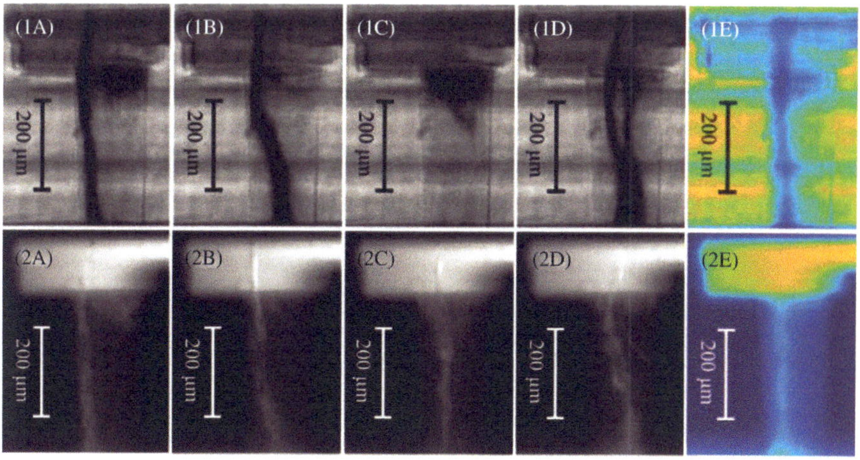

Figure 9.8 Distinct states of cavitation for different injection events visualized by: shadowgraphy (1A-1D) and LIF (2A-2D) for the convergent nozzle. The corresponding false color mean images for the complete measurement are shown for both shadowgraphy (1E) and LIF (2E). *From C. Conrad, S. Bornschlegel, A. Durst, D. Jordan, M. Wensing Influence of the nozzle geometry on in-nozzle cavitation investigated in real-size glass nozzles with shadowgraphy and LIF. 14th Triennial International Conference on Liquid Atomization and Spray Systems, USA (2018) 1–6.*

internal flow. The shadowgraph images provide the position of cavitation formations in the flow. The LIF data are able to reveal the shape of the liquid-vapor interface. This helps in visualization of the cavitation structures, which mostly appear black in the shadowgraph. The flow inside the nozzle exhibits differently from of string and sheet cavitation, and at least one of those two is always present. The shadowgraphy and LIF images for the same event show a lot of differences, though they were taken only 200 ns apart. This shows how fast the flow changes between different states. Even though the fluctuations were very fast, the mean images in false colors show that there still exists a preferred location of the string cavitation.

Walther et al. performed measurements in a single-hole diesel injector nozzle [22]. The focus of the experiment was to obtained the velocity field of the internal flow by particle image velocimetry. The flow was seeded with fluorescent particles, which were tracked and evaluated by a cross-correlation algorithm.

Fig. 9.9 shows two images from the study of Walther et al. [22]. Fig. 9.9A shows the shadowgraph image of the internal flow. It can be seen that cavitation occupies the nozzle hole on the shadowgraph image. Fig. 9.9B shows the exposure image obtained from the light emitted by the fluorescent particles. It can be seen that the fluorescent particles are visible also in the nozzle hole, which provides additional information in comparison to the shadowgraph images.

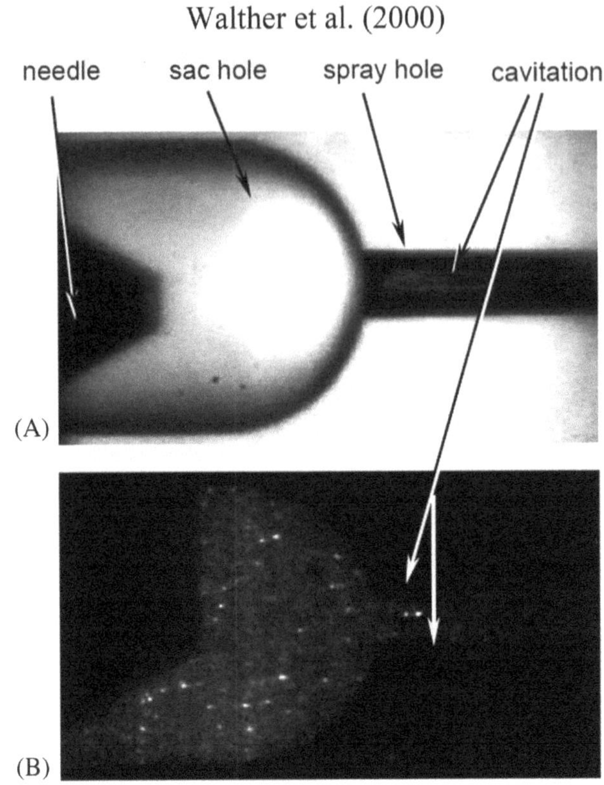

Figure 9.9 (A) Shadowgraphy exposure. (B) Exposure of light emission from fluorescent particles. *From J. Walther, J. K. Schaller, R. Wirth, C. Tropea Investigation of internal flow in transparent diesel injection nozzles using fluorescence particle velocimetry (FPIV). Eighth International Conference on Liquid Atomization and Spray Systems (2000), USA.*

9.1.3.5 X-ray in Nozzles

The subsection 9.1.3.2 discussed the use of transparent nozzles to provide an optical view into the internal nozzle flow. An alternative approach to obtain optical access into the nozzle flow is to use X-ray radiation. Owing to the high energy of the X-rays, a high penetration ability into materials or optically dense multiphase flows can be provided.

X-ray radiation can be used in different ways for investigating internal flow and spray formation. First, the X-ray radiation is being absorbed during the propagation through a flow structure, whether it is nozzle flow or spray, according to the Beer–Lambert law. The degree of absorption can be evaluated and used to retrieve some information about the flow, such as structure or density. Second, the X-ray radiation can produce a phase shift during propagation through an object. Under appropriate conditions the phase shift can produce phase contrast in the image. This technique is

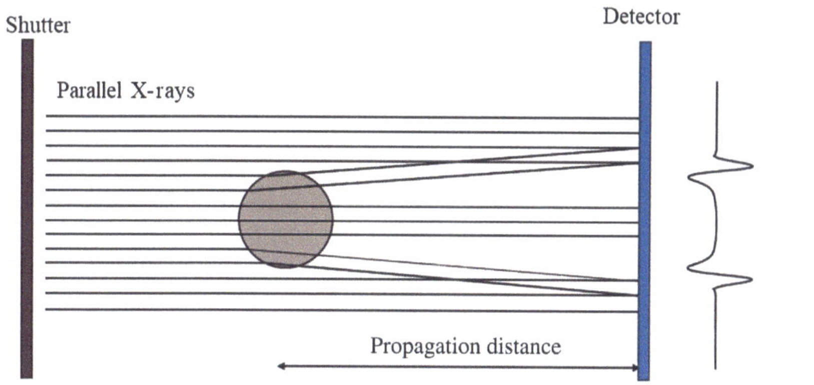

Figure 9.10 Depiction of the formation of phase contrast.

intensively used for the study of internal flow or sprays and is called X-ray phase contrast imaging (XPCI). The principles of this technique will be described, and some experimental studied that were relied on this technique will be discussed.

The development of phase contrast is illustrated in Fig. 9.10, which shows X-rays passing through a liquid droplet based on the assumptions of geometric optics. The fraction of X-rays that pass through the center of the droplet do not deviate from the original direction; however, the intensity of these X-rays is reduced, owing to absorption. The X-rays that pass the droplet on its edges are refracted, and the initial direction of the X-rays is altered. This creates an image on the detector, in which the droplet edges appear with a higher intensity in comparison to the rest of the image. Therefore the contrast of the droplet in the image is improved. It is important to note that the refractive indices of typical liquid fuels and gases are quite similar for X-rays. For instance, the difference in the real part of the refractive indices for *n*-heptane as fuel and nitrogen as a surrounding gas is 1.15×10^{-6} for a photon energy of 11.93 keV. This means that the propagation distance between the object and the detector must be quite long to create a phase contrast in the image plane. The real part of the refractive index of *n*-heptane is less than that of nitrogen. Therefore the brighter intensity appears in the outer area of the image. The darker part of the image appears in the area where the diffracted X-rays would be in the absence of the droplet.

A schematic representation of the experimental setup used at the Argonne National Laboratory is shown in Fig. 9.11, which depicts a technical principle of an injection study by means of XPCI. The synchrotron facility operates in the hybrid-singlet mode, in which X-ray bunches run in a specific pattern in the storage ring. An X-ray beam from the storage ring passes through the safety shutter, which is used to block the X-ray radiation completely when necessary. After that, the X-ray beam passes through the fast shutter, which absorbs 99% of X-rays to avoid damaging the sensitive experimental equipment, such as the scintillator. This shutter is fast and allows the X-ray beam to go

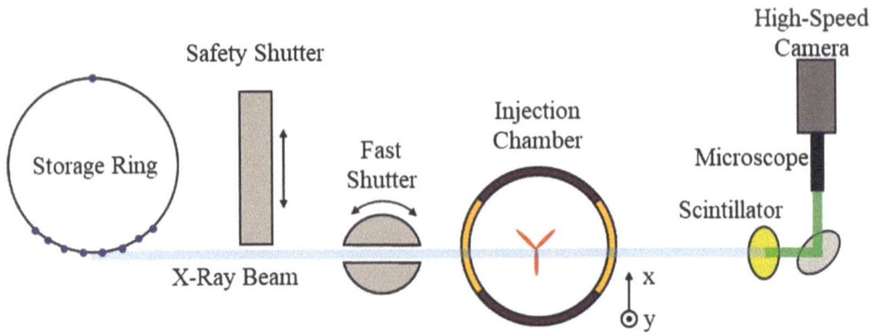

Figure 9.11 Setup for XPCI at the Argonne National Laboratory.

through for the time duration that is necessary for the imaging. Next, the X-ray beam reaches an injection chamber that has Kopton windows to improve the transmittance for X-rays. A fuel injector is mounted in the injection chamber. The chamber can be automatically moved to set imaging positions in the flow. With a limited X-ray beam size of about 1×1 mm, this setup allows successive scanning of a larger region of interest. Afterwards, the X-ray beam propagates further and reaches the scintillator, which converts the X-ray radiation into visible light. This light is deflected by a mirror and recorded by a high-speed camera mounted with an objective lens that defines the optical resolution of the imaging system. In this arrangement the camera is protected against direct exposure from the X-rays, which could damage the camera.

The X-ray radiation is significantly more weakly affected by droplets than by visible light. Therefore the effects of multiple scattering are not an issue, which is quite important in the study of dense sprays. Some pictures of dense sprays obtained by the XPCI technique were shown in Fig. 9.2. Illumination with a single pulse, which is separated from the bunch train in the synchrotron pattern, is typically employed for flow visualization. There are some studies that employed the XPCI technique for investigating the highly transient behavior of cavitation inside injector nozzles. It is important to note that injector nozzles absorb significant X-ray photon energy when the XPCI method is applied. owing to the high absorption coefficient of the nozzle material. One way to overcome this problem is to increase the photon flux, for instance, by setting a longer exposure time. However, this reduces temporal resolution and thereby limits this approach to visualizing and investigating the highly transient cavitation formation inside injector nozzles. Alternatively, instead of increasing the photon flux, the absorption of the X-rays by the nozzle can be diminished if the nozzle is made of a lighter material than is typically the case. It is usually made of stainless steel. Moon et al. performed X-ray imaging of the internal flow of GDI nozzles made of aluminum [25]. The high transmittance of aluminum facilitated higher photon flu; therefore an internal flow and, in particular, transient cavitation structures were visualized with sufficient temporal resolution. Fig. 9.12A shows the

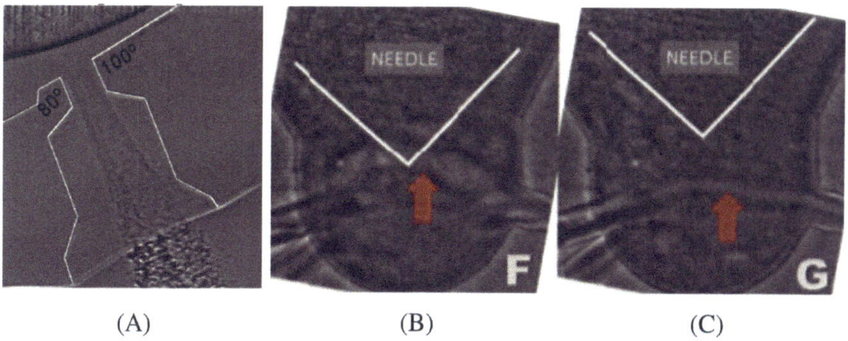

Figure 9.12 Cavitation in fuel injector nozzles observed by XPCI. (A) Geometric-induced cavitation in a multihole GDI nozzle. (B, C) String cavitation in a multihole diesel nozzle. *(a) From S. Moon, K. Komada, Z. Li, J. Wang, K. Kimijima, T. Arima, Y. Maeda High-speed X-ray imaging of in-nozzle cavitation and emerging jet flow of multi-hole GDI injector under practical injection conditions. 13th Triennial International Conference on Liquid Atomization and Spray Systems, Taiwan (2015). (b,c) From S. Moon, W. Huang, J. Wang First observation and characterization of vortex flow in steel micronozzles for high-pressure diesel injection, Experimental Thermal and Fluid Science 105 (2019) 342–348.*

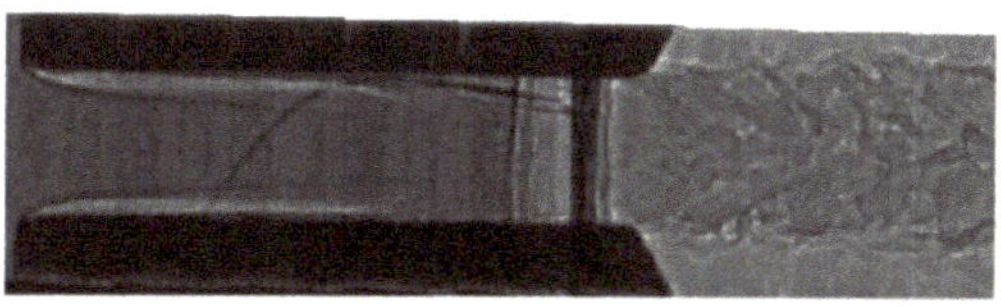

Figure 9.13 XPCI of cavitation and emerging liquid jet on a planar nozzle [12].

geometric-induced cavitation in the GDI nozzle made of aluminum, which was observed in the work of Moon et al. [25]. It can be seen that internal flow and emitting jet can be viewed simultaneously on one image. Another approach that allows visualization of internal flow and cavitation inside nozzles was introduced by Moon et al. [26]. In that sturdy, a steel nozzle was investigated. The flow was seeded with W microparticles, which increased the absorption property of the flow. Therefore the photon flux was increased, but in that case the liquid phase could still be recognized in XPCI images, owing to the higher absorption of the liquid flow. Fig. 9.12B and C show two example images from the work of Moon et al. [26], in which transient behavior of string cavitation was observed in a steel diesel nozzle.

Sou et al. performed visualization of cavitation and emerging jets on planar nozzles using XPCI method, in which various nozzle sizes were tested [12]. Fig. 9.13 is an XPCI image of a planar nozzle used in the study. It shows that this method is also applied in some more fundamental studies on cavitation.

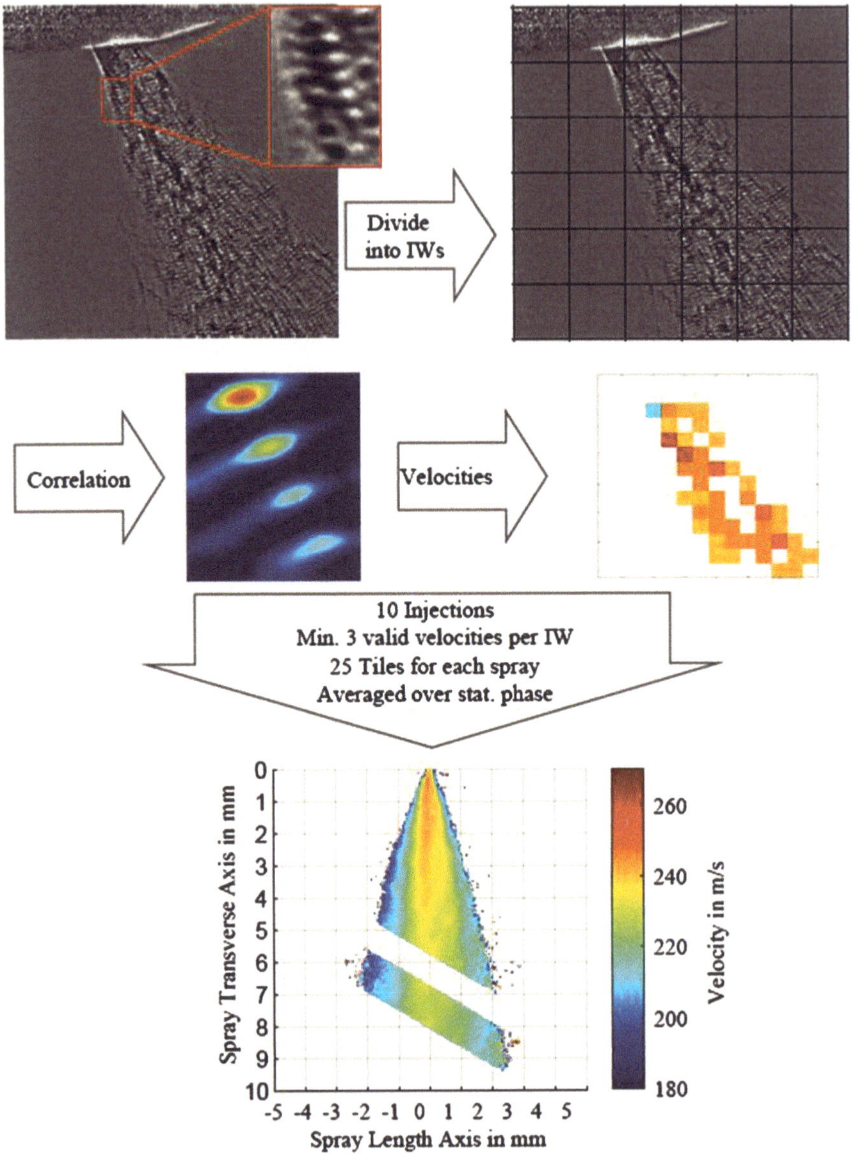

Figure 9.14 Schematic sequence of the computation of the velocity from multiexposed XPCI images.

At the Advanced Photon Source of the Argonne National Laboratory, the bunch train in the storage ring allows the possibility of obtaining multiexposed images of the flow. The multiexposed images can be used to retrieve velocity information by means of the autocorrelation evaluation. The evaluation procedure is shown in Fig. 9.14. The time between two exposures is given as 68 ns, so computing the velocity can be done by determining the distance the spray structures have moved within that time. Therefore the image is first divided into 24 × 24 pixels, the so-called interrogation windows (IW). The normalized 2D autocorrelation is calculated for each of these windows. To reduce the computation time, instead of choosing the whole image as the input image, parts of the image are chosen that are sufficiently large not to crop the results but still much smaller than the original image. The result of the correlation of an IW is shown in the middle left part of Fig. 9.14. The maximum in the upper left corner is the so-called self-correlation point, where the IW is correlated with its original position. In normalized autocorrelation the correlation coefficient at this point is 1. The maxima following to the bottom right result from the further exposures. Not shown is the 180-degree blur of the autocorrelation, which leads to a cutoff in the figure, a symmetrical distribution to the upper left. The distance between the autocorrelation point and the first maximum is then used to determine the velocity vectors. In order to further consider only sufficiently good correlations, the results are first validated by the correlation coefficient of the first maximum. This value is set at 0.2. Despite this filtering, velocity vectors of implausible magnitude and direction are detected, especially in the area of the nozzle tip. Therefore velocities greater than 110% of nozzle exit velocity according to Bernoulli are rejected. A further favorable filter criterion is the deviation of the angle from the spray centerline. Here, an angle of 30 degrees proves to be useful.

Bauer et al. performed X-ray CT imaging of the nozzle, which represents another source of X-ray radiation, to obtain the 3D distribution of time-averaged void structures and internal flow with various cavitation types with large flow rate, such as cloud cavitation, bubble cavitation, and film cavitation [27]. Cross-sectional CT images of cavitation structures in a nozzle hole are shown in Fig. 9.15.

9.1.4 Transient Internal Flow Behavior

In this section the transient behavior of the internal flow is discussed. Mamaikin et al. visualized the transient behavior of cavitation formations in a transparent real-size GDI nozzle [23]. The transparent nozzle was manufactured from acrylic glass. Fig. 9.16 shows shadowgraph images of the internal nozzle flow obtained from a single injection. The experiment was conducted at an injection pressure of 10 MPa and a back pressure of 0.1 MPa. The fuel used in the experiment was a mixture of calibration fluid (ISO 4113) and 1-bromonaphthalene. Although some thermophysical properties of this mixture, like viscosity and vapor pressure, differ significantly from those of gasoline, it allows great optical access into the internal flow due to matching of the refractive indices of this mixture and the acrylic glass. Fig. 9.16A shows an image of the nozzle before the beginning of the injection. Some residual bubbles can be observed that were entrained into the nozzle

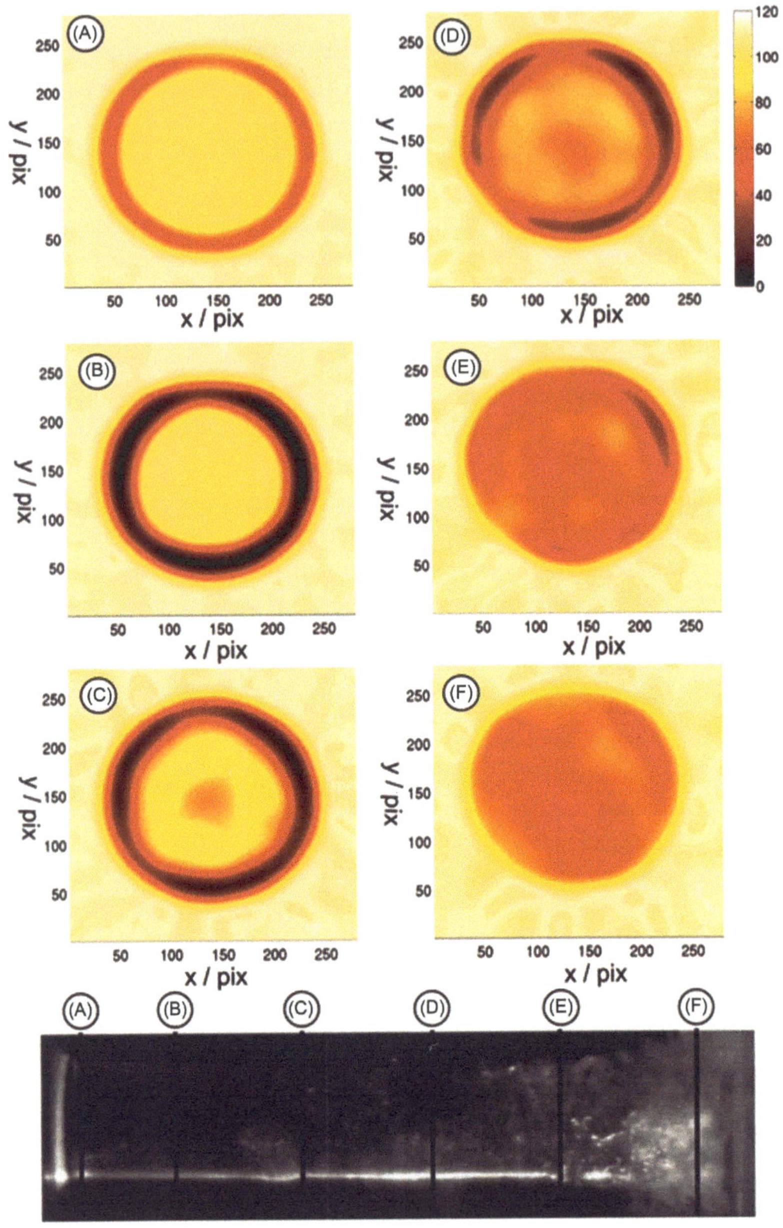

Figure 9.15 Cross-sectional CT images of cavitation structures for different observation planes (A-F) along the hole axis. *From A. Sou, S. Minami, R. Prasetya, R. H. Pratama, S. Moon, Y. Wada, H. Yokohata (2015) X-Ray Visualization of Cavitation in Nozzles with Various Sizes. 13th Triennial International Conference on Atomization and Spray Systems.*

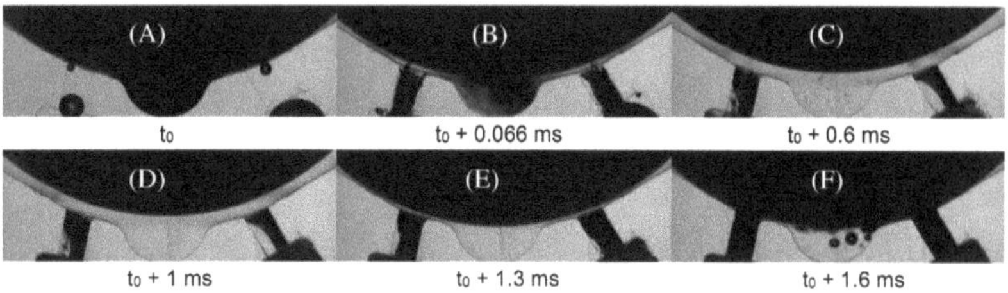

Figure 9.16 In-nozzle transient gas structures during an injection. The images (A-F) correspond to different time points that are specified on each individual image. *From D. Mamaikin, T. Knorsch, P. Rogler, P. Leick, M. Wensing High speed shadowgraphy of transparent nozzles as an evaluation tool for in-nozzle cavitation behavior of GDI injectors. 28th Conference on Liquid Atomization and Spray Systems, Spain (2017) 1027–1034.*

holes and sac from the outside air environment. The start of the injection is shown in Fig. 9.16B, where the bubbles are rapidly flushed away from the nozzle and cavitation structures form in the holes. In this stage, string cavitation is observed in the areas between the inlets of the nozzle holes and the needle. It can be assumed that at this low needle lift, as seen on this image, strong vortices occur inside the nozzles holes, which create a radial pressure drop in the hole from the wall toward the center of the hole. This leads to the formation of string cavitation. Fig. 9.16C shows the flow structure in the steady-state injection phase, where geometric-induced cavitation is observed inside the holes. Fig. 9.16D refers to the steady-state injection phase as well, but to the later injection time. It can be noticed that the geometric-induced cavitation structures observed in the holes maintain their shape in the steady-state injection. It can be seen that the cavitation formation on the right side of each hole originates on the entrance of the hole and extends to the hole exit. This is due to the high deflection of the flow as it enters the hole. The left side of each hole reveals a smoother flow deflection, which results in a shorter cavitation length. It can be observed that smoother flow deflection, like that for the left side of the right hole, leads to shorter cavitation length. Fig. 9.16E shows the internal flow at the end of injection. It can be seen that there is more cavitation occupying the nozzle holes in comparison to the steady-state phase, which is a consequence of increased flow turbulence due to the needle motion. Fig. 9.16F refers to the nozzle flow after the needle has closed. Some gas structures remain in the nozzle holes and are entrained into the nozzle sac, where they appear in the form of large bubbles.

The possibility of evaluating internal flow and cavitation behavior based on the near-nozzle jet parameters is discussed next. The internal flow is evaluated in terms of two parameters: injection rate and degree of hole filling. In the following, the evaluation of these parameters is provided for the case of the diesel jet.

The degree of spray hole filling C_A is defined as the actual mass flow rate $\dot{m}$ over the mass flow rate estimated from the spray velocity at the nozzle exit v_{exit} (inside the nozzle would be even better, but it is experimentally hard to access), the area of the spray hole A, and the density of the fuel ρ. In mathematical form, it can be given as follows:

$$C_A = \frac{\dot{m}_{measured}}{\dot{m}_{estimated}} = \frac{\dot{m}}{v_{exit} \cdot \rho \cdot A}$$

The mass flow rate can be measured, for instance, by using the HDA method, which was described previously. The nozzle exit velocity in this case is estimated with phase contrast velocimetry (PCV) [28]. Although other techniques, such as structural image velocimetry or laser-two-focus velocimetry, can be considered [29,30], PCV is a measurement technique that allows determination of the velocities closest to the nozzle and hence makes it possible to evaluate the outlet velocity very precise. Simpler techniques for velocity determination, such as the penetration speed of the spray front as seen in shadowgraphy or Schlieren images, are not able to measure the nozzle outlet velocity precisely enough.

Fig. 9.17 shows the transient development of a diesel jet, which was obtained by the XPCI method. The images clearly show a high amount of cavitation for

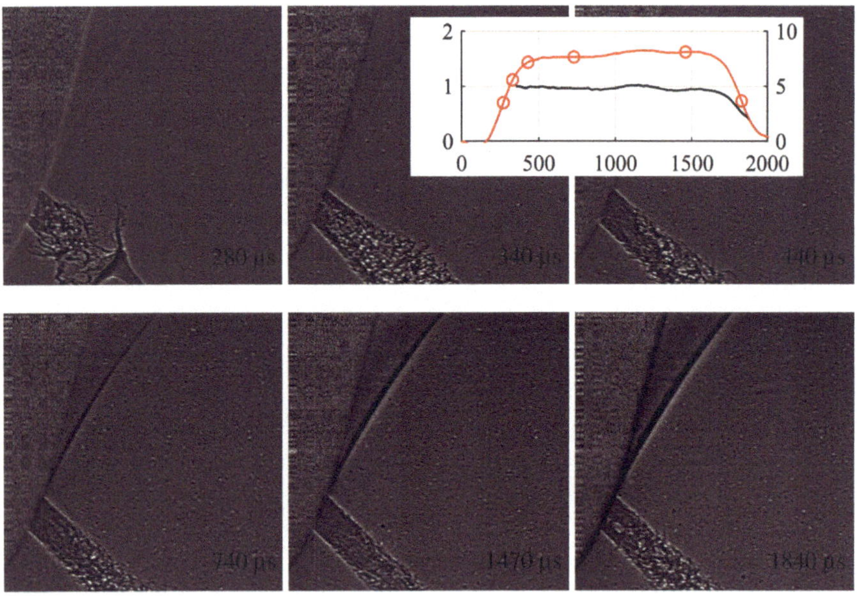

Figure 9.17 Temporal development of a near-nozzle jet injected by a diesel injector. The degree of spray hole filling and injection rate are shown in the plot by the black and red lines, respectively (axes on plot). The fuel was calibration fluid (ISO 4113), and the injection pressure was 50 MPa.

the first two displayed time steps during the opening phase of injection. Owing to the low velocities resulting from the choked flow from the needle opening, it is not possible to evaluate the spray velocities and hence the degree of spray hole filling for this point. For the next three time steps, the flow has stabilized, and no cavitation is present in the jet exiting the nozzle. This shows good correlation, the degree of spray hole filling being very close to 1. In the last time step, which is within the closing phase of the injector, cavitation is present, also indicated by C_A being clearly lower than 1.

It is remarkable that the temporal development of the graph for the degree of spray hole filling is more or less similar to the injection rate. This is because the velocity is proportional to the rate and therefore is relatively constant and that the small modulations observed in velocity have the shape of the injection rate.

Fig. 9.18 shows the degree of transient hole filling that was estimated for the diesel nozzle at three injection pressure levels. The horizontal axis depicts the time after electrical start of injection (ESOI). For all three operating points, the degree of spray hole filling reaches values slightly larger than 1, which is physically not possible. The small deviations are relativized, keeping in mind that the result is computed from two measured values: the mass flow rate and the velocity.

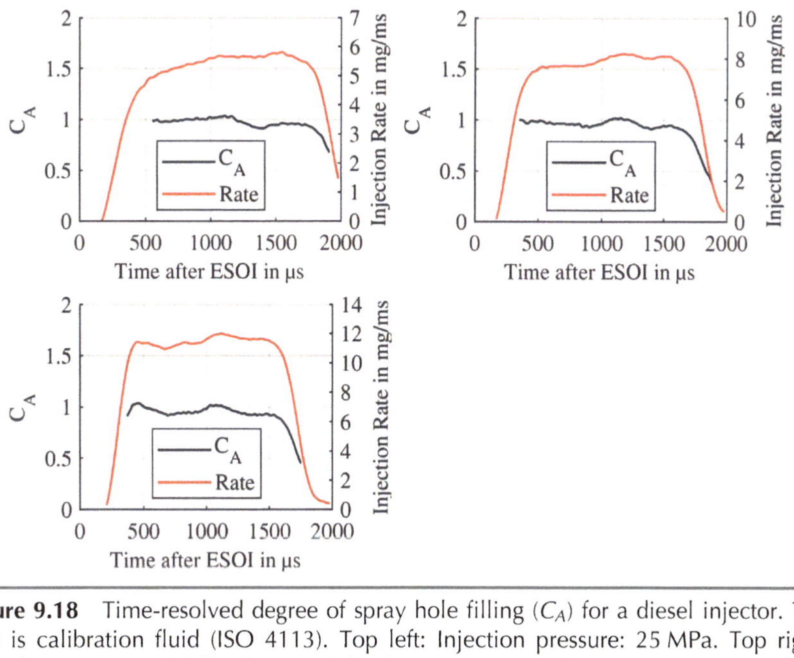

Figure 9.18 Time-resolved degree of spray hole filling (C_A) for a diesel injector. The fuel is calibration fluid (ISO 4113). Top left: Injection pressure: 25 MPa. Top right: Injection pressure: 50 MPa. Bottom: Injection pressure: 100 MPa.

For the gasoline injector, for both investigated pressures, the degree of spray hole filling is slightly above 60%, as can be seen in Fig. 9.19. The high-frequency modulation of the rate and hence C_A result from pressure pulsations in the fuel line.

For both injectors, when the needle closes, it becomes obvious that the degree of spray hole filling decreases drastically, which indicates a high amount of cavitation in the sac volume and the spray hole.

Fig. 9.20 shows the degree of spray hole filling averaged over the stationary phase of the injection for a gasoline injector (20 and 24 MPa) and a diesel injector (25, 50, and 100 MPa). These injectors are the same as selected for the comparison of the liquid jet appearance for GDI and diesel shown in Fig. 9.2. The graph shows clearly the large differences between the two injectors. While the spray hole of the gasoline injector is filled with gas for more than one-third, it is below

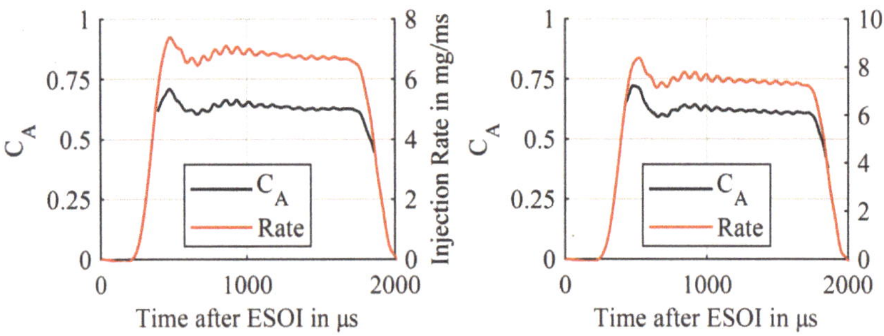

Figure 9.19 Time-resolved degree of spray hole filling (C_A) for a GDI injector. The fuel is *n*-heptane. Left: Injection pressure 10 MPa. Right: Injection pressure 24 MPa.

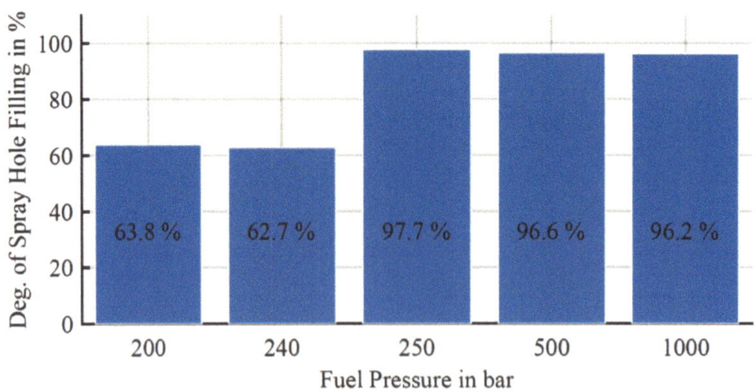

Figure 9.20 Averaged degree of hole filling C_A over the stationary phase for a GDI injector (left two bars) and a diesel injector (right three bars).

4% for the diesel injector. This indicates the very different breakup mechanisms of the both injector types very clearly. While the gasoline injector remarkably promotes cavitation, in order to achieve a good atomization, it has to be suppressed in the diesel injector mostly, in order to avoid damage to the nozzle from erosion. A closer look to the injection pressure for both nozzles indicates, that the degree of spray hole filling is decreasing with increasing pressure. This indicates an increasing amount of cavitation with increasing pressure.

9.2 Influences on Cavitation and Effects for Fuel Injection

Various approaches are used to govern the cavitation behavior in nozzles. For example, nozzle holes are manufactured with the profile of positive or negative conicity. Also, a hole inlet can be rounded to facilitate smooth inflow into the nozzle holes, which is typically achieved by a hydroerosive grinding process, in which a rounding of the hole edges is achieved by flowing an abrasive fluid through the nozzle holes.

9.2.1 Nozzle Geometry

Balewski et al. investigated the influence of the nozzle geometry on the liquid breakup in a noncavitating flow [31]. Various nozzle designs were produced with the intention of changing the inflow geometry before the orifice. The nozzles were manufactured with a single hole, and flow conditions with relatively low Reynolds numbers of 2300 were investigated in this research. The study found that the upstream conditions influence the liquid jet breakup, owing to the difference in the secondary flow field or the turbulence level. That again emphasizes the problem of differentiating the contributions of cavitation and turbulence to liquid jet breakup in the relevant injection conditions, which was mentioned earlier in the chapter. Iin the following sections the geometrical influence on cavitation is discussed.

9.2.1.1 Nozzle Hole Profile

Payri et al. investigated the influence of cavitation in diesel nozzles on sprays [32]. The authors studied nozzles with cylindrical and convergent (conical) holes. The authors used a silicone technique for determination of the nozzle hole geometry. The cavitation was evaluated on the basis of a hydraulic characterization with measurements of the mass flow rate. The authors state that the nozzle with conical holes was not subjected to cavitation for the operating conditions tested, whereas the nozzle with cylindrical holes induced cavitation. The critical cavitation number was defined by means of the collapse of the mass flow rate. This number was found to be affected by the Reynolds number (or injection pressure) and not only by the nozzle geometry. The authors state that cavitation leads to an increase in the spray cone angle. Fig. 9.21A shows a comparison of the tangent of half of the cone angle as a function of cavitation number K for both nozzles. The cavitation number was determined according to the work of Nurick [33] as follows:

$$K = \frac{P_i - P_v}{P_i - P_b}$$

where P_i is the injection pressure, P_b is the back pressure, and P_v is the vapor pressure of the fuel. The evaluation was performed for the time span between 1500 and 3300 μs of the injection duration revealing a more stable behavior of the angle. It can be seen that the spray cone angle reveals an increase of about 15% when cavitation appears for the cylindrical nozzle. Fig. 9.21B shows a comparison of the transient behavior of the cone angle for both nozzles at different injection pressure levels. It can be seen that the cylindrical nozzle is subjected to higher deviation of the cone angle during injection, which can be also linked with the formation of cavitation.

Conrad et al. [24] used 3D single-hole real-size glass nozzles with five variable parameters to investigate the influence of the geometry on the inner flow and spray propagation. A side view of the nozzle and its parameters is shown in Fig. 9.22. Typical nozzle parameters that can be varied in this approach are the bore length L and the inlet and outlet diameters. These three parameters together determine the conicity of the nozzle. In addition, the height of sac hole H and the bore offset from the inlet H_{off} can be varied.

The conicity of the nozzle bore strongly influences cavitation behavior in the nozzle. Conrad et al. investigated the inner nozzle for glass nozzles with convergent, cylindrical, and divergent nozzle designs [24], as illustrated in Fig. 9.23. The results show string cavitation in all three nozzles. The differences are in the amount of sheet cavitation at the nozzle wall. For the convergent nozzle, almost

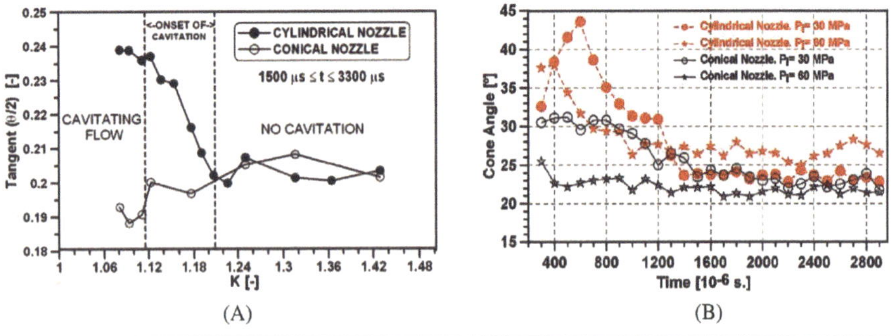

Figure 9.21 Comparison of parameters related to the spray angle for cylindrical and conical holes: (A) tangent of the half of cone angle, (B) cone angle. *From F. Payri, V. Bermudez, R. Payri, F. J. Salvador The influence of cavitation on the internal flow and the spray characteristics in diesel injection nozzles. Fuel 83 (2004) 419–431.*

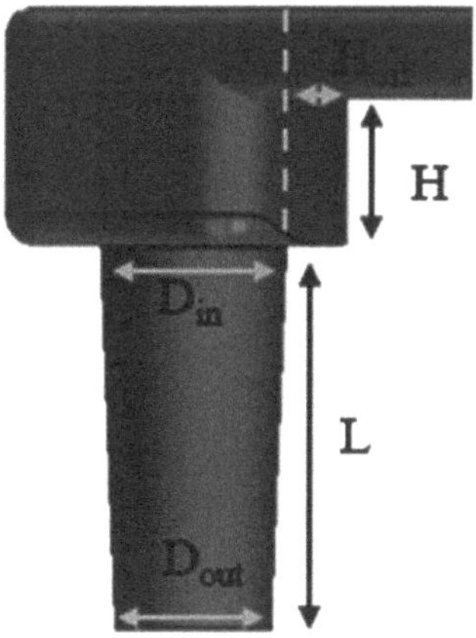

Figure 9.22 Nozzle parameters.

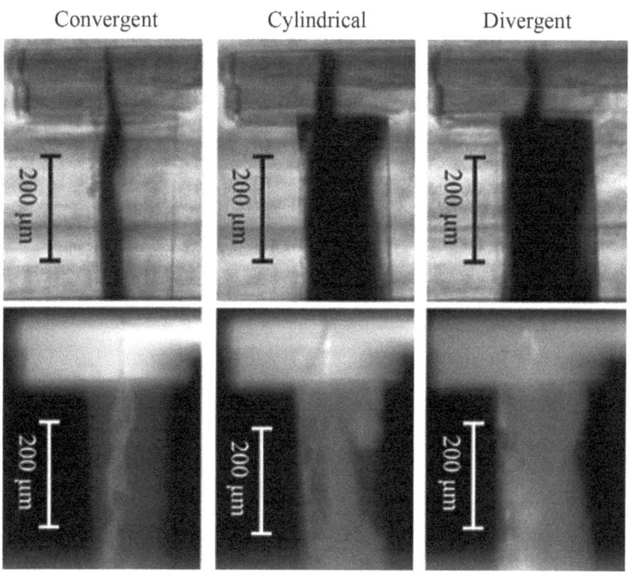

Figure 9.23 Shadowgraph images (top) and LIF images (bottom) for different nozzle geometries [24].

no sheet cavitation forms. The cylindrical and divergent nozzle both exhibit strong sheet cavitation from the inlet of the nozzle bore to the oulet. In the cylindrical nozzle, less sheet cavitation forms on the right side of the nozzle.

In a later study, Conrad et al. investigated the influence of all five parameters of the glass nozzle on the spray propagation. The results show that the relationship of the bore length to the nozzle diameter is the most critical parameter where nozzles with a high *L/D* exhibit high spray penetration and a low cone angle, as can be seen in Figs. 9.24 and 9.25.

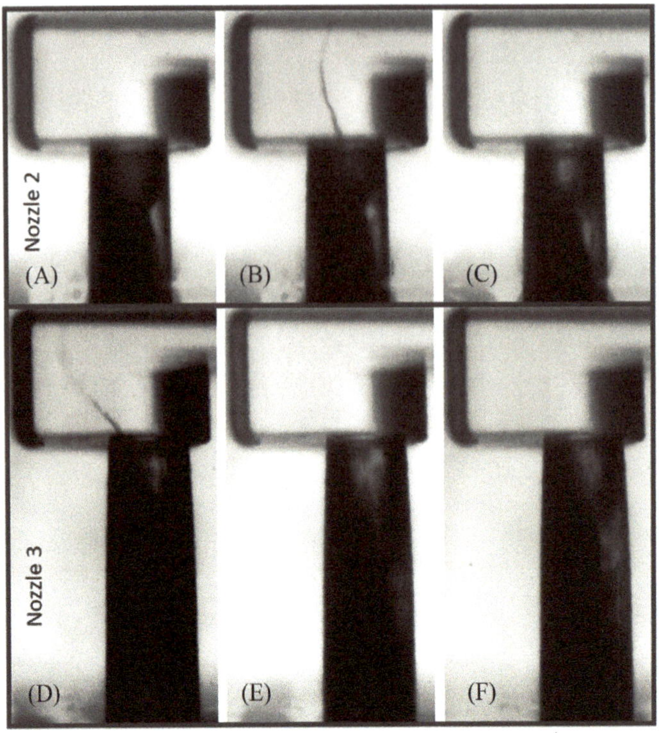

Figure 9.24 Distinct images of the in-nozzle flow taken for different injection events for nozzles 2 (A–C) and 3 (D–F).

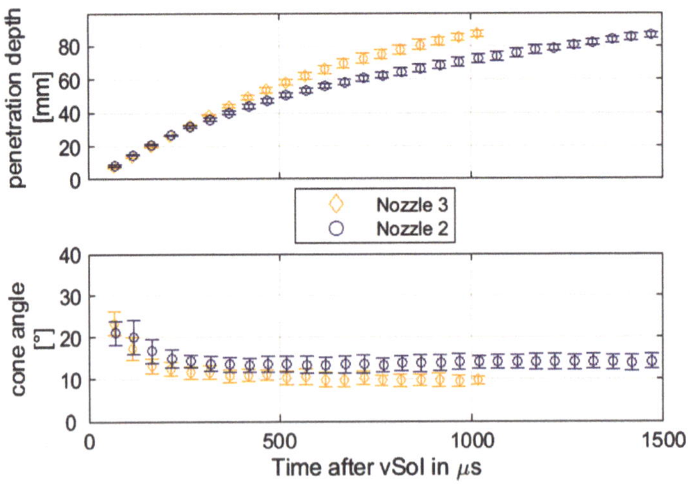

Figure 9.25 Experimental penetration depth and cone angle of nozzles 2 and 3.

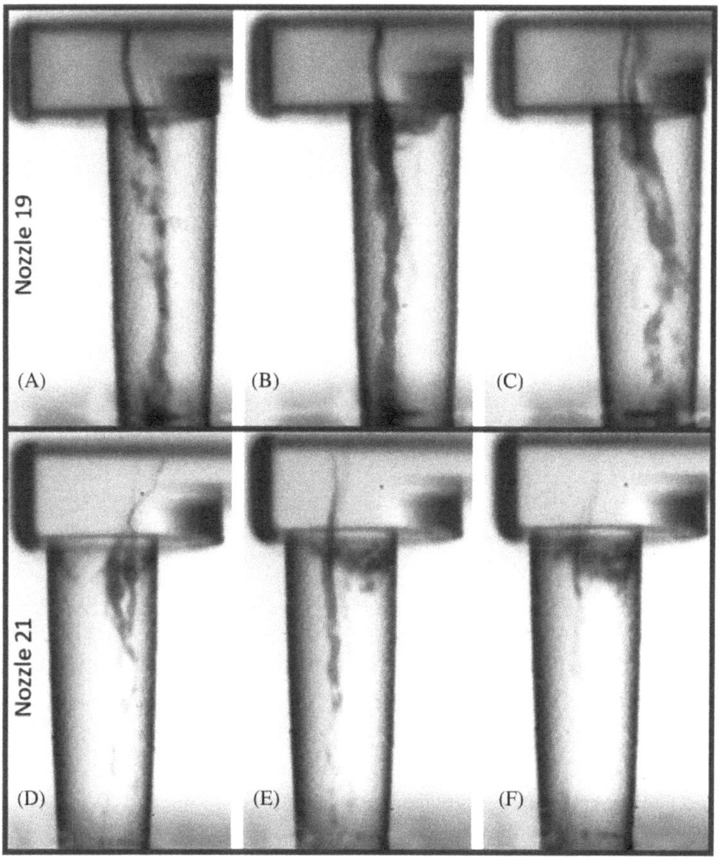

Figure 9.26 Distinct images of the in-nozzle flow taken for different injection events for nozzles 19 (A–C) and 21 (D–F).

Furthermore, the study shows how a small nozzle offset H_{off} increases the string cavitation inside the nozzle, which results in a wider cone angle of the spray. This effect is shown in Figs. 9.26 and 9.27.

9.2.2 Dynamic Factors

Up to this point, the impact of nozzle geometry on cavitation and an emerging liquid jet was discussed. However, it is known that cavitation inception is influenced not only by the geometrical factors, but also by dynamic factors, such as pressure difference, needle lift, and needle eccentricity [5,15]. Furthermore, the needle lift can affect the injection rate and the discharge coefficient [34]. He et al. observed a more

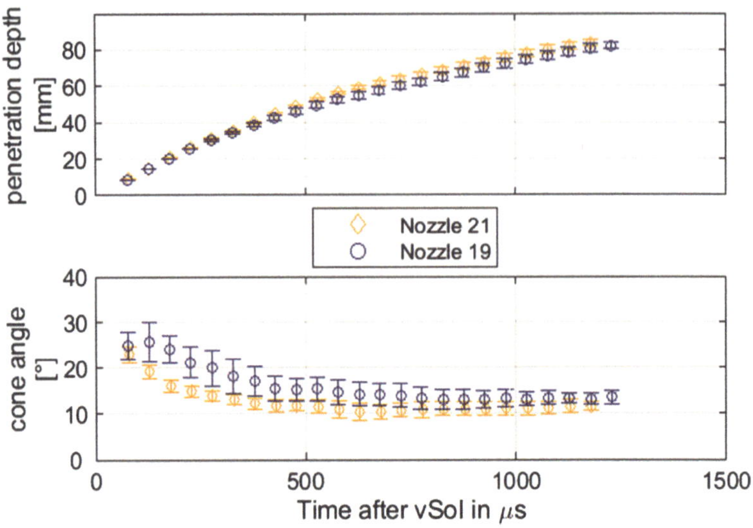

Figure 9.27 Experimental penetration depth and cone angle of nozzles 19 and 21.

obvious effect of the needle eccentricity on the internal flow for a low needle lift case [35]. Gavaises et al. investigated the internal flow and associated spray formation in transparent diesel injector nozzles [36]. The location and probability of string cavitation in a nozzle were found to vary considerably with the needle lift and were less influenced by the Reynolds and cavitation numbers for the fully developed flow conditions. The authors also showed that string cavitation leads to a considerable transient variation in the spray cone angle and the deflection angle, which can contribute to cycle-to-cycle variation in the spray for diesel injectors. The spray cone angle becomes larger when string cavitation is present in a nozzle hole. Mamaikin et al. performed an extensive investigation of the internal flow inside a real-size transparent GDI nozzle [37]. The measurements were conducted with high spatial and temporal resolution for a transient injector operation for a nozzle design that inhibited the formation of geometric-induced cavitation. Visualization of the internal flow in a steady-state injection period did not reveal the appearance of cavitation in the observation time window. However, visualization in the closing phase of the injection revealed the formation of string cavitation. Fig. 9.28 illustrates subsequent images that were recorded during the closing injection phase. The time gap corresponds to 0.2 μs. The first images reveal string cavitation, which occupies the lower part of the hole. The cavitation formation seems to extend upstream along the injection time. At some point, a second origin of string cavitation appears in the area between the needle and the nozzle hole entrance. This formation expands quickly and merges with the already existing string cavitation, which results in a long string cavitation structure that extends from the needle to the hole exit.

Figure 9.28 Development of string cavitation inside a transparent GDI nozzle hole. *From D. Mamaikin, T. Knorsch, P. Rogler, M. Wensing Experimental investigation of flow field and string cavitation inside a transparent real-size GDI nozzle. Experiments in Fluids 61 (154) (2020).*

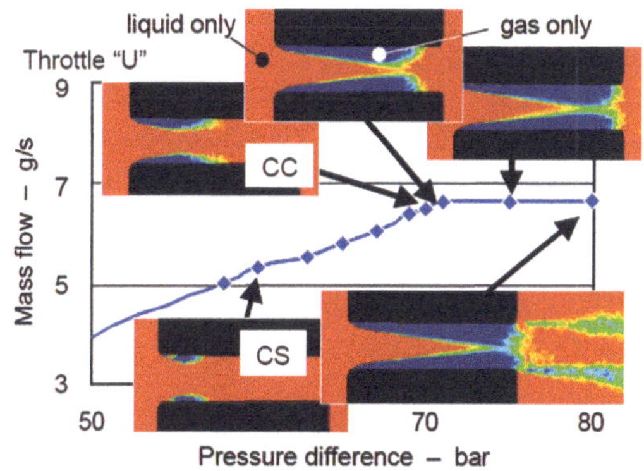

Figure 9.29 The influence of the pressure difference on mass flow for a planar nozzle. *From E. Winklhofer, E. Kull, E. Kelz, A. Morozov Comprehensive hydraulic and flow field documentation in model throttle experiments under cavitation conditions. 17th International Conference on Liquid Atomization and Spray Systems (2001), Switzerland.*

Winklhofer et al. show how the cavitation in a nozzle affects the degree of hole filling and subsequently the mass flow [38]. This was investigated by using a 2D nozzle using mass flow measurements, shadowgraphy, visualization of the pressure field, and particle image velocimetry. For all investigations the inlet pressure was held constant at 10 MPa, and the ambient pressure was changed. Fig. 9.29 shows the influence of pressure difference on mass flow of the nozzle. The investigation showed that the start of cavitation in a nozzle has a small influence on the mass flow. Only when the cavitation reaches a critical point can an effect on the mass flow be observed. This critical cavitation point is defined as the start of critical flow where the mass flow cannot be further increased by decreasing the ambient pressure. This behavior can be used to determine the cavitation tendency of a nozzle by using only hydraulic measurements.

9.2.3 Fuel and Operating Conditions

In addition to the geometrical and dynamic factors that affect the internal flow and cavitation in injector nozzles, the fuel can influence the in-nozzle cavitation dynamics and the emerging jet parameters, owing to the possible variation in the relevant thermophysical properties. Payri et al. investigated the influence of different fuels on internal flow under cavitating conditions, using a transparent nozzle with a cylindrical hole [39]. Four different fuel types were selected for this study: *n*-heptane, *n*-decane, *n*-dodecane, and diesel. The researchers concluded that less viscous fuel tends to cavitate sooner. They link this observation to a higher Reynolds number for the fuel with less viscosity. In addition to the fuel properties, an effect of the operating conditions on the internal flow and emerging jet is essential. Zigan et al. studied the impact of fuel temperature and fuel properties on the internal flow and the spray development [40]. Two fuels with large differences in viscosity and volatility (*n*-decane and *n*-hexane) were investigated at 25°C and 70°C. The researchers found that a reduced viscosity increases cavitation in the nozzle and turbulence, which leads to a narrower spray with a longer penetration. In addition, they stated that the shot-to-shot fluctuations of the sprays were found to originate in the highly unsteady, cavitating nozzle flow.

Aleiferis et al. investigated various fuels, such as commercial gasoline, ethanol, isooctane, *n*-pentane, butanol, and E10 (a mixture of 10% ethanol and 90% gasoline) under different operating conditions [41]. Imaging of the internal flow and emerging liquid jet was conducted on two nozzles made of Perspex. Each nozzle was manufactured with a single hole at an inclination of 60 degrees with respect to the vertical axis. Each nozzle was designed to produce a hole length to diameter ratio of 5, which is a relatively high value for conventional GDI nozzles. However, a greater hole length was desired in this study to improve robustness of the transparent nozzle, which was operated in challenging conditions of high pressure and temperature. The injector needle was set to a fixed open position, providing steady-state injection to be realized in the experiments. The injection pressure was set at 2.3 and 4 MPa for each nozzle individually to match the mass flow rate of a single hole in a real injector considered at 15 MPa. All fuels were observed to appear with similar cavitation patterns and atomization at a fuel temperature of 20°C. The sprays at the fuel temperature of 90°C and the back pressure of 0.05 MPa revealed a wider angle due to the expansion of cavitation bubbles. The researchers concluded that the level of fuel superheating determines the efficiency of atomization. For that, cavitation plays an important role, as it supplies plenty of vapor bubbles, which serve as nucleation sites. The two alcohols were observed in the second wind-induced regime and crossed into the atomization regime only at elevated temperatures. Fig. 9.30 shows the images for gasoline

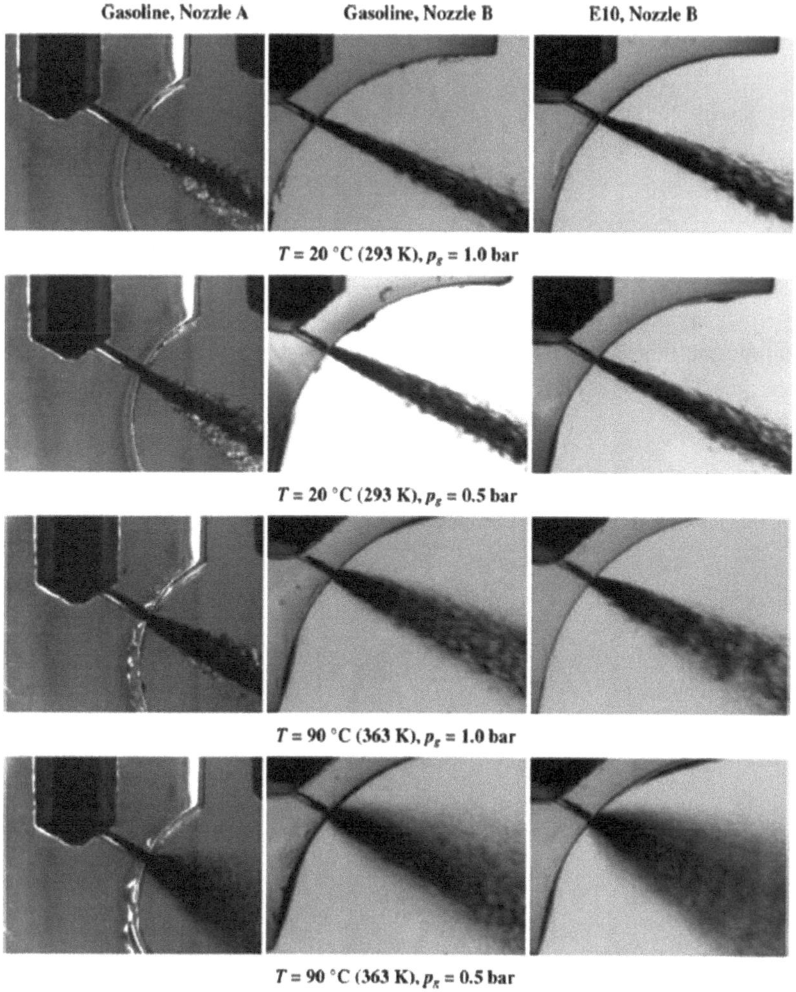

Figure 9.30 Imaging of the internal flow and the emerging jet for gasoline and E10. *From P. G. Aleiferis, J. Serras-Pereira, A. Augoye, T. J. Davies, R. F. Cracknell, D. Richardson Effect of fuel temperature on in-nozzle cavitation and spray formation of liquid hydrocarbons and alcohols from a real-size optical injector for direct-injection spark-ignition engines. International Journal of Heat and Mass Transfer 53 (21–22) (2010) 4588–4606.*

and E10 injected by the two nozzles (nozzles A and B) at different operating conditions. It can be noticed that at 20°C, the effect of back pressure change is small. For the experiments with the fuel temperature of 90°C, the change in the spray appearance is remarkable. As the authors state, the fuel begins to atomize immediately outside the nozzle for this condition. This is due to the increased level of flash boiling. The spray appears to be asymmetric with a deviation toward the side of the cavitation. The cavitation pattern for E10 and gasoline seems to be similar, apart from the case of $T = 90°C$ and $P = 0.05$ MPa, in which it appears to be longer for E10. The E10 sprays are more sensitive to the operating conditions in comparison to gasoline, as can be seen in the comparison based on nozzle B. The authors attribute this observation to the higher vapor pressure of E10 in comparison to gasoline.

Conrad et al. investigated the flow inside their glass nozzles under flash-boiling conditions [24]. The mean images from the study are shown in Fig. 9.31. The authors found only a small effect of the vapor pressure on the nozzle flow, stating that the string cavitation in their nozzle stabilized, but no increase in sheet cavitation could be found. The authors also discussed how their choice of fuel in general has a higher vapor pressure than gasoline or surrogate fuels used in other studies.

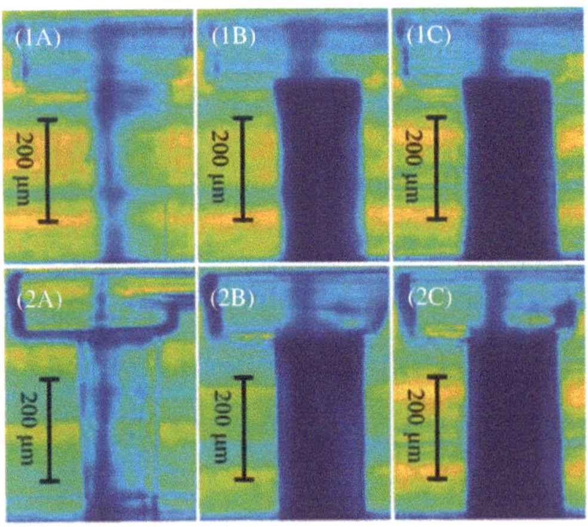

T_{Fuel}	30 °C
$p_{ambient}$	0.1 MPa
p_{vapor}(toluene)	0.0049 MPa
p_{vapor}(n-decane)	0.0003 MPa
T_{Fuel}	90 °C
$p_{ambient}$	0.03 MPa
p_{vapor}(toluene)	0.0540 MPa
p_{vapor}(n-decane)	0.0063 MPa

Figure 9.31 Shadowgraph false color mean images for cold conditions: (1A) convergent, (1B) cylindrical, (1C) divergent nozzles and for flash-boiling conditions: (2A) convergent, (2B) cylindrical, (2C) divergent nozzles. Vapor pressures are from the REFPROP database. *From C. Conrad, S. Bornschlegel, A. Durst, D. Jordan, M. Wensing Influence of the nozzle geometry on in-nozzle cavitation investigated in real-size glass nozzles with shadowgraphy and LIF. 14th Triennial International Conference on Liquid Atomization and Spray Systems, USA (2018) 1–6.*

9.2.4 Erosion Damage

The destructive action of cavitation is attributed to two mechanisms:

- Shock waves influenced by the collapse of vapor bubbles.
- Generation of a high-speed jet that performs as a water hammer effect on the surface.

The collapse of cavitation bubbles is accompanied by a high temperature and pressure rise inside the bubbles, which is attributed to the high compressibility of the gas-vapor medium and inertia of the liquid. The high local temperatures and the high-frequency pressure waves lead to erosive wear of the surface of the injector nozzle. Erosion occurs when the pressure waves destroy the oxide layer on the material surface. The air in the liquid creates a new oxide layer, which is significantly thinner. The interrelation of cavitation and erosion damage in injector nozzles was investigated by Gavaises et al. [43]. Enlarged transparent nozzles were used to investigate the internal flow of diesel injector nozzles and corresponding erosion damages. The researchers observed surface damage on those nozzle areas and the formation and development of cavitation in the flow during the injection. This was confirmed by the numerical simulations of the flow performed in the framework of this study.

9.3 Conclusion

Cavitation plays an important role in the operation of modern fuel injection systems. In particular, internal flow in high-pressure fuel injectors is affected by this phenomenon. On the one hand, cavitation can improve atomization performance, owing to the additional microturbulence caused by collapse of the gas bubbles. On the other hand, cavitation can lead to the fast deterioration of injector nozzles as a consequence of an erosion process. Therefore an adequate understanding and prediction of the cavitation behavior in injector nozzles are necessary for further improvements of fuel injectors.

Various experimental techniques have been developed to study cavitation behavior and its effect on the spray formation for fuel injection systems. Hydraulic methods, such as HDA, are commonly used to characterize the internal flow of the fuel injectors, which can also estimate the degree of cavitation inside the injector nozzles. Different optical approaches involve glass nozzles of various scales that enable visualization of the internal flow. These methods are essential, as they provide a nonintrusive approach to evaluate the impact of various parameters, such as injector nozzle design, fuel properties, and experimental conditions, on the internal flow behavior. Furthermore, advanced approaches are currently available that employ fast

X-ray imaging to investigate the internal flow and cavitation inside metal injector nozzles. This experimental approach is also used to study dense spray regions.

Cavitation can appear in both gasoline and diesel injectors, especially considering relevant operating conditions. However, it plays a more relevant role for gasoline injectors. The experimental observations obtained from XPCI show that the gasoline can leave the nozzle already as spray; in other words, the first ligaments are formed at the exit of the nozzle. By contrast, it was shown that the diesel spray can leave the injector as an intact core. The liquid breakup process for the diesel injectors is substantially influenced by the high flow turbulence and aerodynamic forces due to significantly higher injection pressure. The extent of cavitation appearing in the nozzle hole is closely related to the degree of hole filling. During the stationary spray phase with the nozzle needle fully open, the degree of hole filling is close to 100% in many diesel injection cases and about two-thirds or even less in GDI cases. Nevertheless, cavitation is also extremely relevant in the diesel sector. Cavitation causes more damage in the diesel sector, owing to the higher injection pressures, and even slight damage can influence the spray and strongly change the mixture formation and thus ultimately the formation of pollutants in the engine process.

References

[1] A. Ferrari, Modelling approaches to acoustic cavitation in transmission pipelines, International Journal of Heat and Mass Transfer 53 (2010) 4193–4203.

[2] F.J. Salvador, J. De la Morena, M. Crialesi-Esposito, J. Martinez-Lopez, Comparative study of the internal flow in diesel injection nozzles at cavitating conditions at different needle lifts with steady and transient simulation approaches, Proceedings of the Institution of Mechanical Engineers, Part D: Journal of Automobile Engineering 232 (8) (2018) 1060–1078.

[3] F. Köpple, D. Seboldt, P. Jochmann, A. Hettinger, A. Kufferath, M. Bargende, Experimental investigation of fuel impingement and spray-cooling on the piston of a GDI engine via instantaneous surface temperature measurements, SAE International Journal of Engines 7 (3) (2014) 1178–1194.

[4] R. Payri, J. Gimeno, J. De la Morena, P.A. Battiston, A. Wadhwa, R. Straub, Study of new prototype pintle injectors for diesel engine application, Energy Conversion and Management 122 (2016) 419–427.

[5] I. Gilles-Birth, M. Rechs, U. Spicher, S. Bernhardt, A study of the in-nozzle flow characteristic of valve covered orifice nozzles for gasoline direct injection. SAE Technical Paper, 2005-01-3684, 2005.

[6] B. Janetzky Messtechnik für einspritzsysteme, in: K. Mollenhauer, H. Tschöke (Eds.). 3. Auflage, Springer, Kapitel 5.4, 2007.

[7] Moehwald GmbH, Hydraulischer Druckanstiegs-Analysator. Technical brochure. Injection rate and quantity measurement. <https://moehwald.de>.

[8] T. Knorsch, P. Rogler, M. Miller, W. Wiese, On the evaluation methods for systematic further development of direct-injection nozzles, SAE Technical Paper 2016-01-2200, 2016.

[9] T. Mederer, M. Wensing, A. Leipertz, Investigation of the interaction of charge motion and residual gas concentration in an optically accessible SI engine, SAE Technical Paper 2013-01-0558, 2013.

[10] R. Welss, S. Bornschlegel, M. Wensing, LIF laser scanning to investigate different spray characteristics and its effect on the mixture formation in an DISI engine, in: Proceedings of the 13th Triennial International Conference on Liquid Atomization and Spray Systems, Taiwan, 2015.

[11] A. Sou, S. Hosokawa, A. Tomiyama, Effects of cavitation in a nozzle on liquid jet atomization, International Journal of Heat and Mass Transfer 50 (2007) 3575–3582.

[12] A. Sou, S. Minami, R. Prasetya, R.H. Pratama, S. Moon, Y. Wada, et al. X-ray visualization of cavitation in nozzles with various sizes, in: Proceedings of the 13th Triennial International Conference on Liquid Atomization and Spray Systems, Taiwan, 2015.

[13] D.P. Schmidt, C.J. Rutland, M.L. Corradini, P. Roosen, O. Genge, Cavitation in Two-Dimensional Asymmetric Nozzles, SAE Technical Paper (1999) 1999-01-0518.

[14] M. Nouri, N. Mitroglou, Y. Yan, C. Arcoumanis, Internal flow and cavitation in a multi-hole injector for gasoline direct-injection engines, SAE Technical Paper, 2007-011405, 2007.

[15] C. Arcoumanis, M. Gavaises, J.M. Nouri, E. Abdul-Wahab, R.W. Horrocks, Analysis of the flow in the nozzle of a vertical multi hole diesel engine injector, SAE Transactions Journals of Engines 107 (1998) 980811.

[16] P.G. Aleiferis, Y. Hardalupas, D. Kolokotronis, A. Taylor, A.M.K.P. Arioka, M. Saito, Experimental investigation of the internal flow field of a gasoline model injector using micro-PIV, SAE Technical Paper Series, 2006-01-3374, 2006.

[17] A. Andriotis, M. Gavaises, C. Arcoumanis, Vortex flow and cavitation in diesel injector nozzles, Journal of Fluid Mechanics 610 (2008) 195–215.

[18] H. Afzal, C. Arcoumanis, M. Gavaises, N. Kampanis, Internal flow in diesel injector nozzles – modelling and experiments, in: Proceedings of IMechE Seminar on Fuel Injection Systems, London, Paper S492/S2/99, 1999.

[19] B.A. Reid, M. Gavaises, N. Mitroglou, G.K. Hargrave, C.P. Garner, E.J. Long, et al., On the formation of string cavitation inside fuel injectors, Experiments in Fluids 55 (2014) 1662.

[20] C. Arcoumanis, H. Flora, M. Gavaises, M. Badami, Cavitation in real-size multi-hole diesel injector nozzles. SAE transactions (109) section 3, Journal of Engines (2000) 1485–1500.

[21] C. Badock, R. Wirth, A. Fath, A. Leipertz, Investigation of cavitation in real size diesel injection nozzles, International Journal of Heat and Fluid Flow 20 (5) (1999) 538–544.

[22] J. Walther, J.K. Schaller, R. Wirth, C. Tropea, Investigation of internal flow in transparent diesel injection nozzles using fluorescence particle velocimetry (FPIV), in: Proceedings of the Eighth International Conference on Liquid Atomization and Spray Systems, USA, 2000.

[23] D. Mamaikin, T. Knorsch, P. Rogler, P. Leick, M. Wensing High speed shadowgraphy of transparent nozzles as an evaluation tool for in-nozzle cavitation behavior of GDI injectors, in: Proceedings of the 28th Conference on Liquid Atomization and Spray Systems, Spain, 2017, pp. 1027–1034.

[24] C. Conrad, S. Bornschlegel, A. Durst, D. Jordan, M. Wensing Influence of the nozzle geometry on in-nozzle cavitation investigated in real-size glass nozzles with shadowgraphy and LIF, in: Proceedings of the 14th Triennial International Conference on Liquid Atomization and Spray Systems, USA, 2018, pp. 1–6.

[25] S. Moon, K. Komada, Z. Li, J. Wang, K. Kimijima, T. Arima, et al., High-speed X-ray imaging of in-nozzle cavitation and emerging jet flow of multi-hole GDI injector under practical injection conditions, in: Proceedings of the 13th Triennial International Conference on Liquid Atomization and Spray Systems, Taiwan, 2015.

[26] S. Moon, W. Huang, J. Wang, First observation and characterization of vortex flow in steel micronozzles for high-pressure diesel injection, Experimental Thermal and Fluid Science 105 (2019) 342–348.

[27] D. Bauer, H. Chaves, C. Arcoumanis, Measurements of void fraction distribution in cavitating pipe flow using X-ray CT, Measurement Science and Technology 23 (5) (2012).

[28] A. Neubauer, Messtechnische Erfassung primärer Strahlstrukturen von Benzin- und Dieselsprays. Dissertation, Friedrich-Alexander-Universität Erlangen-Nürnberg, 2020.

[29] N. Kawaharada, K. Gröger, H. Ueki, F. Dinkelacker, Laser 2-focus velocimeter for droplet characterization in dense sprays, in: Proceedings of the 19th International Symposium on the Application of Laser and Imaging Techniques to Fluid Mechanics, Portugal, 2018.

[30] T. Dageförde, K. Gröger, N. Kawaharada, F. Dinkelacker, Velocity field measurements with high speed structural image velocimetry in the primary atomization region of future diesel fuels, SAE Technical Paper 2020-01-2112, 2020.

[31] B. Balewski, B. Heine, C. Tropea, Experimental investigation of the correlation between nozzle flow and spray using laser Doppler velocimeter, phase Doppler system, high-speed photography and X-ray radiography, Atomization and Sprays 20 (1) (2010) 57–70.

[32] F. Payri, V. Bermudez, R. Payri, F.J. Salvador, The influence of cavitation on the internal flow and the spray characteristics in diesel injection nozzles, Fuel 83 (2004) 419–431.

[33] W.H. Nurick, Orifice cavitation and its effect on spray mixing, Journal of Fluids Engineering 98 (4) (1976) 681–687.

[34] P. Viera, R. Payri, A.B. Swantek, D.J. Duke, N. Sovis, A.L. Kastengren, et al., Powell Linking instantaneous rate of injection to X-ray needle lift measurements for a direct-acting piezoelectric injector, Energy Conversion and Management 112 (2016) 350–358.

[35] Z. He, W. Zhong, Q. Wang, Z. Jiang, Z. Shao, Effect of nozzle geometrical and dynamic factors on cavitating and turbulent flow in a diesel multi-hole injector nozzle, International Journal of Thermal Sciences 70 (2013) 132–143.

[36] M. Gavaises, A. Andriotis, Cavitation inside multi-hole injectors for large diesel engines and its effect on the near-nozzle spray structure. SAE Technical Paper 2006-01-1114, 2006.

[37] D. Mamaikin, T. Knorsch, P. Rogler, M. Wensing, Experimental investigation of flow field and string cavitation inside a transparent real-size GDI nozzle, Experiments in Fluids 61 (2020) 154.

[38] E. Winklhofer, E. Kull, E. Kelz, A. Morozov, Comprehensive hydraulic and flow field documentation in model throttle experiments under cavitation conditions, in: Proceedings of the 17th International Conference on Liquid Atomization and Spray Systems, Switzerland, 2001.

[39] R. Payri, F.J. Salvador, J. Gimeno, O. Venegas, Study of cavitation phenomenon using different fuels in a transparent nozzle by hydraulic characterization and visualization, Experimental Thermal and Fluid Science 44 (2013) 235–244.

[40] L. Zigan, J.-M. Shi, I. Krotow, I. Schmitz, M. Wensing, A. Leipertz, Fuel property and fuel temperature effects on internal nozzle flow, atomization and cyclic spray fluctuations of a direct injection spark ignition-injector, International Journal of Engine Research 14 (6) (2013) 543–556.

[41] P.G. Aleiferis, J. Serras-Pereira, A. Augoye, T.J. Davies, R.F. Cracknell, D. Richardson, Effect of fuel temperature on in-nozzle cavitation and spray formation of liquid hydrocarbons and alcohols from a real-size optical injector for direct-injection spark-ignition engines, International Journal of Heat and Mass Transfer 53 (21-22) (2010) 4588–4606.

[42] E. Lemmon, NIST reference fluid thermodynamic and transport properties database: Version 9.0, NIST Standard Reference Database 23.

[43] M. Gavaises, D. Papoulias, A. Andriotis, E. Giannadakis, A. Theodorakakos, Link between cavitation development and erosion damage in diesel injector nozzles. SAE Technical Paper 2007-01-0246, 2007.

Chapter 10

Cavitation in Positive Displacement Pumps

Phoevos Koukouvinis, Mithun Murali-Girija, Ioannis K. Karathanassis and Manolis Gavaises
Department of Mechanical Engineering and Aeronautics, School of Mathematics, Computer Science and Engineering, CITY, University of London, London, United Kingdom

10.1 Introduction

Pumps are devices that are used to increase the static pressure of liquids by converting mechanical energy to a high-pressure liquid flow. Pumps are used extensively in mechanical applications ranging from the automotive and aeronautical sectors to chemical and bioengineering fields. A rough classification of pumps is presented in Fig. 10.1. As shown, pumps can be classified as dynamic or rotodynamic pumps, which operate on the basis of the continuous interaction of a rotating impeller with the working liquid (e.g., a centrifugal pump), and positive displacement pumps, which will be the focus of this chapter. Positive displacement pumps are a subcategory of pumps that operate by trapping an amount of fluid in a cavity of varying geometry and volume and forcing it through the discuarge port.

Positive displacement pumps can further be classified, depending on the type of motion of the pumping volume boundaries, as reciprocating or rotating. The best possible examples of the two categories are piston pumps and gear pumps, respectively; see also Fig. 10.2. These pumps will be further examined as the

Cavitation and Bubble Dynamics. DOI: https://doi.org/10.1016/B978-0-12-823388-7.00010-2

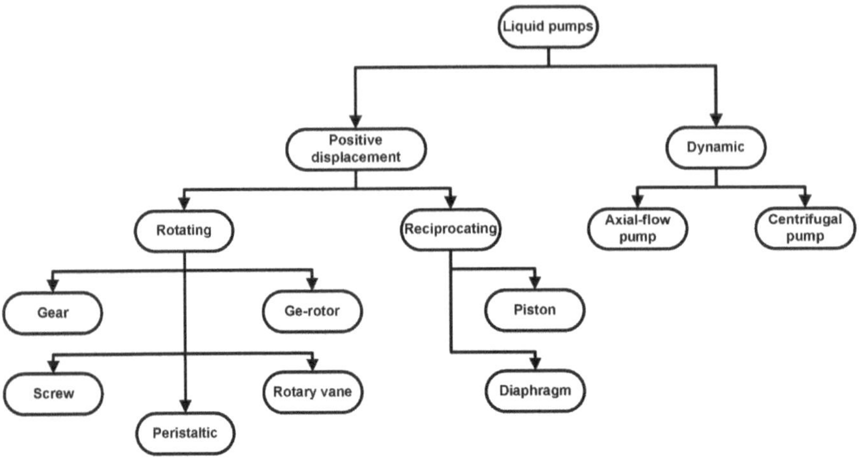

Figure 10.1 Classification of pump types as positive displacement pumps and dynamic pumps.

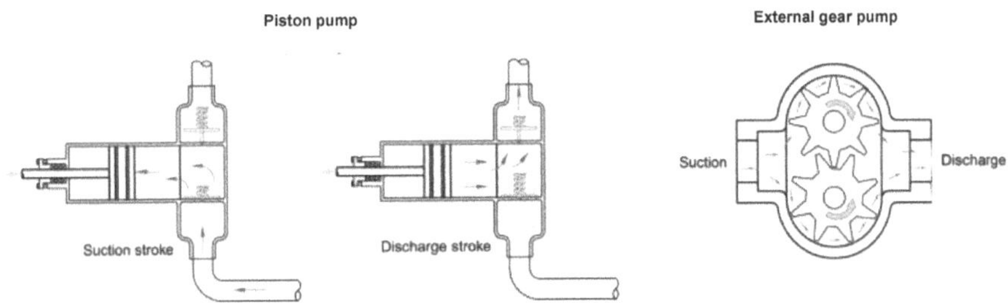

Figure 10.2 Diagrams of positive displacement pumps. Left: A piston pump. Right: An external gear pump. *Adapted from M. Stewart, Pump fundamentals, in: Surface Production Operations, Elsevier, 2019, pp. 7–59 [3], with permission.*

most representative of their category. Both reciprocating and rotating pumps find extensive use in fuel applications. For example, the high-pressure systems of modern diesel engines operate with high-pressure piston pumps, able to reach pressure ratios of more than 2000 bar [1]. On the other hand, rotating pumps and in particular external gear pumps are operating at lower pressure ratios, of 10–100 [2] and are more commonly found in aircraft fuel systems, but also for applications involving hydraulic oils or highly viscous fluids.

An important characteristic of piston pumps is the fact that they operate in discrete cycles. As shown in Fig. 10.2, during the suction cycle, the piston recedes from the cylinder, increasing the available volume for the liquid, hence creating suction. A valve opens, allowing liquid to enter the pump cylinder. After the piston has reached the maximum volume, it advances inside the cylinder, pushing the liquid and increasing its pressure. At the same time, the discharge valve opens, allowing the liquid to be expelled from the cylinder. For that reason it is common for piston pumps to be operated in arrays of multiple cylinders at different crank angles so that when one pump is ingesting liquid, the other is discharging, thus improving flow uniformity. By contrast, gear pump operation does not require valves, as the liquid is trapped in the volumes formed between the gear teeth and the casing (in the case of an external gear pump) or the two gears (in the case of an internal gear pump). Hence even if there is some pulsation of the flow, gear pump operation resembles more a continuous process.

Furthermore, an inherent aspect of positive displacement pumps, especially relevant in high-pressure applications and piston pumps, is the pressurization-induced heating of the liquid. Indeed, in considering an adiabatic and reversible (hence isentropic) liquid compression, the simplistic assumption of incompressible liquid implies that there is no temperature change, since entropy, s, is only a function of temperature, T, as follows:

$$s_1 - s_{\text{ref}} = c ln\left(\frac{T_1}{T_{\text{ref}}}\right) \tag{10.1}$$

where ref corresponds to an arbitrarily defined reference state. However, at large pressure ratios, density changes of liquids become considerable. Hence an isentropic process implies temperature changes as well. In particular, the general formula for entropy for a compressible medium is [4]:

$$s_1 - s_{\text{ref}} = \int_{\text{ref}}^{1} \frac{c_p}{T} dT - \int_{\text{ref}}^{1} \frac{1}{\rho^2}\left(\frac{\partial \rho}{\partial \text{T}}\right)_p dp \tag{10.2}$$

Thus an isentropic compression will induce heating. The amount of heating depends on the thermodynamic properties. As an example, to illustrate the effect, assuming properties of dodecane [5], which is commonly used as a single-component surrogate for diesel fuel, Table 10.1 lists temperature changes for dodecane.

As has been demonstrated, compression heating is nonnegligible; for example, at a pressure ratio of 2000, a temperature increase of 24 K is expected. The aforementioned discussion is applicable for gear pumps as well; however, since the pressure ratios are smaller, approximately 100–200, the temperature increase is only about 2–3.5 K. It is important to note that the aforementioned estimations are based on the assumption of isentropic compression, which is inherently an idealization; hence it represents the minimum possible achievable heating under the assumption of adiabatic conditions. In reality, the actual compression heating

TABLE 10.1 Pressurization-induced heating for dodecane. Isentropic compression starting from 300 K and 1 bar.

Temperature (K)	Pressure (bar)	Density (kg/m^3)	Pressure ratio	ΔT (K)
300	1	744.36	1	0
300.17	10	744.91	10	0.17
300.88	50	747.32	50	0.88
301.74	100	750.21	100	1.74
307.83	500	770.15	500	7.83
314.09	1000	789.86	1000	14.09
324.07	2000	820.08	2000	24.07
332.06	3000	843.41	3000	32.06
340.05	*4000*	*866.74*	*4000*	*40.05*
348.04	*5000*	*890.07*	*5000*	*48.04*

Italicized entries imply extrapolated conditions.

will be higher, owing to nonidealities, such as friction, and can be estimated through the introduction of an isentropic efficiency, η, as follows:

$$\eta_C = \frac{h_{2,s} - h_1}{h_2 - h_1} \tag{10.3}$$

where h represents the enthalpy, index 1 represents the suction of the pump, index 2 represents the discharge of the pump, and 2,s represents the ideal/isentropic discharge of the pump. Isentropic efficiency is a manufacturing parameter of the pump and is in the range of 90%. Similarly, the work consumed during the compression is the mass flow rate times the enthalpy difference from discharge to suction, that is,

$$W = \dot{m}(h_2 - h_1) \tag{10.4}$$

For a general compressible fluid, enthalpy is calculated as follows:

$$h_1 - h_{\text{ref}} = \int_{\text{ref}}^{1} c_p dT + \int_{\text{ref}}^{1} \left[\frac{1}{\rho} + \frac{T}{\rho^2} \left(\frac{\partial \rho}{\partial T} \right)_p \right] dp \tag{10.5}$$

where ref is an arbitrarily defined reference state. When density is assumed to be constant, the better-known relation of enthalpy for incompressible materials is obtained:

$$h_1 - h_{\text{ref}} = \int_{\text{ref}}^{1} c_p dT + \frac{p_1 - p_{\text{ref}}}{\rho} \tag{10.6}$$

10.2 Simulation Methods

A particular complexity of positive displacement pumps in general is associated with their method of operation, implying moving boundaries and changing chamber geometries. This fact renders the simulation of such devices difficult, since

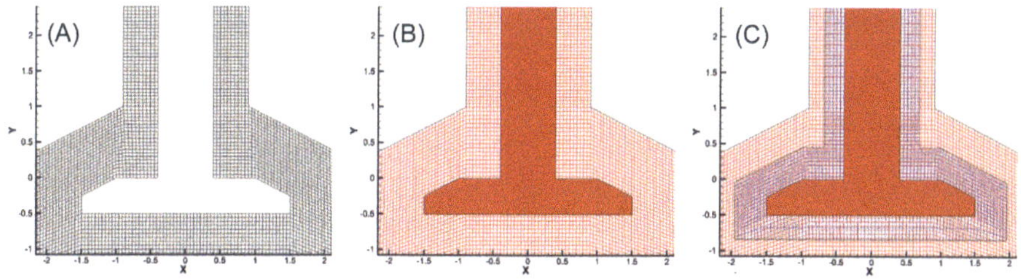

Figure 10.3 An indicative illustration of different mesh types for moving zones, demonstrated for a valve. The valve is the inverted "T"-structure at the middle of each figure. (A) Body-fitted mesh. (B) Immersed boundary. (C) Overset mesh. The red colored region implies "deactivated" background cells (indicated by the *red* mesh). The overset mesh is indicated by the blue mesh and is attached on the moving needle.

common computational methods such as finite volume require a computational mesh, which has to adjust dynamically to the geometry changes. Whereas mesh deformation can be handled with the space conservation law [6], the operation of positive displacement pumps often comes with changing geometry topology; examples are opening and closing of valves for piston pumps and the formation of successive chambers between gear teeth for gear pumps. Such features may be approached in different ways (see also Fig. 10.3):

- By using deforming body-fitted meshes. In this case the mesh deforms, following the boundary geometry. In cases in which the mesh resolution becomes too coarse or the mesh quality is too low, introduction of additional mesh layers (known as layering), remeshing, and interpolation of fields may be employed. Changing the mesh topology can be achieved by disconnecting adjacent regions, disabling cell zones and creating artificial wall boundaries. Although this approach is doable, it requires considerable user intervention and scripting to be performed successfully.
- By employing immersed boundary (IB) techniques. In this case the computational mesh remains fixed in space and time. However, the solid geometry is mapped onto the computational grid by appropriately marking cells occupied by the solid and by enforcing appropriate conditions. This approach was initially utilized for biological applications [7] with flexible geometries. Later, this approach was modified and used for rigid body simulations by other researchers [8,9]. There are many variations of the IB technique; the interested reader is referred to [10,11]. Here, a simple application of the continuous forcing approach [7,11] in which the forcing term (f_{IB}) is added to the continuous form of momentum equation. The term f_{IB} is the source term that forces the flow to follow the boundaries of the immersed body. This force is proportional to the difference in flow velocity ($\bar{U}$) and target body velocity ($\bar{U}_{IB}$, which is the velocity of the immersed body). The formulation of the forcing term for

a transient flow is given in Eq. (10.7):

$$f_{\mathrm{IB}} = -\frac{\rho_m C}{dt}\alpha_{\mathrm{IB}}\left(\bar{\mathrm{U}} - \bar{\mathrm{U}}_{\mathrm{IB}}\right) \tag{10.7}$$

In Eq. (10.7), C is an arbitrary coefficient that can be tuned to adjust the strength of the immersed body velocity, dt is the time step, and α_{IB} is the masking function, which takes a value of 1 if the cell is inside the immersed body (solid) and 0 if the cell is outside (fluid). Any value between 0 and 1 refers to a cell that is partially occupied by the wall (see Fig. 10.4). Further to the momentum equation, source terms are also included for the transport equations for turbulent quantities and energy (if applicable). The main advantage of IB methods is their flexibility and relative simplicity. Their biggest disadvantage is the near-wall region description; in that region, gradients become very large and therefore cannot be described accurately with the uniform meshes that are commonly employed as a background in IB techniques. Although finer resolution can be locally employed by using adaptive meshing, for example, such an effort is less efficient in terms of both accuracy and computational cost in comparison to body-fitted deforming meshes.

- By employing overset meshes. With this technique, two sets of disconnected meshes describing the fluid are created; one is attached on the moving region, and the other is a background mesh, attached on the stationary zone. The two meshes are required to overlap, and interpolations are carried out in this overlap region to enforce the connectivity between the two zones. The greatest advantage of overset meshes is their flexibility and accuracy; computational meshes can be refined near walls and cells can be dynamically enabled or disabled as the moving body moves around. In a way, it can be viewed as a hybrid of the body-fitted and IB methods. However, its disadvantage is related to the necessity of an overlap region between the moving and stationary zones, which needs a thickness of around 4–5 cells. This cannot always be achieved, especially in cases of thin gaps, such as those in gear pumps, causing convergence issues.

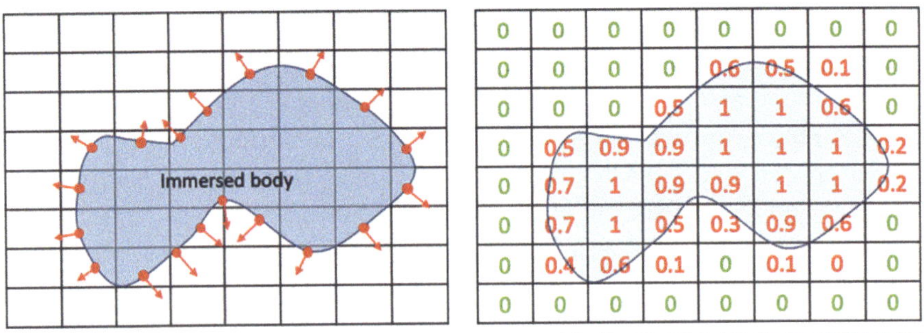

Figure 10.4 Immersed body representation using Lagrangian points (left) to mask function in an Eulerian frame (right).

10.3 Cavitation in Positive Displacement Pumps

In the case of positive displacement pumps, cavitation can occur as a result of throttling effects at the suction of the pump, induced by the geometrical layout and the conditions of operation. Places that are susceptible to cavitation are sharp geometrical features or narrow openings, even on the high-pressure side, for example, formed at the thin gap between the valve and the valve seat when the former opens. At such regions, flow acceleration is so strong that pressure may drop locally below saturation, forming cavitation pockets. Furthermore, pulsation induced by piston pumps for liquid delivery results in complex wave dynamics due to the waves passing through the fluid and the solid components, potentially leading to cavitation formation due to rarefaction.

Once formed, cavitation can have the following major side effects:

- Reduction in the delivered flow rate due to reduction of the available cross section for the liquid and/or flow choking. Although in general this is a nuisance, as it reduces the efficiency of the pump, it can also be exploited, for example, in piston pumps in the case of part-load operation, to limit the delivered mass flow rate without changing the stroke length of the piston [1].
- Collapse of cavitation bubbles poses a threat to the structural integrity of the device itself as well as to other components of the common rail system. Cavitation-induced erosion may be attributed to two primary mechanisms: impingement of pressure waves emitted during the bubble implosion on material surfaces or the impulse momentum of a liquid microjet created as a result of the asymmetrical bubble collapse that impinges on the solid surface. Along with cavitation erosion, additional damage may be caused by the introduction of metallic debris in the flow, which can increase erosion, especially in the tight passages of positive displacement devices.
- A further side effect is the increase in vibrations due to the sudden temporal changes in the fluid compressibility. The pure liquid is very stiff, whereas the cavitating mixture is highly compressible.

10.4 Cavitation in Piston Pumps

10.4.1 Short Literature Review on Existing Studies

Only limited studies are available in the literature that discuss the development of two-phase flow in positive displacement pumps bearing a piston plunger and a separate pressurization chamber, where cavitation can also emerge during the phases of volume expansion. Geometrical modifications to mitigate the effect of cavitation within axial piston pumps have been proposed by Wang [12] and Kumar and Bergada [13]. Petzold et al. [14] visualized the transient, two-phase flow within a metering pump employing a closed chamber pressurized by a moving piston. Two variations of the configuration were considered, having different porthole designs, to verify which one achieved higher dosage accuracy, which was determined by the extent of bubble growth in the pump chamber. Giuffrida et al. [15] investigated, in terms of liquid delivery efficiency, the operation of a radial positive displacement pump made up of three separate pressurization chambers and three plungers connected through a common cam ring. It was verified that the delivered flow rate decreased for increasing values of the pump driving speed and constant delivery pressure. This trend was attributed to the onset of cavitation during the pump suction period at high cam rotational speed.

From a numerical modeling point of view, the studies available so far that refer to positive displacement pumps adopt a series of simplifications regarding either the prevailing flow conditions or the moving components of the geometrical layout. Iannetti et al. [16] evaluated the effect of the noncondensable air-mass fraction on the cavitation phenomena occurring within a positive displacement water pump using a two-phase mixture model. The two-phase field arising within an axial piston oil pump, where the flow enters the compression chamber through a V-shaped slit, was investigated by Tsukiji et al. [17], using both flow visualization and a numerical model. The comparison of flow visualization data with computational fluid dynamics (CFD) predictions revealed that the extent of the vapor cloud emerging at the location of geometrical constriction can be accurately captured by the numerical model. However, only the case of a static piston close to its upper dead state was considered. Roemer et al. [18] performed a dynamic mesh simulation to predict the compressible flow emerging within a chamber connected to high- and low-pressure manifolds, a layout that corresponds to a simplified unit cell of a digital displacement pump. Husmeier et al. [19] used an Eulerian-Eulerian multiphase model along with a moving mesh technique to predict the two-phase field that emerges inside a fuel delivery pump, where the inlet manifold and discharge nozzle are obliquely located on top of a pressurization chamber.

10.4.2 Application: Cavitation in a Piston Pump

The high-pressure diesel fuel pump that is considered in the present investigation is depicted in Fig. 10.5. The pump in question is of the positive displacement type, comprising a manifold section and a sealed chamber connected through a moving valve. The liquid is pressurized within the chamber through the translational movement of a piston plunger. The pump receives diesel fuel from a low-pressure pump, not examined here, at a pressure of 6.5×10^5 Pa and a temperature of approximately 318 K. The duration of the pumping cycle for the purposes of the present

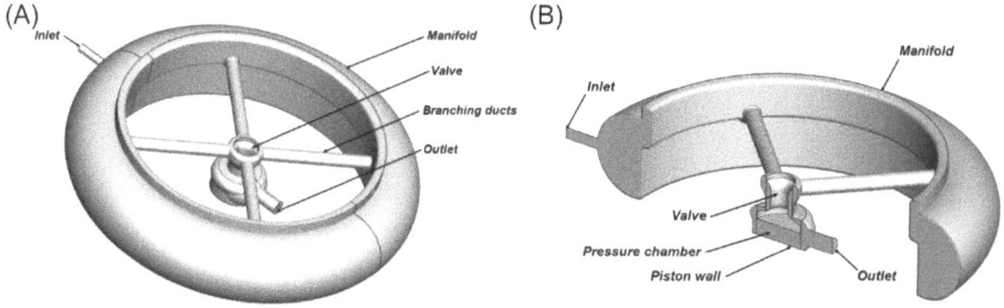

Figure 10.5 (A) Full view and (B) section view of the fluid volume inside the pump, showing the main pump components.

investigation was taken to be equal to 12×10^{-3} s, the main interest being to elucidate the development of the cavitation topology during the suction period of the pump operation and mainly at part-load conditions, where throttling is applied in the pump inlet, to limit the maximum delivered fuel.

The main geometrical components comprising the pump layout are also indicated in Fig. 10.5. Note that the pump geometry is symmetrical with respect to the *xz*-plane. The pump inlet that directs fuel into the manifold volume can be seen in the upper left part of Fig. 10.5A. The cross section of the inlet pipe has been adjusted according to the throttling applied in order to simulate the part-load condition effect. The manifold volume has a ring shape with a semicircular cross section and acts as a buffer volume to absorb sharp transient effects. Four circular pipes, placed in a cross pattern, connect the manifold volume with the pump cylinder entrance. A mushroom-shaped valve, as illustrated in the section view in Fig. 10.5B, is located at the cylinder entrance to prevent liquid from counterflowing toward the inlet during the compression phase. In the instance shown in Fig. 10.5 the valve lift is 1.0 mm, while the maximum valve lift is approximately 1.6 mm. A moving piston wall located at the bottom of the cylinder volume, which can also be seen in the lower part of Fig. 10.5B, represents the pump plunger. The piston is depicted at its top dead-center point, corresponding to the minimum cylinder volume. The piston displacement during the pump stroke follows a sinusoidal motion with 5-mm amplitude. Finally, the pressurized liquid exits the pump through a cylindrical duct. In the actual device, a check valve is employed at the duct outlet that opens when the desired pressure is reached, that is, at approximately 2000 bar (not shown here).

In the actual mechanical configuration the valve on top of the pressure cylinder moves as a result of the pressure and spring forces acting on it; however, to simplify the simulation, the valve was assumed to move following a prescribed motion, estimated through simplified one-dimensional (1D) analysis performed beforehand, which provided the profile shown in Fig. 10.6. Likewise, a sinusoidal movement profile was imposed for the movement of the plunger to replicate the force of the crankshaft. An additional line is evident on the graph in Fig. 10.6, referred to as

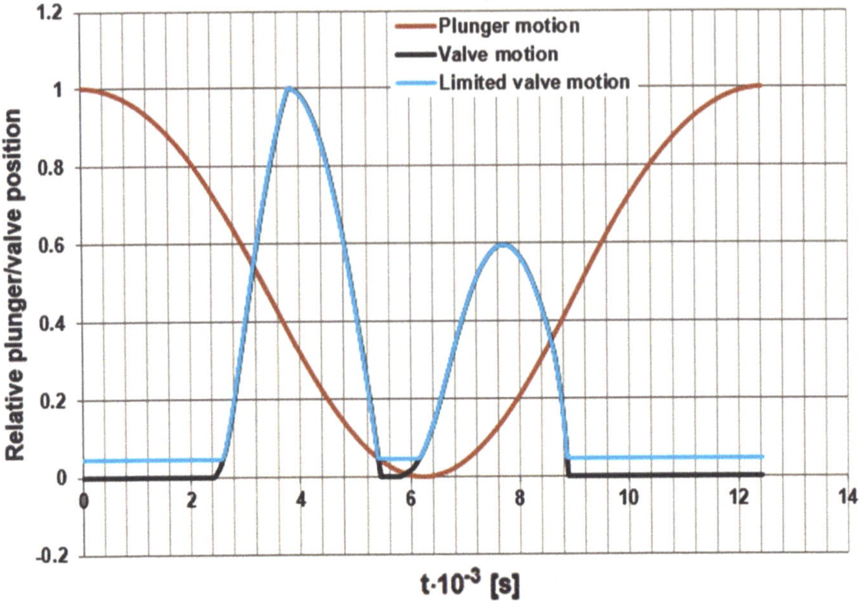

Figure 10.6 The pump plunger and valve motion profiles (normalized by the maximum displacement in each case).

limited valve motion, since the valve movement had to be limited to a minimum lift of 75 μm for purely numerical (grid topology) purposes. Once the valve was considered fully closed, the interfaces corresponding to the clearance of 75 μm between the valve and valve seat were switched to walls to prevent liquid from flowing into the chamber. By contrast, when the valve lift was assigned values larger than 75 μm topology, the interfaces were switched to a permeable condition.

A cycle of pump operation commences with the downward movement of the plunger, while the valve is kept closed, thus resulting in expansion of the liquid within the cylinder. At 2.6×10^{-3} s the valve opens, signifying the start of the suction period, during which fuel enters the chamber. The valve closes once again at 5.4×10^{-3} s, and the piston reaches its lower dead-center point at 6.24×10^{-3} s and subsequently starts its ascending motion. From 6.2 to 8.9×10^{-3} s the valve is open with the piston moving upwards; hence surplus liquid is pushed toward the manifold. At approximately 8.9×10^{-3} s the valve seals the chamber, and the last part of the pump stroke simulation is characterized by constant liquid pressurization up to 2000×10^5 Pa. In the real application, at that pressure, the pressurized fuel unseats the outlet check valve and enters the common rail system.

To resolve the flow inside the pump, it is necessary to discretize the fluid volume in a set of computational elements over which the conservation equations are solved. By taking advantage of the existing symmetry, the computational domain was truncated to half the original geometrical layout in order to decrease the computational cost. The governing equations of the flow are continuity and momentum equations; the energy equation was omitted, as vaporization within the pump is governed exclusively by inertial forces [20]. Turbulence effects were accounted for with the k-omega shear stress transport (SST) model, as the Reynolds number can reach values up to 38,000 depending on the phase of the operation cycle and the characteristic length scale of each specific subvolume. Since there is evidence in the literature that turbulence has an influence on cavitation structures [21,22] and, moreover, that it is not always possible for this interaction to be captured with traditional Reynolds-averaged Navier–Stokes models, the SAS variant, capable of capturing the larger transient turbulent structures, was employed. Cavitation was modeled with the use of a novel barotropic, homogenous equilibrium model (HEM), which aims to replicate phase change without introducing an additional partial differential equation for vapor fraction transport, since the mass transfer rate is considered infinite, and consequently, density is linked directly to pressure (see also Chapter 1: Review of Numerical Methodologies for Modeling Cavitation), solving only continuity and momentum equations. Besides, the barotropic model is capable of predicting the Rayleigh collapse of vapor structures inherently, contrary to two-phase, mass transfer models, making it suitable for erosion prediction [21,23].

According to the formulation of the HEM, an appropriate equation of state (EoS) is required that corresponds to the phase change of the liquid to the liquid–vapor mixture. In this work, the influence of thermal effects has been omitted, since the diesel temperature and thus vapor pressure increase is very small [24] with respect to the extreme pressurization (i.e., 2000×10^5 Pa), which is much higher than the pressure prevailing in the pump chamber for the entire flow time apart from the very last time steps, where all cavitation has already collapsed. Such an approximation is often employed in the simulation of cavitating flows in engineering applications [25–27], and in conjunction with the objective of tracking the probability of erosion formation [28], the use of the barotropic model is justified. The form of the EoS used is shown in Fig. 10.7.

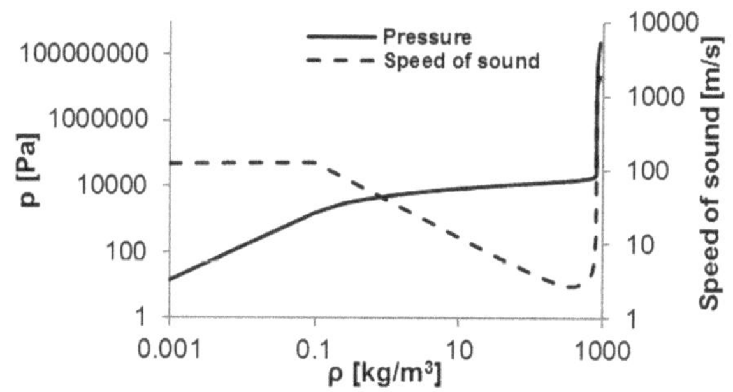

Figure 10.7 Pressure variation with density as produced by the barotropic HEM EoS and respective variation in the speed of sound.

Selected instances of the simulated pumping cycle, shown in Fig. 10.8, are suitable for elucidating the development and topology of the cavitating flow that arises within the geometrical layout during the various stages of pump operation. In the first stages of the cycle ($t < 2.6 \times 10^{-3}$ s), while the valve is still closed, pressure drops inside the cylinder only, with the simultaneous appearance of vaporous structures (indicated as a low-density mixture), whereas the manifold remains constant. As expected, the depressurization of the chamber in the first stages of operation causes the fuel to cavitate, and a mixture of liquid and vapor forms, the lowest density being detected in the vicinity of the moving piston wall. After opening of the valve, which occurs at $t = 2.6 \times 10^{-3}$ s, a rarefaction wave travels upstream toward the inlet, and fuel flows inside the manifold and consequently the cylinder. At the same time, cavitation forms between the valve and the valve seat as a result of the acceleration of fuel in the vicinity of the valve passage. The low pressures prevailing in that region cause further vapor formation, and the density locally reaches values as low as 48 kg/m^3. Once the valve closes for a second time, high pressures can be detected in the region upstream of it (see the instance at 5.4 ms in Fig. 10.8); this effect has been well documented in the literature and termed the water hammer effect [29], which is caused by fluid deceleration in that region. Finally, the valve opens again at 6.2×10^{-3} s, and all the flow processes already mentioned are repeated. Although the piston is reducing the cylinder volume during the second valve opening, pressure in the cylinder is still low even at 8.8 ms, owing to the existence of vaporous structures. Actually, the liquid pressurization within the chamber commences after a time period of 9.2×10^{-3} s. The emergence of cavitation inside the pump geometry is illustrated clearly through contour plots of the density field presented in Fig. 10.8.

Another notable feature characterizing the pump suction phase is the formation of cavitation rings that collapse on the manifold wall opposite the inlet duct. These vaporous structures set in due to the low-pressure regions existing on each side of the fuel jet that enters the manifold volume (see also Fig. 10.8 at $t = 3.9 \times 10^{-3}$ s) and are in fact cavitating vortices attributed to the shear flow induced by the jet penetration into the manifold volume. At the time instances of 5.4 and 5.8×10^{-3} s, vaporous structures still occupy a significant part of the cylindrical chamber since the downward displacement of the piston is still forcing the liquid in the cylinder to expand. The cavitation topology is designated by the complex flow field arising within the cylinder. The variation of the liquid volume and liquid volume fraction inside the pump cylinder is depicted in Fig. 10.9. The minimum predicted liquid volume fraction is approximately 58%, a value that has been confirmed by actual observations, since it was found that at part-load conditions, the pump delivers liquid of approximately half the swept volume.

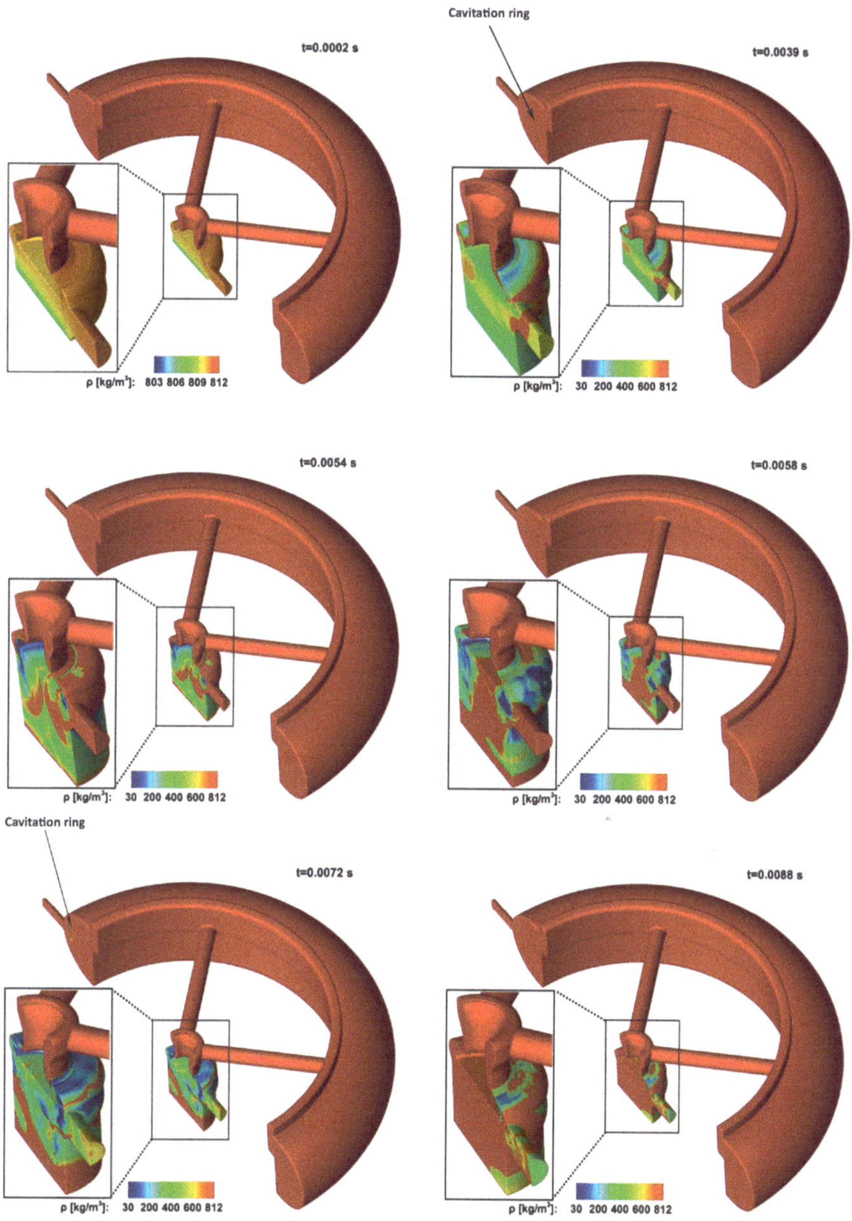

Figure 10.8 Density distribution inside the piston pump and manifold.

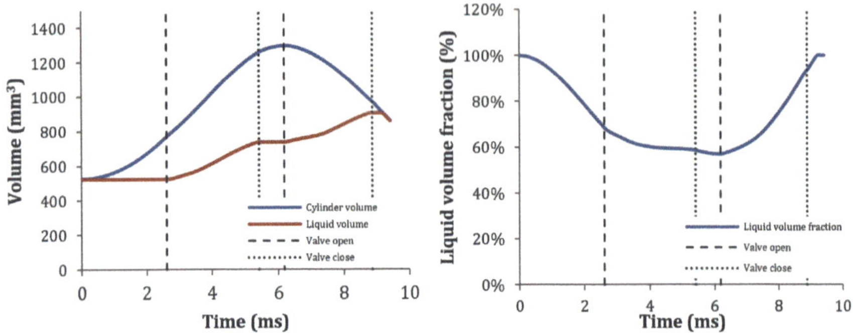

Figure 10.9 Variation in the liquid volume (left) and the liquid volume fraction (right) inside the cylinder.

10.5 Cavitation in Gear Pumps

10.5.1 Short Literature Review on Existing Studies

In the last few decades, visualization techniques such as strobe photography and particle image velocimetry have become popular among researchers for studying flow characteristics inside gear pumps. A study of the visualization of flow inside the gear pump using strobe photography was conducted by Eaton et al. [30]. They captured averaged images of cavitation formation in the gear-meshing region over several meshing cycles, observing cavitation even when the fluid was highly saturated with air content. The time-resolved particle image velocimetry (TRPIV) technique was used to study the turbulence inside the gear pump by Castilla et al. [31], utilizing air bubbles as the tracer material to avoid damaging the gears. A similar approach was used to investigate the flow inside the suction chamber of an external gear pump [32] and to investigate geometry features, such as relief groves, and the effect of parameters such as rotational speed, inlet pressure, and oil temperature [33]. More recently, Antoniak and Stryczek [34] presented a study that made the visualization of the flow in all characteristic areas of the pump possible by constructing the body of their pump from a transparent material. On the basis of the operating conditions, they observed three forms of cavitation occurring independently inside the pump; cavitation bubble (*B*), cavitation veil (*Ve*), and cavitation jet (*J*), as shown in Fig. 10.10A–C, respectively. Depending on the operating condition, these forms of cavitation can take extreme forms. By significantly lowering the inlet pressure the cavitation, bubbles (*B*) will change into foam (*F*) (Fig. 10.10D). If the rotation speed of the pump increases, a

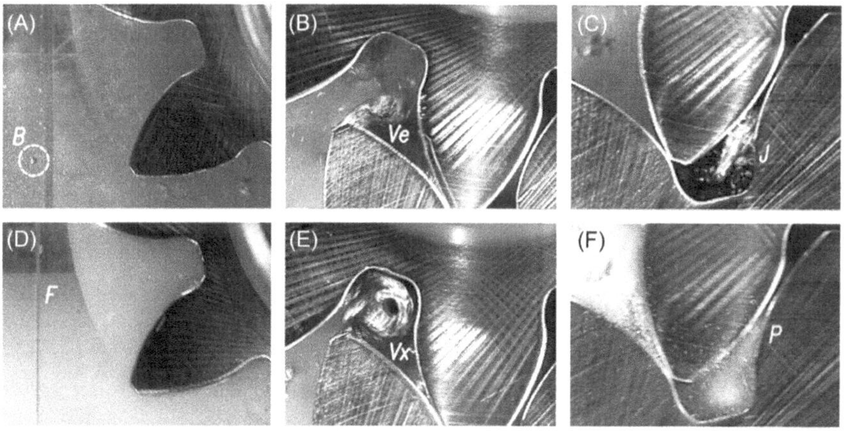

Figure 10.10 Characteristic forms of cavitation observed in gear pumps: (A) Bubbles, (B) veil, (C) jet; (D) foam; (E) vortex and (F) plume. *Reproduced with permission from P. Antoniak, J. Stryczek, Visualization study of the flow processes and phenomena in the external gear pump, Archives of Civil and Mechanical Engineering, 18 (4) (2018) 1103–1115 [34].*

veil (*Ve*) will be transformed into a cavitation vortex (*Vx*) (Fig. 10.10E). Lowering the inlet pressure as well as increasing the pump speed will also affect the appearance of the jet (*J*). The reduced pressure in the chamber will facilitate the transformation of the bubbles (*B*) into cavitation foam (*F*), and eventually this accompanying foam (*F*) will make the jet (*J*) look less clear and resemble a plume (*P*), as shown in Fig. 10.10F. A detailed description of the flow streams at inlet and outlet chambers as well as at the inlet and outlet bridges are presented by Antoniak and Stryczekk [34]. Their study was limited to low- and medium-pressure ratios (in the range of 10 bar) because of the structural limitations of the transparent material.

Despite the pump's simple design, modeling of such a pump is not so trivial. Initial studies were based on graphical [35] or theoretical approaches [36,37]. With the recent advances in engineering and computational technologies, more advanced models capable of performing complex CFD simulations have been developed. One popular but simple approach to modeling gear pumps using lumped parameters can be found in Refs. [38,39]. In their model (HYGESim), the fluid dynamics and the mechanical gear motion are modeled by using AMESim, the lateral gaps are modeled by using CFD, and these models are coupled together with many other submodels to include cavitation, material erosion, and the lateral motion of the gears.

A complete CFD simulation of the gear pump poses a major challenge concerning numerical meshing of gears with small clearances and solid-to-solid contact between the gears. Furthermore, the modeling requires a dynamic mesh that can be adapted to the rotation of the gears without losing the mesh quality. These issues in mesh handling have been a research focus for the past few decades. A superposition method in which the stationary and moving regions are meshed separately and the data are interpolated between these meshes during the simulation was employed by Bruce et al. [40]. The interpolation between the mesh was a critical step in their method. An arbitrary Lagrangian-Eulerian (ALE) approach using a pseudopressure operator for describing the nodal velocity is presented by Riemslagh et al. [41]; it was applied to incompressible flow in a lobe pump and an external gear pump. One limitation of this model, according to Houzeaux and Codina [42], was the limited control over the mesh in the clearance region due to the automatic remeshing. An alternative method called FMALE (fixed mesh ALE) was proposed by Houzeaux and Codina [42], in which the evolving geometry is meshed a priori instead of remeshing in the classical ALE. This approach ensured that the mesh quality was maintained; however, it required considerable computational effort. In another study, a dynamic mesh approach using an unstructured mesh that deforms and remeshes to accommodate the gear motion, as implemented in Ansys Fluent, was employed by Strasser [43]. This approach can be expensive if the mesh count is large and smaller time-step values are used, which is typically the case for gear pumps, as frequent adaption of the mesh is required to maintain the mesh quality. A hybrid between the deform/remesh approach of Strasser [43] and the mesh replacement approach of Houzeaux and Codina [42] was employed by Castilla et al. [44]. Their study was focused on the flow field in the inlet chamber, and they concluded that the modeling of contact between the teeth and a proper choice of turbulence model are essential to capture the flow fluctuations that are responsible for the pressure ripples, noise, and vibration. A three-dimensional (3D) simulation of the external gear pump using the OpenFOAM toolkit was recently reported by Castilla et al. [45]. Since this study was dedicated to mesh manipulation and modeling gear contacts, cavitation and turbulence were not taken into account. Another custom-made approach that is focused on flow simulations in compressors can be found in Refs. [46–48]. The numerical algorithm that was developed is implemented into a meshing solver called SCORG, which has the capability of working alongside many commercial CFD solvers.

Despite its importance, none of the works described above considered the effect of cavitation. One of the first CFD works to consider cavitation in a gear pump was reported by del Campo et al. [49], examining the influence of different parameters, such as suction chamber geometry and gear revolutions per minute (rpm), on cavitation and volumetric efficiency. In a follow-up work, the effect of operating pressure on pump performance in the presence of cavitation was studied [50]. The researchers reported that the impact of suction side cavitation on the outlet flow and pressure ripples observed at the outlet for low pressure (10 bar) disappear when a mean pressure jump of 100 bar is applied. However, no significant change in the inlet flow and morphology of cavitation is observed with the increase in operating pressure. In their study, unlike the approach of Castilla et al. [44], in which a tiny wall was defined to model the contact between the gears that will deform with the gear rotation until the new mesh replaces it, del Campo et al. [50] modeled the gear contact by increasing the dynamic viscosity to a higher value at the contact location. According to del Campo et al. [50], the approach of Castilla et al. [44] produces numerical errors when cavitation is considered. CFD simulations using 1D and 3D gear pumps for improving the porting design and for reducing cavitation erosion in an external gear pump were conducted by King et al. [51]. A recent study considered cavitation in the numerical analysis of a 3D gear pump [52]. The researchers used the commercial code PumpLinx for their simulation and validated their results against experimental data from the pump manufacturer. Their cavitation model was based on the work of Singhal et al. [53], which considers the effect of a finite amount of noncondensable gas. To better predict the volumetric efficiency, the authors considered all the leakage paths and grooves in their model but omitted the contact between the gears. Recently, the SCORG solver was combined with the pressure-based algorithm implemented in Star-CCM+ for predicting cavitation in twin-screw pumps by Yan et al. [54]. They employed a combination of volume-of-fluid method and the Scherr-Sauer model for predicting cavitation. Their meshing strategy allowed the modeling of small clearance and, hence were able to predict the leakage flows. They reported a slight increase in volumetric efficiency due to the reduction in leakage when cavitation is present in the clearances.

It has been pointed out in the literature by many authors, for example, del Campo et al. [49], Yoon et al. [55] that the interpolations that area involved in using a traditional remeshing/deforming approach or a mesh replacement approach could lead to numerical errors, especially in solving for multiphase flows and with high gear rpm. Moreover, these approaches are computationally expensive, owing to the small time-step requirements for maintaining the mesh quality in small clearances. To overcome these complexities, approaches such as overset mesh, which uses a background mesh and many component meshes, have been employed in the past [56] for simulating turbomachinery. Even though this approach ensures better mesh quality, the numerical errors that arise from the interpolation of data between the base and component mesh cannot be avoided. From the personal experience of the authors, modeling clearance between the gears is also challenging, since the data interpolation requires at least four overlapping cells between the background and the component mesh.

An alternative approach that is becoming increasingly popular for such simulations is the IB method. In a recent study, the immersed solid approach implemented in Ansys CFX was employed by Yoon et al. [55] for modeling a 3D gear pump operating at very high speed (10,000 rpm). The authors claimed that the use of this method simplified their numerical simulation setup and significantly reduced the computation time. Their model was validated against the experimental data for flow rate measurements. They concluded that the dominant geometrical parameters that influence the pump flow rate are the tip and the lateral clearances. The limitation of their work is that it was not applied in multiphase or variable-density flows. More recently, a numerical algorithm for reducing the computational cost while finding the optimum parameter for cavitation optimization in a double-suction centrifugal pump was conducted by Pei et al. [57]. They proposed an optimization algorithm that claims to reduce the time for NPSHr by at least 44.4%. They used the Zwart–Gerber–Belamri model implemented in CFX for modeling cavitation with a frozen rotor approach for modeling the rotor motion.

10.5.2 Application: Cavitation in a Gear Pump

To mimic the complex motion of the geometries, an IB approach was employed [2]. The IB approach makes the modeling of wall-to-wall contact (e.g., to model the needle closure, the contact between two gears in a gear pump) and the complex motion of the geometry easier where the canonical body-fitted grids with remeshing/deforming approaches can be computationally expensive and inefficient.

Simulations were first performed for a 2D gear pump; the pump geometry was publicly shared by Roger Ull [58], as shown in Fig. 10.11A. The only difference in geometry between [58] and the present study is that the curved shoulders at the inlet and outlet are replaced with sharp ones in this work. The most difficult features in a gear pump simulation, as was explained earlier, are the modeling of small clearances and the contact between the gears while maintaining a reasonable mesh quality. In the present simulation the application of the IB method has made the implementation of these tasks more manageable. To model the contact between the gears, one of the gears is rotated by a small angular position to make the gear surfaces touch each other before exporting the coordinate positions of the gears into IB code. In Fig. 10.11 the clearances between the gears and between the gears and the casing along with the contact between gears are highlighted by using IB points superimposed on the background mesh. An additional refinement for the numerical mesh is provided at the location where the two gears come into contact, as highlighted in Fig. 10.11. This refinement ensures sufficient grid

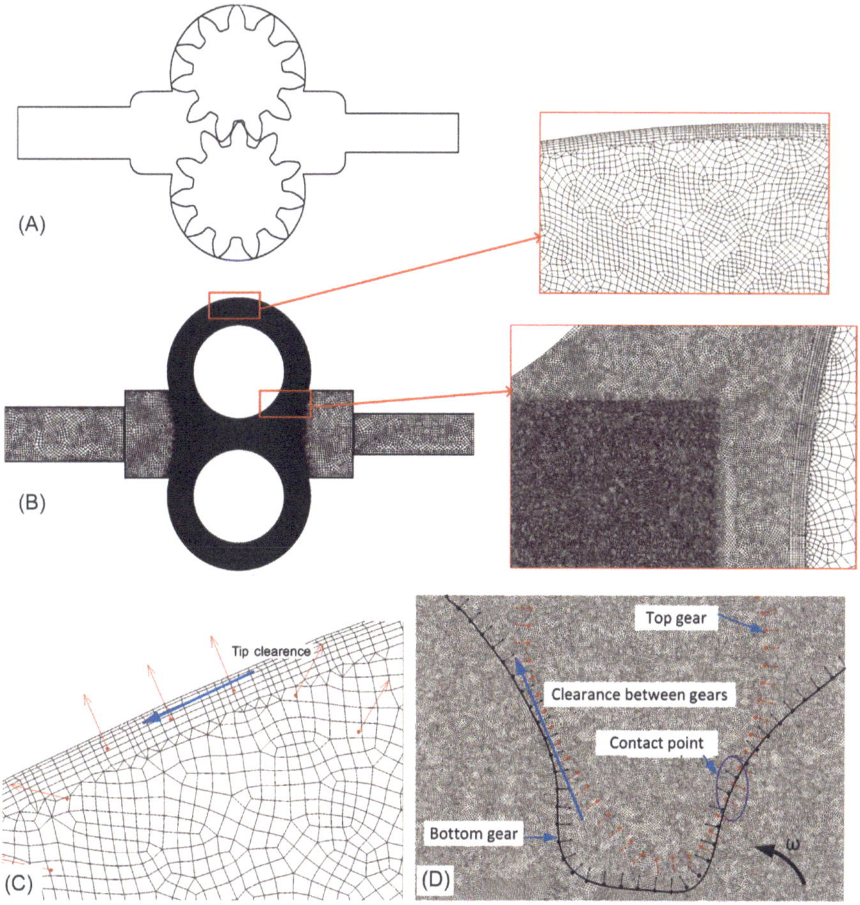

Figure 10.11 (A) Line diagram of the gear pump. (B) The computational mesh for immersed boundary simulation. Modeling of (C) tip clearances and (D) clearance and the point of contact between the gears. *Part figure (A) taken from A. Roger Ull, Study of mesh deformation features of an open source CFD package and application to a gear pump simulation, 2012 [58].*

elements in the clearances to capture the leakage flow between the gears. With this refinement and the addition of boundary layer refinement to limit the y^+-values below 5 in most of the domain for the extreme condition considered, the total mesh count reaches up to 0.5 million cells. Constant absolute pressure boundary conditions have been applied to the inlet and outlet boundaries. The multiphase flow model used in this study is fully compressible and utilizes the $k-\omega-$SST turbulence model with the Reboud et al. correction [59] for the

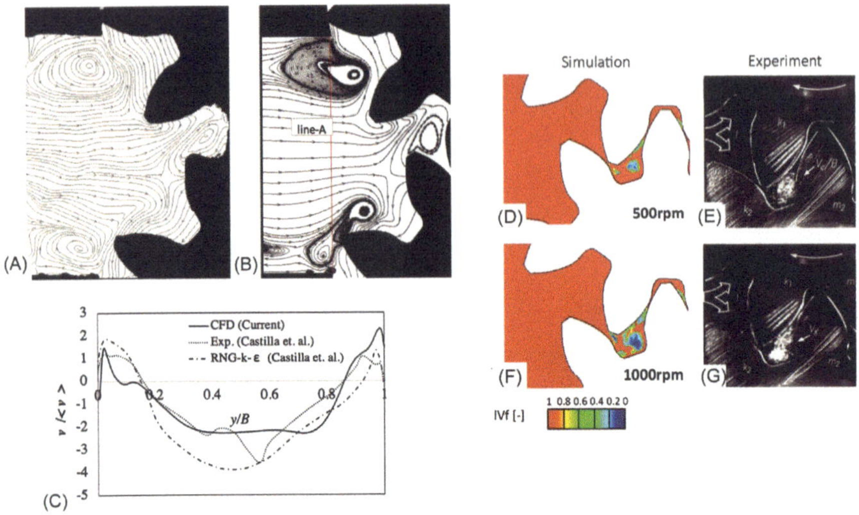

Figure 10.12 Left: Comparison of streamlines between (A) experiments by Castilla et al. [44] using TRPIV (B) simulations. (C) Indicative comparison of the velocity distribution, along line A. Right: Comparison of cavitation for two different speeds. (D, F) Liquid volume fraction distribution and (E, G) experiment [33].

turbulence viscosity. In all the cases considered in this study, a relatively low-pressure jump is applied between the inlet and the outlet ($\sim$10 bar), similar to the numerical study of Roger Ull [58].

In Fig. 10.12A and B the streamlines observed by Castilla et al. [44] using TRPIV are compared with the simulation discussed above. The flow field and the velocity profile plotted in Fig. 10.12C show good correlation with the experimental values. In addition, a qualitative comparison is made for the cavitation predictions at two different gear rpms, as shown in Fig. 10.12D–G. It can be observed that the locations of the cavitation predicted from the simulations correlate very well with the vortex core observed in the gap between the teeth at the suction chamber observed in the experimental work of Stryczek et al. [33] for both the cases being considered.

The effect of modeling contact between the gears on cavitation and pump performance is studied by comparing two cases, one with a clearance of 74 μm [58] between the gears and the other condition when gears are in contact. Unlike the unitary contact ratio (a ratio that defines the average number of teeth that are in contact with the mating gear during a period in which a tooth goes into and out of contact) studied in the simulations of del Campo et al. [49], in the present study, the contact between the gears occurs at two locations; that is, before the first contact separates, the next pair of gears come into contact. This creates a region isolated from the inlet and outlet where very high pressures are expected in a 2D

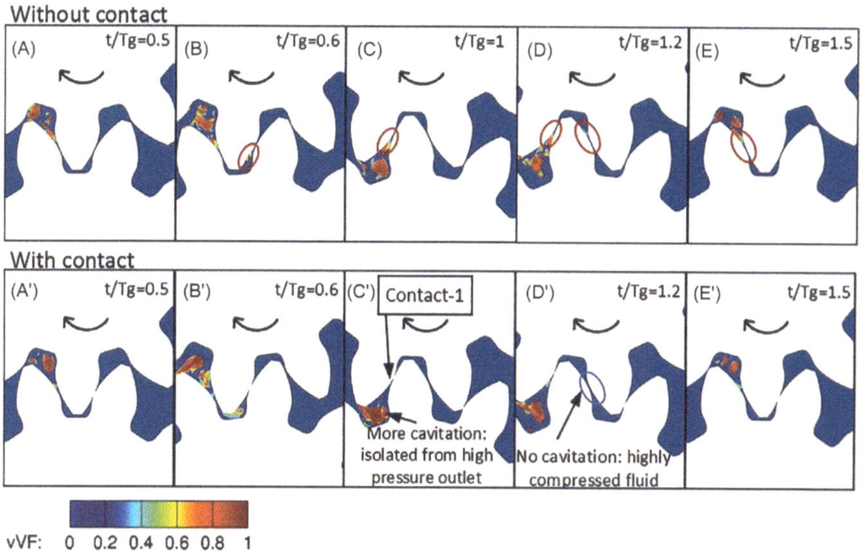

Figure 10.13 Comparison of vapor volume fraction over a gearing period with (A–E) clearance between the gears and (A′–E′) with contact between gears. The time interval between the contours from A to E is not uniform; they are chosen to highlight specific events over a period. Time is nondimensionalized with *Tg,* defined as *Tg* = 11 rpm/60.

simulation. As the gear rotates, the fluid inside the pockets is compressed, and the liquid on the high-pressure side leaks toward the inlet through small clearances. The acceleration of this leakage flow through clearance creates a low-pressure zone and results in the formation of cavitation, as highlighted by the red circles in Fig. 10.13A–E. This cavity acts as a fictitious contact point that separates the inlet chamber from the outlet, as also observed by del Campo et al. [49]. On the other hand, when the contact between the gears are modeled, the first contact point (Contact-1) act as the barrier between the inlet and outlet chambers, and the solid–solid contact does not allow flow across it; hence no cavitation is observed at contact location (Fig. 10.13B′ and C′). When the gears cross half-pitch rotation, the second contact is established, trapping a volume of fluid between the two contacts. Further rotation of the gear raises the pressure in the trapped volume up to 30–40 times the inlet pressure. This region is of particular interest in many studies (e.g., Refs. [30,33,45]), as this pressure rise can produce noise and vibration in the pump. To overcome this, the gear pumps are provided with relief grooves to drain this trapped fluid to the inlet chamber. Owing to the pressurized fluid in the trapped volume, cavitation does not appear in this region (Fig. 10.13D′) when the gears are in contact. While this is true for 2D simulations, for a real gear pump with relief grooves, the high-velocity jet from the trapped volume toward the inlet chamber through grooves can lead to erosion due to jetting.

An overall increase in cavitation is observed when the gears are in contact. This increase is attributed mainly to the increase in cavitation just ahead of the contact point, where the pressure is much lower, owing to the isolation of the of high-pressure side from the inlet by the contact point compared to the case without contact, in which relatively higher pressures are realized, owing to the continuous connection between the inlet and the outlet chamber. This can be confirmed by comparing the total vapor generation over time for the two cases; indeed, the mean value of vapor volume fraction is 0.19% for gear contact against 0.14% without contact. The mean value of the mass flow rate also increases by approximately 4.5% when the gears are in contact. Such an increase in flow rate with contact between the gears was also acknowledged by Castilla et al. [44] for a unity contact ratio. They also reported that the flow fluctuation increases with the contact, owing to the water hammer effects caused by the sudden opening and closing of the inner teeth chamber to both the outlet and inlet.

The effect of noncondensable gas (NCG) on cavitation is studied by varying its content from 0% to 1% by volume at a constant rotational speed of 2000 rpm. In the absence of noncondensable gas in the fluid, a reduction in local pressure below vapor pressure immediately causes the liquid to change its phase to vapor phase. On the other hand, in a similar situation, if a certain amount of noncondensable gas is present in the liquid, the reduction in local pressure is recovered to a certain extent by the expansion of the highly compressible gas and hence limiting cavitation. The formation of cavitation and the expansion of the NCG at the inlet chamber under different conditions are shown in Fig. 10.14. The reduction in cavitation with increasing NCG can be appreciated from this figure. In addition, a comparison of vapor contours at 0%, 0.2%, and 1% NCG reveals that the core of cavitation remains the same, which is at the core of the vortices, while the spread of the vapor cavity around the core is limited by the expansion of the gas.

10.6 Conclusions

In this chapter a short overview of physical mechanisms and cavitation formation in positive displacement pumps is discussed. Positive displacement pumps are an integral part in many high-pressure applications in the fuel and petrochemical industry and for highly viscous fluids. However, their inherent method of operation renders them rather tricky to resolve numerically. Several alternative methods for the computational geometry manipulation exist, such as deformable meshes, immersed boundaries, and overset meshing, However, they all come with limitations and have to be examined per application.

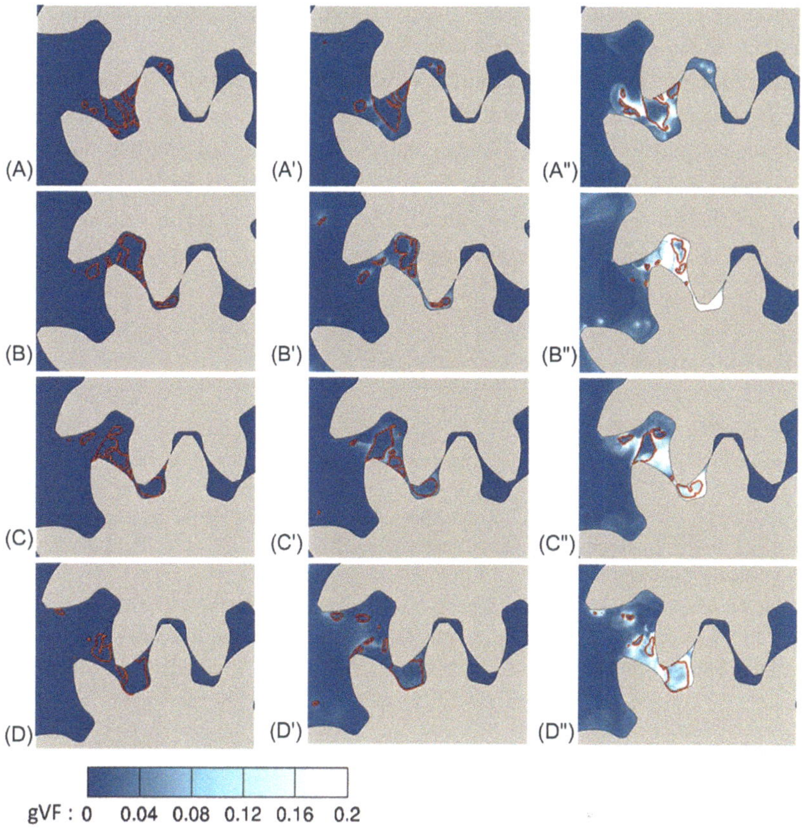

Figure 10.14 Evolution of NCG volume fraction and cavitation; (A–D) for 0% NCG, (a′–D′) for 0.2% and (A″–D″) for 1% NCG content. The isolines of 50% vapor volume fraction are shown in red. The instances are chosen randomly to highlight main features over a gearing period.

It was demonstrated that thermodynamic effects during compression are not insignificant. Especially when high-pressure diesel fuel pumps are considered, constant properties cannot be used, and liquid heating can be important. Moreover, such effects will be more pronounced in new generations of fuel injection systems, given experimental configurations operating at 5000 bar [60]. Despite their application on high-pressure ratios, positive displacement pumps are still susceptible to cavitation, mainly at the suction side, where pressure is lower. Additional regions where cavitation may develop are narrow passages, as those for example that are formed between valve and valve seat, during valve opening in piston pumps. Something similar applies for gear pumps, in which cavitation may develop in the clearance between gears, owing to liquid leaking from the high-pressure side to the low-pressure side. Cavitation is also expected to be more pronounced as the rpm of the pump increase, since the flow rate through passages and local pressure drops will consequently increase. Finally, the effect of noncondensable gas on cavitation occurring in gears pump was studied for the first time by varying the gas content in the flowing liquid from 0% to 1% by volume. The results suggest that the presence of noncondensable gas reduces the formation of cavitation without having much impact on the average outlet flow rates. However, an increase in fluctuation in the outlet flow is observed with the increase in gas content.

References

[1] P. Koukouvinis, I.K. Karathanassis, M. Gavaises, Prediction of cavitation and induced erosion inside a high-pressure fuel pump, International Journal of Engine Research 19 (3) (2018) 360–373.

[2] M.-G. Mithun, P. Koukouvinis, I.K. Karathanassis, M. Gavaises, Numerical simulation of three-phase flow in an external gear pump using immersed boundary approach, Applied Mathematical Model 72 (2019) 682–699.

[3] M. Stewart, Pump fundamentals, Surface Production Operations, Elsevier, 2019, pp. 7–59.

[4] D.E. Winterbone, A. Turan, General thermodynamic relationships, Advanced Thermodynamics for Engineers, Elsevier, 2015, pp. 141–162.

[5] E.W. Lemmon, M.L. Huber, Thermodynamic properties of n-dodecane, Energy and Fuels 18 (4) (2004) 960–967.

[6] I. Demirdžić, M. Perić, Space conservation law in finite volume calculations of fluid flow, International Journal of Numerical Methods in Fluids 8 (9) (1988) 1037–1050.

[7] C.S. Peskin, Flow patterns around heart valves: a numerical method, Journal of Computational Physics 10 (2) (1972) 252–271.

[8] D. Goldstein, R. Handler, L. Sirovich, Modeling a no-slip flow boundary with an external force field, Journal of Computational Physics 105 (2) (1993) 354–366.

[9] G. Iaccarino, G. Kalitzin, C. Elkins, Numerical and Experimental Investigation of the Turbulent Flow in a Ribbed Serpentine Passage, Stanford, CA, 2003.

[10] M.D. de Tullio, P. De Palma, G. Iaccarino, G. Pascazio, M. Napolitano, Immersed boundary technique for compressible flow simulations on semi-structured grids, BT – Computational Fluid Dynamics 2006, 2009, pp. 365–370.

[11] R. Mittal, G. Iaccarino, Immersed boundary methods, Annual Review of Fluid Mechanics 37 (1) (2005) 239–261.

[12] S. Wang, The analysis of cavitation problems in the axial piston pump, Journal Of Fluids Engineering 132 (7) (2010) 074502.

[13] S. Kumar, J.M. Bergada, The effect of piston grooves performance in an axial piston pumps via CFD analysis, International Journal of Mechanical Sciences 66 (2013) 168–179.

[14] M. Petzold, J. Weber, E. Dautry, O. Ohligschläger, A. Müller, Visualization and analysis of the multiphase flow in an electromagnetically driven dosing pump, in: ASME/BATH 2013 Symposium on Fluid Power and Motion Control, 2013.

[15] A. Giuffrida, A. Ficarella, D. Laforgia, Study of the delivery behaviour of a pump for common rail fuel injection equipment, Proceedings of the Institution of Mechanical Engineers, Part I: Journal of Systems and Control Engineering 223 (4) (2009) 521–535.

[16] A. Iannetti, M.T. Stickland, W.M. Dempster, An advanced CFD model to study the effect of non-condensable gas on cavitation in positive displacement pumps, Open Engineering 5 (1) (2015).

[17] T. Tsukiji, K. Nakayama, K. Saito, S. Yakabe, Study on the cavitating flow in an oil hydraulic pump, in: Proceedings of 2011 International Conference on Fluid Power and Mechatronics, 2011, pp. 253–258.

[18] D.B. Roemer, P. Johansen, H.C. Pedersen, T.O. Andersen, Simulation of dynamic behaviour of a digital displacement motor using transient 3D computational fluid dynamics analysis, in: ASME/BATH 2013 Symposium on Fluid Power and Motion Control, 2013.

[19] F. Husmeier, D. Greif, P. Sampl, J. Strucl, W. Edelbauer, Numerical simulation of compressible multi-phase flow in high pressure fuel pump, in: Vol. 2, Fora: Cavitation and Multiphase Flow; Fluid Measurements and Instrumentation; Microfluidics; Multiphase Flows: Work in Progress; Fluid-Particle Interactions in Turbulence, 2014.

[20] C. Brennen, Cavitation and Bubble Dynamics, Oxford University Press, 1995.

[21] M. Dular, O. Coutier-Delgosha, Numerical modelling of cavitation erosion, International Journal of Numerical Methods in Fluids 61 (12) (2009) 1388–1410.

[22] O. Coutier-Delgosha, J.L. Reboud, Y. Delannoy, Numerical simulation of the unsteady behaviour of cavitating flows, International Journal of Numerical Methods in Fluids 42 (5) (2003) 527–548.

[23] C. Egerer, S. Hickel, S. Schmidt, N. Adams, Large-eddy simulation of turbulent cavitating flow in a micro channel, Physics of Fluids 26 (2014).

[24] N.I. Kolev, N.I. Kolev, Thermodynamic and transport properties of diesel fuel, Multiphase Flow Dynamics 4 (2011) 293–327.

[25] H. Naseri, P. Koukouvinis, M. Gavaises, Evaluation of turbulence models performance in predicting incipient cavitation in an enlarged step-nozzle, Journal of Physics: Conference Series 656 (1) (2015).

[26] P. Koukouvinis, M. Gavaises, Simulation of throttle flow with two phase and single phase homogenous equilibrium model, Journal of Physics: Conference Series 656 (1) (2015).

[27] P. Limbach, T. Müller, R. Skoda, Application of a compressible flow solver and barotropic cavitation model for the evaluation of the suction head in a low specific speed centrifugal pump impeller channel, Journal of Physics: Conference Series 656 (2015) 012065.

[28] M.S. Mihatsch, S.J. Schmidt, N.A. Adams, Cavitation erosion prediction based on analysis of flow dynamics and impact load spectra, Physics of Fluids 27 (10) (2015) 103302.

[29] G.K. Batchelor, An Introduction to Fluid Dynamics, Cambridge University Press, 2000.

[30] M. Eaton, P.S. Keogh, K.A. Edge, The modelling, prediction, and experimental evaluation of gear pump meshing pressures with particular reference to aero-engine fuel pumps, Proceedings of the Institution of Mechanical Engineers, Part I: Journal of Systems and Control Engineering 220 (5) (2006) 365–379.

[31] R. Castilla, J. Wojciechowski, P.J. Gamez-Montero, A. Vernet, E. Codina, Analysis of the turbulence in the suction chamber of an external gear pump using time resolved particle image velocimetry, Flow Measurement and Instrumentation 19 (6) (2008) 377–384.

[32] N. Ertürk, A. Vernet, R. Castilla, P.J. Gamez-Montero, J.A. Ferre, Experimental analysis of the flow dynamics in the suction chamber of an external gear pump, International Journal of Mechanical Sciences 53 (2) (2011) 135–144.

[33] J. Stryczek, et al., Visualisation research of the flow processes in the outlet chamber–outlet bridge–inlet chamber zone of the gear pumps, Archives of Civil and Mechanical Engineering 15 (1) (2015) 95–108.

[34] P. Antoniak, J. Stryczek, Visualization study of the flow processes and phenomena in the external gear pump, Archives of Civil and Mechanical Engineering 18 (4) (2018) 1103–1115.

[35] T.E. Beacham, High-pressure gear pumps, Proceedings of the Institution of Mechanical Engineers 155 (1) (1946) 417–452.
[36] N.D. Manring, S.B. Kasaragadda, The theoretical flow ripple of an external gear pump, Journal of Dynamic Systems, Measurement, and Control 125 (3) (2003) 396.
[37] M. Borghi, B. Zardin, E. Specchia, External Gear Pump Volumetric Efficiency: Numerical and Experimental Analysis, 2009.
[38] A. Vacca, M. Guidetti, Modelling and experimental validation of external spur gear machines for fluid power applications, Simulation Modelling Practice and Theory 19 (9) (2011) 2007–2031.
[39] J. Zhou, A. Vacca, P. Casoli, A novel approach for predicting the operation of external gear pumps under cavitating conditions, Simulation Modelling Practice and Theory 45 (2014) 35–49.
[40] D. Bruce, M. Wilson, S. Generalis, Flow field analysis of both the trilobal element and mixing disc zones within a closely intermeshing, co-rotating twin-screw extruder, International Polymer Processing 12 (4) (1997) 323–330.
[41] K. Riemslagh, J. Vierendeels, E. Dick, An arbitrary Lagrangian–Eulerian finite-volume method for the simulation of rotary displacement pump flow, Applied Numerical Mathematics 32 (4) (2000) 419–433.
[42] G. Houzeaux, R. Codina, A finite element method for the solution of rotary pumps, Computer & Fluids 36 (4) (2007) 667–679.
[43] W. Strasser, CFD investigation of gear pump mixing using deforming/agglomerating mesh, Journal of Fluids Engineering 129 (4) (2006) 476–484.
[44] R. Castilla, P.J. Gamez-Montero, N. Ertürk, A. Vernet, M. Coussirat, E. Codina, Numerical simulation of turbulent flow in the suction chamber of a gearpump using deforming mesh and mesh replacement, International Journal of Mechanical Sciences 52 (10) (2010) 1334–1342.
[45] R. Castilla, P.J. Gamez-Montero, D. del Campo, G. Raush, M. Garcia-Vilchez, E. Codina, Three-dimensional numerical simulation of an external gear pump with decompression slot and meshing contact point, Journal of Fluids Engineering 137 (4) (2015) 041105.
[46] A. Kovacevic, N. Stosic, I.K. Smith, Three dimensional numerical analysis of screw compressor performance, Journal of Computational Methods in Sciences and Engineering 3 (2) (2003) 259–284.
[47] G. Bianchi, S. Rane, A. Kovacevic, R. Cipollone, S. Murgia, G. Contaldi, Grid generation methodology and CFD simulations in sliding vane compressors and expanders, IOP Conference Series: Materials Science and Engineering 232 (2017) 012053.
[48] G. Singh, S. Sun, A. Kovacevic, Q. Li, C. Bruecker, Transient flow analysis in a roots blower: experimental and numerical investigations, Mechanical Systems and Signal Processing 134 (2019) 106305.
[49] D. del Campo, R. Castilla, G.A. Raush, P.J. Gamez Montero, E. Codina, Numerical analysis of external gear pumps including cavitation, Journal of Fluids Engineering 134 (8) (2012) 081105.
[50] D. del Campo, R. Castilla, G. Raush, P. Gamez-Montero, E. Codina, Pressure effects on the performance of external gear pumps under cavitation, Proceedings of the Institution of Mechanical Engineers, Part C: Journal of Mechanical Engineering Science 228 (16) (2014) 2925–2937.
[51] A. King, R. Plant, A. Kovacevic, Improved design of gear pump porting through CFD simulation, in: International Conference on Compressors and Their Systems, 2009, pp. 387–394.
[52] E. Frosina, A. Senatore, M. Rigosi, Study of a high-pressure external gear pump with a computational fluid dynamic modeling approach, Energies 10 (8) (2017).
[53] A.K. Singhal, M.M. Athavale, H. Li, Y. Jiang, Mathematical basis and validation of the full cavitation model, Journal of Fluids Engineering 124 (3) (2002) 617.
[54] D. Yan, A. Kovacevic, Q. Tang, S. Rane, Numerical investigation of cavitation in twin-screw pumps, Proceedings of the Institution of Mechanical Engineers, Part C: Journal of Mechanical Engineering Science 232 (20) (2018) 3733–3750.

[55] Y. Yoon, B.H. Park, J. Shim, Y.O. Han, B.J. Hong, S.H. Yun, Numerical simulation of three-dimensional external gear pump using immersed solid method, Applied Thermal Engineering 118 (2017) 539–550.
[56] C. Kato, M. Kaiho, A. Manabe, An overset finite-element large-eddy simulation method with applications to turbomachinery and aeroacoustics, Journal of Applied Mechanics 70 (1) (2003) 32.
[57] J. Pei, M.K. Osman, W. Wang, D. Appiah, T. Yin, Q. Deng, A practical method for speeding up the cavitation prediction in an industrial double-suction centrifugal pump, Energies 12 (11) (2019) 2088.
[58] A. Roger Ull, Study of mesh deformation features of an open source CFD package and application to a gear pump simulation, 2012.
[59] J.-L. Reboud, B. Stutz, O. Coutier, Two-phase flow structure of cavitation: experiment and modelling of unsteady effects, Thirds International Symposium Cavitation 2017 (April) (1998) 203–208.
[60] W. Vera-Tudela, R. Haefeli, C. Barro, B. Schneider, K. Boulouchos, An experimental study of a very high-pressure diesel injector (up to 5000 bar) by means of optical diagnostics, Fuel 275 (2020) 117933.

Index

Note: Page numbers followed by "*f*" and "*t*" refer to figures and tables, respectively.

D

FSC
www.fsc.org
MIX
Papier | Fördert gute Waldnutzung
FSC® C083411

Kolibri 360 GmbH
Ferdinand-Jühlke-Straße 7
99095 Erfurt, Deutschland
produktsicherheit@kolibri360.de